Artificial Intelligence Enabled Real Time Environmental Monitoring

Binota Thokchom • Rishemjit Kaur
Pardeep Singh • Pengpeng Qiu
Editors

Artificial Intelligence Enabled Real Time Environmental Monitoring

Springer

Editors
Binota Thokchom
Department of Environmental Science
Dhanamanjuri University
Imphal, Manipur, India

Pardeep Singh
Department of Environmental Studies
PGDAV College, University of Delhi
New Delhi, Delhi, India

Rishemjit Kaur
CSIR – Central Scientific Instruments Organisation
Chandigarh, Chandigarh, India

Pengpeng Qiu
State Key Laboratory of Advanced Fiber Materials, Department of Material Science and Technology
Donghua University
Shanghai, Shanghai, China

ISBN 978-981-95-3766-2 ISBN 978-981-95-3767-9 (eBook)
https://doi.org/10.1007/978-981-95-3767-9

This Springer imprint is published by the registered company Springer Nature Singapore Pte Ltd.
The registered company address is: 152 Beach Road, #21-01/04 Gateway East, Singapore 189721, Singapore

Preface

The present book explores the intricate applicability of Artificial Intelligence (AI) in environmental challenges. The contents of this book have been designed to congregate quality information and innovative methodologies for a better understanding, assessing, analyzing, interpreting, or comprehending environmental problems. It will further help in deciphering various environmental obstacles. Internet of Things (IoT)-based problem-solving approaches make the quantification of all these monitoring, analysis, prediction, mitigation, etc., amenable and producible and have better empirical significance. With the intervention of features like classified automated data, it is going to be an exclusively thought-provoking contribution while addressing the issues. Precisely, comprehensive knowledge will be intensified in traversing the various functionalities of environmental ecosystems.

This volume brings together research and insight on the important environmental sectors such as air and water quality monitoring, biomedical waste management, environmental decision support systems, environmental sustainability, sustainable agriculture, urban planning and building, smart cities, climate change prediction, seafood adaptations, etc. Many experts have contributed their remarkable knowledge on different models of AI to bring out the incredible solutions to shape a better ecosystem. Finally, this book will be a source of transformative sketches for researchers, professionals, policymakers, and many who have a keen aspiration and belief that AI can metamorphose the panorama of environmental sustainability. It is a call to action, innovation, and collaboration to harness the full potential of AI in conserving and protecting this beautiful Earth.

Imphal, Manipur, India — Binota Thokchom
Chandigarh, India — Rishemjit Kaur
New Delhi, India — Pardeep Singh
Shanghai, China — Pengpeng Qiu

Contents

Editors and Contributors

About the Editors

Binota Thokchom is an Assistant Professor at Dhanamanjuri University, Manipur. She has served as DST-Inspire Faculty at Manipur University and IIT-Guwahati. She had also worked at the Analytical Instrumentation Division, NEERI, Nagpur. Her interest covers several environmental issues. She is developing water-treating technologies and nanoparticles, which are also under review for patent certifications.

Rishemjit Kaur is a Principal Scientist at CSIR-Central Scientific Instruments Organisation, Chandigarh, and an Associate Professor at the Academy of Scientific and Innovative Research (AcSIR), specializing in artificial intelligence. Her research focuses on the development of Artificial Intelligence and natural language processing for social good, particularly in agriculture, nutrition, and food. She is developing large language models for Indian farmers in local languages.

Pardeep Singh is a Senior Assistant Professor in the Department of Environmental Studies at PGDAV College, University of Delhi, India. Focusing on pressing environmental issues and sustainable management strategies, his research interests encompass global environmental change, climate change, and its multifaceted impacts on the Indian Himalayan Region. He has organized numerous international expeditions across the Indian Himalayan Region and has represented India at academic platforms.

Pengpeng Qiu is an Associate Professor in the State Key Laboratory of Advanced Fiber Materials, Department of Material Science and Technology at Donghua University. The scope of his research is the design and synthesis of functional mesoporous materials for various environmental-related applications. Till now, he has published more than 60 papers in international SCI journals in the fields of materials science and engineering.

Contributors

Geeta Aggarwal PGDAV College, University of Delhi, New Delhi, India

Seema Aggarwal Miranda House, University of Delhi, New Delhi, India

Inga Aleksandrova HSE University, Institute for Statistical Studies and Economics of Knowledge, Moscow, Russia

Bayartungalag Batsaikhan University of Mongolian Academy of Sciences, Ulaanbaatar, Mongolia

Megha Chauhan LNJN-NICFS, National Forensic Sciences University (NFSU), Delhi, Delhi, India

Sujatro Ray Chowdhuri Central Ground Water Board, CHQ, Faridabad, India

Shikha Daga Department of Commerce, PGDAV College, University of Delhi, New Delhi, India

Alina Demekhina HSE University, Institute for Statistical Studies and Economics of Knowledge, Moscow, Russia

Daria Gribkova HSE University, Institute for Statistical Studies and Economics of Knowledge, Moscow, Russia

Bashabi Gupta Miranda House, University of Delhi, New Delhi, India

Nimisha Jadon School of Studies in Environmental Chemistry, Jiwaji University, Gwalior, India

Milu Maria Jose Miranda House, University of Delhi, New Delhi, India

Jipson Joseph School of Law, Christ University, Bangalore, Karnataka, India

Rituraj Kashyap Department of Computer Science & Engineering, Faculty of Engineering & Technology, Rama University, Kanpur, Uttar Pradesh, India

Atul Kumar PGDAV College, University of Delhi, New Delhi, India

Yulia Milshina HSE University, Institute for Statistical Studies and Economics of Knowledge, Moscow, Russia

Achyutananda Mishra School of Law, Christ University, Bangalore, Karnataka, India

Kriti Mishra Central Ground Water Board, SUO, New Delhi, India

Ananya Pandey School of Law, Christ University, Bangalore, Karnataka, India

Mayuri Pandey Banaras Hindu University, Varanasi, India

Pragya Pathak Department of Computer Science & Engineering, Faculty of Engineering & Technology, Rama University, Kanpur, Uttar Pradesh, India

Keisham Radhapyari Regional Chemical Laboratory, Central Ground Water Board, North Eastern Region, Guwahati, Assam, India

Swati Shrivastava School of Studies in Environmental Chemistry, Jiwaji University, Gwalior, India

Madhukar Singh Central Ground Water Board, SUO, New Delhi, India

Pardeep Singh Department of Environmental Studies, PGDAV College, University of Delhi, New Delhi, India

Shashi Kant Singh Central Ground Water Board, CHQ, Faridabad, Haryana, India

Sonal Thukral Department of Commerce, Delhi School of Economics, University of Delhi, New Delhi, India

Shaily Varshney Applied Chemistry and Chemicals Technology Department, SGSITS, Indore, India

Kiran Yadav Department of Commerce, PGDAV College, University of Delhi, New Delhi, India

Monika Yadav Department of Applied Sciences & Humanities, Faculty of Engineering & Technology, Rama University, Kanpur, Uttar Pradesh, India

Durmus Cagrı Yıldırım Tekirdag Namık Kemal University, Tekirdag, Turkey

Seda Yıldırım Tekirdag Namık Kemal University, Tekirdag, Turkey

[illegible] Chemical Laboratory, Central Ground Water Board, North Eastern Region, Guwahati, Assam, India

[illegible]

[illegible]

[illegible]

[illegible]

Optimizing Collaborative Learning for Sustainability

1

Megha Chauhan and Atul Kumar

Abstract

In the era that is identified as the era of artificial intelligence (AI), this chapter attempts to explore the conjunction of transformative potential of green artificial intelligence (green AI) in fostering collaborative learning. This chapter highlights how AI techniques can expedite group learning by emphasizing on inclusivity, adaptability, and environmental consciousness, when it is programmed and applied in a responsible manner. The ideology of green AI can be traced to sustainable digital education that aims not only to individualized learning experience but also the reduction of ecological and environmental footprints of educational technologies.

Using interdisciplinary approach, the chapter draws the evolution of collaborative learning from conventional face-to-face models to high-tech AI-powered ecosystem that focuses in intelligent teaching systems, virtual tutors, adaptive forums, data analytics, etc. With the help of real-world case studies from institutions like Stanford and MIT, the authors illustrate how AI-powered tools can hone skills like teamwork, problem-solving, critical thinking, and analyzing while aligning with United Nation Sustainable Development Goals.

The chapter also highlights the ethical, technological, and pedagogical constraints for the implementation of AI in education. It can be said that the balanced approach is the need of the hour where technologist, academicians, and policymakers cocreate a framework that ensures the protection of data, equity in access, inclusivity, etc. Finally, this chapters positions green AI as a catalytic force that

M. Chauhan (✉)
LNJN-NICFS, National Forensic Sciences University (NFSU), Delhi, Delhi, India

A. Kumar
PGDAV College, University of Delhi, New Delhi, India

B. Thokchom et al. (eds.), *Artificial Intelligence Enabled Real Time Environmental Monitoring*, https://doi.org/10.1007/978-981-95-3767-9_1

can revolutionize collaborative learning for a more responsible, responsive, and sustainable education in the future.

Keywords

Green AI · Collaborative learning · Sustainable education · Digital literacy · Educational technology · Sustainable Development Goals (SDGs)

Green artificial intelligence (AI) in the education system can be defined as the applicability of AI tech that is ecofriendly and energy efficient for enhancing the collaborative learning process while keeping sustainability as one of the primary goals. Majorly it aims at augmenting e-resources, reducing carbon footprints, and enhancing sustainable digitalized education pedagogy (Schwartz et al. 2020). Collaborative learning is the wave of the future in a world where the education landscape has been altered quickly and significantly (Soller et al. 2005). This encourages students to work together, share information, and advance their essential traits such as critical thinking, communication skills, and group participation, which is a powerful way. However, group work can be challenging for anyone who has attempted to facilitate it. Challenging group dynamics bring different learning styles, personalities, and skill levels together, challenging managing the class (Michaelsen et al. 2023). That is when the disruptive power of AI and new ways to augment and optimize collaborative and sustainable learning come into play.

1.1 Concept of Green AI in Education

Green AI in education can be defined as the convergence of sustainable AI tools and practices within the education system. This approach primarily focuses on the development and application of AI-driven tools and technologies that are sustainable and energy efficient, reduction of ecological wastage, and conservation of environmental assets ultimately aiming for long-term sustainability. The possibility of green AI in education comprehends the generation of AI models that are energy efficient, promoting sustainable digital practices and advocating green consciousness among the stakeholders of education. By aligning AI with SDG, green AI aims to strike a balance between technology, environment, and sociological responsibility, thereby ensuring that the use of AI shouldn't be done at the cost of the environment.

Think about the classroom where AI will study each student to produce a balanced, impactful mix of children. It can constantly monitor group interactions and offer immediate suggestions to students and teachers. This can ensure that everyone participates and aids in the learning process (because who learns anything when no one does homework or prepares for a presentation?).

AI can help create intelligent collaboration platforms that adjust according to the student group's needs (Zhai et al. 2021). These platforms will not only be able to make tailor-made recommendations, resolve disputes, and suggest ways by which group dynamics can work but also help increase productivity—by adding teaching with AI in collaborative environments to overcome traditional gaps, enabling a dynamic and equitable approach (Pedro et al. 2019).

It equips students for the workforce—not just in terms of collaboration and digital literacy but also by enhancing the learning experience. AI can replicate live examples and help students develop team problem-solving and decision-making competence (Markauskaite et al. 2022). Working directly with technology promotes application over theory in a way that lets students develop job skills during lecture hours and prepare themselves to excel from day 1 once they hit the workforce.

Another great benefit of AI in collaborative and sustainable learning is inclusivity. AI ensures that all students can partake and excel in collaborative activities by recognizing the varying needs of its diverse population (Sanusi et al. 2022). The tailored approach can help narrow the achievement gap and create a more inclusive and supportive space.

Entering the digital era, the tandem of AI and collaboration is a powerful tool capable of changing the whole approach toward education. Considering the contribution of AI to human-computer interactions, the AI response to the challenges of group dynamics ensures educators can focus on significantly making the environment for their students more supportive and collaborative. This chapter addresses the ethical concerns brought about by AI's implications. It gives examples of the practical implementation and the pathways for AI development in education to help educators. Decision-makers have a vision of integrating AI into a collaboration to support its social presence in the future. To reach that goal, we must contribute to creating a fair, inclusive, and open educational system suitable for everyone. By converging technology-driven tools and technology like inclusive and flaccid learning environments, customized tutoring systems, and virtual classrooms, green AI makes sure that the advancement in educational pedagogy is in coherence with the sustainability goals (Luccioni, Schmidt, Vardanyan, and Bengio (2021)).

Key Principles of Green AI

1. *Energy efficiency*: AI algorithms and models should be built in such a manner that they use less computational power, which would make them more energy efficient and reduce carbon emissions. They may involve optimizing codes, employing ecological data processing tools and techniques, and selecting hardware that functions optimally on less energy (van Wynsberghe 2021).
2. *Sustainable infrastructure*: This refers to the creation of infrastructures that use renewable sources of energy and support ecologically friendly practices, for example, creating data centers powered by renewable energy sources, minimiz-

ing wastage and optimizing utilization of resources, and promoting energy-efficient devices in educational settings.[1]

3. *Responsible development and disposition*: This includes surety that the technologies that are being designed and used are in consonance with the environment. This principle supports the revisit of environmental footprints during the initial phase of development and espouses the practices that alleviate the negative impact of technology on the environment (van Wynsberghe 2021).

1.2 The Evolution of Green Learning: From Traditional Methods to Collaborative AI Integration

Collaborative learning has long been considered a teaching component with the belief that students learn best when actively interacting with their peers. Traditional collaborative learning methods, such as group projects, peer tutoring, and classroom discussions, have been utilized for a period to promote comprehension, critical thinking, and social skills (Panitz 1999). These approaches are rooted in the social constructivist theory that emphasizes knowledge construction through interactions and shared experiences (Vygotsky 1978). Studies have shown that students who participate in learning tend to achieve levels of success and retention compared to those who study independently (Johnson and Johnson 1989).

During the early stages, collaborative learning mainly depended on face-to-face interactions and having everyone there in person. Teachers gave group assignments and guidance, but how well it worked mostly came down to how the group worked together and each person's input. This way of learning was helpful, but it faced issues like not everyone participating equally, some voices being louder than others, and struggles with matching schedules and resources (Dillenbourg 1999). When digital tech came around, collaborative learning changed a lot. Online forums, virtual classrooms, and tools like Google Docs and Slack let students work together from far away, breaking distance barriers and making learning settings more flexible (Barkley et al. 2005). These tools allowed students to communicate at different times, making independent collaboration easier. However, even with these new things, regular digital tools couldn't adjust to each person's learning needs or how the group worked in real time.

Interestingly, technology is advancing so rapidly that educational paradigms are being transformed and address the limitations of conventional and digital collaborative learning by involving the participation of the individual, adaptive, and knowledge-based technology as a solution to the problem (Rane et al. 2023). AI platforms can also use the data gathered from students' behavior, preferences, and performance to create diverse groups with various techniques and knowledge. They also can give real-time feedback to children and promote group engagement and inclusivity in the learning environment (Rosé et al. 2008). Woolf et al.'s (2009) study

[1] See Green AI: Revolutionizing Sustainability in Artificial Intelligence for a Greener Future | E-SPIN Group.

found that E-tutoring powered by AI systems considerably assists students' learning through personalized feedback and framework. Dillenbourg and Fischer (2007) also demonstrate AI's potential to intervene in the case of interactional problems within a group by offering guidance to the students. The studies conducted in implementing virtual agents and chatbots demonstrated that they can produce learners with more enthusiasm and positive feedback through instant assistance and active participation (Muldner et al. 2011).

Contemplating the probable advantages and challenges facing AI in the field of education is crucial for the assessment of its worth. It is necessary to cope with ethical issues such as data privacy and bias in algorithms that are concerned with fairness and equal access to AI use in schools. Moreover, educators need the tools to promote a holistic, vibrant, productive, and knowledgeable environment. Moreover, this chapter provides more details on how AI has been used as a tool in the education sector and how it is not simply another means of achievement. This chapter goes a bit further and sheds light on what we need to understand about the existing traditional pedagogies, as those have been the special ones for developing new ways to include AI in the education system. Understanding the history and development of technology makes it easier for us to perceive just how far AI has changed collaborative learning and how much it could alter our future pedagogy of teaching and learning.

1.3 Implementing AI in Collaborative Learning Environments: Opportunities and Challenges

As we further gauge AI's role in education, the pros and cons should be considered. Data privacy and algorithm prejudice must be considered prudently to guarantee the fair use of AI in institutions.

Educators must learn to use AI tools correctly to improve their teaching, including making learning more exciting, fair, and valuable. This part of the chapter will go deep into how people work together in learning, looking at how things used to be done and how AI is changing the game in schools. Understanding history and tech progress helps us see the significant changes AI can make in collaborative learning and how it can shake up education later (Shah 2023).

Also, AI helps watch and understand how groups talk and act together right away, such as spotting how people communicate and seeing if everyone is taking part equally or if there are problems (Burgess 2017). Educators can step in quickly to help and ensure everyone is included (Dillenbourg and Fischer 2007). Plus, AI-powered chatbots and virtual helpers improve group chats by giving quick feedback and making talks easier. These fancy tools help with active learning and thinking skills so that every student can share helpful something (Muldner et al. 2011).

When bringing AI into teamwork class, there are some issues we need to think about. Ethics, data privacy, and ensuring AI in education are important (Holmes and Porayska-Pomsta 2023). It's critical to be transparent and fair with AI programs so everyone can learn without discrimination risks. Teachers need to learn new stuff to use AI well. Through training and skill-building programs, educators must correctly make the most of AI tools (Blikstein 2018). Teachers play a significant role in helping students work with AI systems and making sure technology adds to teamwork learning instead of taking over.

As AI gets better, we need more research to see how it affects how students get involved, how well they do in school, and how they develop modern skills. Looking at results over time can show us if AI works well, helping us use the best strategies when including AI in schools (Rosé et al. 2008). While putting AI in teamwork classes has an excellent chance of improving our teaching, it's essential to be careful and intelligent about it.

1.3.1 Real-World Examples of Green AI in Teaching and Learning

Stanford University

Stanford University has taken the remarkable stand to converge artificial intelligence with energy efficiency, in educational setup. The institution AI+ Education initiative in collaboration with Stanford Institute of Human-Centered Artificial Intelligence (HAI) is proactively working on the creation of digitalized learning tools that are efficient and environment-friendly. This collaboration aims to work on minimizing the environmental footprints of educational technologies.[2]

In addition to the educational platforms, the university is also exploring hardware that is more energy efficient and supports AI. Advancements in this region are very vital for the reduction of energy consumption associated with AI applications, thereby falling in line with green AI.[3]

The Massachusetts Institute of Technology (MIT) Climate Machine

MIT has developed "Climate Machine," which is an AI-driven algorithm tool that holds the capability of predicting the weather. This tool is designed to read huge datasets so that the environmentalist can understand the complex climatic phenomenon. This machine aims to develop strategies to lessen the harm caused by natural calamities by managing the ecology by the prediction.[4]

Green AI-Powered Sustainability Education Programs

Institutions across the globe are actively working toward the amalgamation of AI into their system. Institutes are increasingly deploying AI-driven climate predictors into their curricula to promote learners' understanding of climate change. These models make use of machine learning algorithms to create a stimulating environment for students to explore and learn about varied environmental policies and their implementation (Chen et al. 2023).

---

[2] See AI + Education • Stanford Accelerator for Learning.

[3] See New material opens the door for energy-efficient computing | Stanford University School of Engineering.

[4] See Overview: Climate Intelligence—MIT Media Lab.

Another new age AI-driven setup being adopted by educational institutions is *Virtual Laboratories.* These virtual labs simulate real-world environmental issues, which let the students conduct experiments and investigations in a virtual setup. This technique reduces the environmental footprint to a much greater extent parallelly letting students develop skills, knowledge, and experience on sustainable practices and environmental issues (Chen et al. 2023).

Furthermore, the incorporation of virtual laboratories is in line with the United Nations Sustainable Development Goals (SDGs), specifically SDG4 (Quality Education) and SDG 10 (Reduced Inequalities). Virtual labs are paving the way to fill in the existing educational gaps irrespective of geographical location and diversification (Poo et al. 2023). In addition, the introduction of virtual laboratories in the education domain has proven to provide personalized feedback by adapting to the algorithm of a learner's need, which in turn enhances the satisfaction and learning experience of a student; it can also be equated to personal mentoring (Zhang and Liu 2024).

Hence it can be said that AI-driven virtual laboratories are a depiction and an operative approach toward sustainability, ecological awareness quality, and impactful education.

1.4 Role of Green AI in Promoting Sustainable Learning

Artificial intelligence (AI) technologies are transforming collaborative learning environments by enhancing student engagement, interaction, and learning outcomes through countless groundbreaking tools and applications.

Intelligent tutoring systems (ITS) Intelligent tutoring systems (ITS) use AI algorithms to give students personalized learning experiences in accordance with their needs. These systems observe how students are doing, what they're good at, and how they learn, to give them specific feedback and study material to learn (Woolf et al. 2013). Acting like a real human tutor, ITS helps students work together in groups by ensuring each student gets the right help and guidance (Falchikov 2003).

Chatbots and virtual agents AI-powered chatbots and virtual agents are crucial for getting students to talk and work with each other in online classes. These cool tools can help students immediately by answering questions, explaining things, and leading discussions (Muldner et al. 2011). Chatbots that know how we talk can make it easy to chat with them, ensuring we're interested and joining group talks (Jain et al. 2018).

Data analytics and learning analytics Data analytics and learning analytics use AI to look at all the data students make while working together. They find patterns in how students talk, work together, and do in class to help teachers know what actions to take (Rosé et al. 2008). By watching how groups change and what each student does, learning analytics let teachers pick the best groups, change how they teach, and step in when needed so everyone learns better.

Augmented reality (AR) and virtual reality (VR) Augmented reality (AR) and virtual reality (VR) tech offer innovative ways to learn by putting you right into a different world using AI (Akçayır and Akçayır 2017). With these unique apps using AI tech, students can move stuff around in virtual worlds with friends while solving complex problems like a team. This helps us think about space better, work with others, and be creative as part of a group study.

Adaptive learning platforms Adaptive learning platforms also use AI to make learning more personal for each student based on what they like doing and how they like to learn. In group work, these platforms pick out the best things for each student to learn from others so that everyone gets involved (Rosé et al. 2008). Since adaptive learning works for everyone differently, it helps ensure all friends can participate in group activities under every circumstance (Garmston and Wellman 2016).

Real-world applications and benefits Using AI tech in group learning has shown significant benefits for schools. Studies say these tools make students want to learn more, do better in classwork, and think more deeply about different subjects (Woolf et al. 2013; Muldner et al. 2011). By giving customized support from beginning to end, making conversations natural between classmates easier than ever, and making groups work well together, AI tech is helping teachers build fun classes that are fair for everyone (Nazaretsky et al. 2022).

Although utilizing AI technology in group learning looks excellent, there are some hard aspects. Protecting data is a big issue and being just while using computer programs that might have biased views. The stakeholders also need to be trained on how to use these tools, and computers should be working correctly to enable the best understanding of all people (Blikstein 2018).

The use of artificial intelligence technologies has the potential to revolutionize collaborative learning environments by supporting individualized instruction, encouraging dialogue, and ensuring effective teamwork. Thoughtfully using this tool, therefore, enables teachers to create innovative, inclusive, student-centered learning experiences.

1.5 Pedagogical Strategies for Integrating Green AI in Learning

To deploy AI in the classroom, we should find the right way to do this, which requires an intelligent facilitator to use it most effectively and to handle the challenges efficiently. Advances and constraints of using AI tools are widespread among educators in their classroom instruction courses to facilitate student interaction, cultivation of teamwork, and expansion covering learning gains.

Designing collaborative activities with AI support Teaching strategies must be aimed at making activities that help teamwork by using AI tools to make students communicate well and interact more. They can employ intelligent tutoring systems (ITS) to make sure that every student learns the way they want in a team, so students receive the proper knowledge that is identified as their needs (Woolf et al. 2013). Moreover, by including AI feedback and adaptive resources, teachers can encourage students to engage and hold meaningful conversations with their peers actively (Chen et al. 2020).

Promoting reflective practice and metacognition AI's applications can support learners in their reflective process and group efforts (Markauskaite et al. 2022). Tools like data analytics can also let educators know how groups work, the contribution of each person in the group, and the learning outcomes (Rosé et al. 2008). Instructors show the generated data to the learners, make them understand the areas that have scope for improvement, and set future goals at an individual and group level. This output will certainly build teamwork skills, self-reflected learning skills, thinking skills, etc.

Cultivating digital literacy and ethical awareness Students need to possess skills in digital literacy and understand the right way to use AI (Wan 2012). Educators should discuss the ethical side of using AI in school. This includes data safety and fair use of AI tools. By encouraging students to make moral choices, teachers help them navigate technology-rich teamwork areas well. They also foster responsible online behavior.

Facilitating seamless integration of AI tools Teaching pedagogies should not aim at introducing AI as a tool used in different exclusive tasks but rather in merging the AI tools into normal teamwork tasks. Muldner et al. (2011) observed that educators should train students to use chatbots and virtual helpers to talk better and share their ideas with other people. An increase in knowledge about changes in technologies and new ideas, which is stimulated by teachers, allows students to use AI tools effectively to achieve the learning objectives as well as with other students (Kim et al. 2022).

Emphasizing continuous assessment and feedback AI enables educators to always monitor the result of group work through feedback systems. The analysis of data in real time to understand what is not being taught and where various groups are challenged helps monitor progress and intervene if a team is struggling (Woolf et al. 2013). As for coaching and assessments during the process, the use of artificial intelligence is applied, and teachers create an environment where learners can continue to improve their teamwork skills and grades (Shi 2023).

Real-world applications and benefits Technique approaches to implementing AI in collaborative learning have shown several advantages in learning environments. The findings suggest that these strategies increase students' engagement, improve academic outcomes, and promote the skills that are deemed valuable in the twenty-first century, including communication and collaboration (Woolf et al. 2013; Rosé et al. 2008). AI technologies, therefore, should be adopted in learning environments in a manner that supports the learning objectives as well as the educational benefits of the users.

Challenges and considerations When incorporating AI into cooperative learning, there are difficulties associated with technology implementation and use, ethical questions, and technology access. These issues need to be met in the context of tiered professional learning, professional ethics, and equal opportunities regarding AI technologies (Blikstein 2018). In this way, by being aware of these considerations in advance, educators can link the possibilities of AI to contribute to improving collaborative approaches to teaching and learning for any learner.

In conclusion, the outlined instructional approaches to integrate AI in collaborative learning mostly focus on the following principles: There should be a purposeful design of activities, metacognition should be encouraged, digital literacy, and the option should always be available to integrate technology into the lesson and assessment emphasized. Thus, by implementing the above approaches and dealing with related issues, tutors are enabled to leverage AI tools to foster innovative, comprehensive, and productive co-participation learning environments.

1.6 Evaluation and Assessment of Green AI Technologies in Collaborative Learning

The incorporation of AI in group learning calls for proper ways of assessing how it is effective in achieving learning and whether it is compatible with the education system. Self-assessment of the outcomes of AI focuses on the aspect of increasing the students' engagement, improving their interactions, and enhancing individualization.

Measuring student engagement and participation One can observe students' engagement in group activities to check AI in group learning. Watching how students interact and participate in tasks gives clues about how AI helps them communicate and be active in groups (Rosé et al. 2008). Educators can use this info to see if AI leads to real student talks and gets students involved.

Assessing learning outcomes and academic achievement With AI tools, educators can track how well students learn and do in real-time group tasks. Smart Tutor Systems can follow student progress, work on group jobs, and if they hit the learning goals (Woolf et al. 2013). This data helps teachers see how individual students and groups do, find areas to improve, and change teaching ways to boost group learning.

Evaluating the impact on collaborative skills and critical thinking Observing how AI affects skills like talking well, solving problems, and adding value to discussions is critical. AI-powered chatbots help students talk with peers and practice these skills (Muldner et al. 2011). Educators can test these skills with both words and numbers to check if students get better at necessary abilities through AI tools.

Addressing ethical and social implications AI tools must also consider the moral and social sides, like data safety, biased information from algorithms, and fair access for all (Cheng et al. 2021). It's vital to ensure that AI used in educational institutions follows fairness rules, is clear about what it does, and is used responsibly. Teachers should discuss key moral points when checking AI tools to ensure that using AI is safe and avoids tech troubles.

Studies show the various uses of AI tech in group learning. Some work proves that AI data can help groups work better together (Rosé et al. 2008). Other studies discuss using Smart Tutor Systems to tailor learning for each student and aid peer teamwork (Woolf et al. 2013).

Though there are good things about trying out AI in teaching groups, there are troubles, too. There should be standard ways to check how well AI works across schools; solid facts should be there for checking; teachers should keep learning more about using tech well (Blikstein 2018). Future research needs to look at how well AI helps students learn over time in class settings.

Thus, it can be stated that the practical assessment of AI technologies in collaborative learning is crucial for the evaluation of their influence, improvements in teaching methodologies, as well as the optimization of the learning outcomes. This chapter has highlighted how through strict assessment criteria and the consideration of ethical issues AI can help educators achieve the potential goal of inclusive, engaging, and efficient collaborative learning environments (Khreisat et al. 2024).

1.7 Best Practices for Implementing Green AI in Collaborative Learning

Using artificial intelligence (AI) in collaborative learning is a must to get things right and make sure it helps students learn better. Teachers can use these to make students more involved, get them working together, and create personalized learning.

Align AI technologies with pedagogical goals In integrating AI into collaborative learning, the first step is to match AI with what teachers want to accomplish as well as what they would like students to acquire. Teachers need to explain the intended learning outcomes, which teamwork competencies they wish to develop, and the content knowledge students should acquire with the help of AI (Blikstein 2018). This way, the teachers can choose the proper tools and methods in artificial intelligence and support the chosen goals providing the learners with productive educational experiences (Pedro et al. 2019).

Provide professional development for educators Teachers are some of the most important agents in getting AI into learning teams. The implication enables the acquisition of the know-how and skills to use AI tools appropriately in teaching for those teachers with good training chances (Chounta et al. 2022). Thus, teachers can think about the applications of AI at events like workshops, online seminars, and working in teams, discuss the most effective moves in the field, and introduce new teaching methodologies that employ AI to enhance students' performance.

Foster collaboration and peer learning New beneficial aspects of effective interaction among learners are offered by AI tools, as learners get opportunities to unite and share their experiences and knowledge as well as complete activities together (Markauskaite et. al 2022). In certain areas where AI supports learning, teachers require an approach that demands the students to engage more with their peers and use good communication skills besides sharing knowledge (Rosé et al. 2008). Incorporated in the teamwork spirit and fairness during joining in of activities, teachers foster learners to work well in a team and apply AI tools in accomplishing their teamwork tasks.

Personalized learning experiences with AI AI-backed adaptive learning places offer custom learning by fitting content, resources, and teaching plans to each person's needs (Woolf et al. 2013). Educators should use AI tools that offer personal advice and paths to learning that are better suited for individuals within teams (Woolf et al. 2013). Working with different learner types and abilities using personal touches from AI techs helps lift learners' interest, gives a deeper understanding of lessons, and improves group work.

Monitor and evaluate AI implementation Continuous assessments are paramount to ascertain the efficiency of AI-related technologies in interactive learning environments (Muldner et al. 2011). The performance metrics and figures using these tools are used to distinguish which parts are better or worse and which are correctable while some teaching changes can be adapted in the future. Thus, educators gain insight into AI application strengths, weaknesses, and opportunities for pedagogical innovation, avoiding the pitfalls of duplication and maximizing the system's efficiency (Muldner et al. 2011). Sources show different capabilities of AI systems through empirical data and modeling of the learning process as possible success stories in collaborative learning environments. For example, Woolf et al. (2013) report AI's influence in improving academic achievement and increasing student participation. Also, Rosé et al. (2008) analyze the influence of AI analytical methods on group interaction and collaborative skills in online learning.

Using artificial intelligence in collaborative learning is not only about the benefits but also about technology integration, the moral dilemma, and the need for professional development (Blikstein 2018), which are some of the educators' significant concerns. Such difficulties should be approached by setting objectives, ethical guidelines, and constant backing to use AI-educational technologies.

It can also be stated that the best practices for implementing AI in collaborative learning are the alignment of technologies with pedals, in-service training for teachers, facilitating cooperation among students, individualizing student-based education, and overseeing adoption through careful assessment. To encourage educators in active, extensive, and influential collaborative learning practice, they can utilize these techniques and incorporate AI technologies in the classroom.

1.8 Future Trends and Innovations in Green AI-Enhanced Collaborative Learning

Artificial intelligence (AI) integration in collaborative learning continues to progress, promising transformative innovations that enhance student engagement, personalize learning experiences, and foster collaborative skills. As technology grows and educational paradigms shift, several key trends and innovations evolve in AI-enhanced collaborative learning environments.

1. Advancements in Adaptive Learning Systems
   The tendencies of AI-enhanced collaborative learning in the not-so-distant future will improve adaptive learning systems, which provide individually tailored educational experiences for each student. AI algorithms power these systems, analyzing student performance data, learning preferences, and level of engagement in real-time (Woolf et al. 2013). Adaptive learning systems help tailor content delivery, feedback mechanisms, and collaborative group formations through dynamic changing to optimize the outcomes of student learning needs—by helping promote individualized pathways for each learner (Muñoz et al. 2022).
2. Enhanced Natural Language Processing (NLP) and Communication Technologies
   Natural language processing (NLP) technologies facilitate seamless communication and collaboration among students in AI-enhanced learning environments. Future innovations in NLP will enable AI systems to understand better and respond to natural language inputs, fostering more interactive and engaging educational interactions (Rosé et al. 2008). Virtual agents and chatbots with advanced NLP capabilities will support students in collaborative tasks, provide real-time assistance, and enhance peer-to-peer communication within digital learning spaces (Kuhail et al. 2023).
3. Integration of Augmented Reality (AR) and Virtual Reality (VR)
   Integrating augmented reality (AR) and virtual reality (VR) technologies hold promises for immersive and interactive collaborative learning experiences. AI-powered AR/VR applications will enable students to engage in virtual collaborative projects, simulations, and experiential learning activities (Dillenbourg and Jermann 2007). These technologies provide realistic environments where students can collaborate, problem-solve, and explore complex concepts, transcending physical limitations and enhancing learning through interactive, multisensory experiences.

4. Data-Driven Insights for Continuous Improvement
AI-generated analytical learning will enable educators to receive guidelines on improving student performance, group collaboration, and learning outcomes. One of the next-generation advancements shall contemplate predictive analytics, which truly foresight student's needs, pinpoint learning gaps, and recommend personalized interventions that happen during the class (Siemens and Gasevic 2012). Data-driven community-building and artificial intelligence are two of the most effective teaching methods. Educators can make smarter decisions thanks to big data and AI algorithms and use them to create enhanced collaborative learning environments and targets for educational goals.
5. Ethical Considerations and Responsible AI Adoption
Given that AI has been continuing to influence education, ethical and responsible AI adoption will be the main topic of discussion. Inevitable developments will focus on the formation of ethical rules, the protection of data privacy, the correction of algorithmic bias, and the provision of fair access to AI-enabled learning materials. Educators, policymakers, and technologists will work together to formulate models that maintain ethical standards and be the basis of a secure, uniform teaching system for each student. There is nothing else to say about the fact that future developments and revolutions in AI-enhanced collaborative learning seem to be the only hope for the future of education. Personalized learning, cooperative skills development, and a rise in the quality of education in general are the significant benefits of these technologies. By deploying these innovations and confronting the problems that come with them, educators can implement creative, captivating, and multicultural learning environments that equip students with relevant skills in the fast-changing digital world.

1.9 Conclusion

Throughout this chapter, we have explored the transformative impact of artificial intelligence (AI) on collaborative learning environments, specifically focusing on optimizing group dynamics and interaction. AI technologies have transformed conventional collaborative learning methods by offering personalized, adaptive, and data-driven solutions that endorse profound engagement and more influential group interactions, thereby creating a holistic and inclusive environment for learners.

The platforms led by AI evaluate the learner's behavior, preferences, and performance to build customized study groups, thereby ensuring a perfectly balanced heterogeneity of skills and knowledge. Such technologies provide immediate feedback and suggestions to enhance learning (Rosé et al. 2008).

The fruition of collaborative learning can be traced back to social constructivist theory. The evolution has progressed from physical interactions to virtual and digital collaboration environments to AI-empowered setups. Initial challenges like fluctuating participation levels and coordination have been well taken care of by AI's ability to accustom personalized learning requirements and group dynamics (Dillenbourg 1999).

As we embrace AI in education, ethical considerations regarding data privacy, algorithmic bias, and equitable access must be carefully navigated. Educators play a crucial role in integrating AI tools effectively, leveraging their potential to create more engaging, inclusive, and productive learning experiences.

Progressive trends in AI-enhanced sustainable learning comprise advancements in adaptive learning systems, enhanced communication technologies, and augmented and virtual reality convergence. These novelties promise to personalize learning experiences further, optimize collaborative skills, and prepare students for success in a digitally driven world (Siemens and Gasevic 2012).

It can be concluded that integrating green AI into learning represents a paradigm shift in educational practices, offering unprecedented opportunities to enhance group dynamics, optimize interaction, and foster collaborative skills among students. By embracing these technologies responsibly and innovatively, educators can pave the way for a more sustainable, dynamic, and inclusive educational landscape.

References

Akçayır M, Akçayır G (2017) Advantages and challenges associated with augmented reality for education: a systematic review of literature. Educ Res Rev 20:1–11

Barkley EF, Cross KP, Major CH (2005) Collaborative learning techniques: a handbook for college faculty. Jossey-Bass

Blikstein P (2018) AI in education: the present and the future. Int J Artif Intell Educ 28(1):1–14

Burgess A (2017) The executive guide to artificial intelligence: how to identify and implement applications for AI in your organization. Springer

Chen L, Chen P, Lin Z (2020) Artificial intelligence in education: a review. IEEE Access 8:75264–75278

Chen L, Chen Z, Zhang Y et al (2023) Artificial intelligence-based solutions for climate change: a review. Environ Chem Lett 21:2525–2557. https://doi.org/10.1007/s10311-023-01617-y

Cheng L, Varshney KR, Liu H (2021) Socially responsible AI algorithms: issues, purposes, and challenges. J Artif Intell Res 71:1137–1181

Chounta IA, Bardone E, Raudsep A, Pedaste M (2022) Exploring teachers' perceptions of artificial intelligence as a tool to support their practice in Estonian K-12 education. Int J Artif Intell Educ 32(3):725–755

Dillenbourg P (1999) What do you mean by collaborative learning? In: Dillenbourg P (ed) Collaborative-learning: cognitive and computational approaches. Elsevier, Oxford, pp 1–19

Dillenbourg P, Fischer F (2007) Basics of computer-supported collaborative learning. Zeitschrift für Berufs- und Wirtschaftspädagogik 21:111–130

Dillenbourg P, Jermann P (2007) Designing integrative scripts. In: Fischer F, Kollar I, Mandl H, Haake JM (eds) Scripting computer-supported collaborative learning: cognitive, computational, and educational perspectives. Springer, pp 275–301

Falchikov N (2003) Learning together: peer tutoring in higher education. Routledge

Garmston RJ, Wellman BM (2016) The adaptive school: a sourcebook for developing collaborative groups. Rowman & Littlefield

Holmes W, Porayska-Pomsta K (2023) The ethics of artificial intelligence in education. Routledge Taylor

Jain M, Kumar P, Kota R, Patel SN (2018) Evaluating and informing the design of chatbots. In: Proceedings of the 2018 designing interactive systems conference. ACM, pp 895–906

Johnson DW, Johnson RT (1989) Cooperation and competition: theory and research. Interaction Book Company

Khreisat MN, Khilani D, Rusho MA, Karkkulainen EA, Tabuena AC, Uberas AD (2024) Ethical implications of AI integration in educational decision making: systematic review. Educ Adm Theory Pract 30(5):8521–8527

Kim J, Lee H, Cho YH (2022) Learning design to support student-AI collaboration: perspectives of leading teachers for AI in education. Educ Inf Technol 27(5):6069–6104

Kuhail MA, Alturki N, Alramlawi S, Alhejori K (2023) Interacting with educational chatbots: a systematic review. Educ Inf Technol 28(1):973–1018

Luccioni A, Schmidt V, Vardanyan V, Bengio Y (2021) Using artificial intelligence to visualize the impacts of climate change. IEEE Computer Graphics and Applications, 41(1):8–14. https://doi.org/10.1109/MCG.2020.3025425

Markauskaite L, Marrone R, Poquet O, Knight S, Martinez-Maldonado R, Howard S et al (2022) Rethinking the entwining between artificial intelligence and human learning: what capabilities do learners need for a world with AI? Comput Educ Artif Intell 3:100056

Michaelsen LK, Knight AB, Fink LD (eds) (2023) Team-based learning: a transformative use of small groups in college teaching. Taylor & Francis

Muldner K, Burleson W, VanLehn K (2011) An analysis of students' gaming behaviors in an intelligent tutoring system: predictors and impacts. User Model User-Adap Inter 21(1):99–135

Muñoz JLR, Ojeda FM, Jurado DLA, Peña PFP, Carranza CPM, Berríos HQ et al (2022) Systematic review of adaptive learning technology for learning in higher education. Eurasian J Educ Res 98(98):221–233

Nazaretsky T, Ariely M, Cukurova M, Alexandron G (2022) Teachers' trust in AI-powered educational technology and a professional development program to improve it. Br J Educ Technol 53(4):914–931

Panitz T (1999) The case for student-centered instruction via collaborative learning paradigms. ERIC

Pedro F, Subosa M, Rivas A, Valverde P (2019) Artificial intelligence in education: challenges and opportunities for sustainable development. UNESCO

Poo MC-P, Lau Y-Y, Chen Q (2023) Are virtual laboratories and remote laboratories enhancing the quality of sustainability education? Educ Sci 13(11):1110. https://doi.org/10.3390/educsci13111110

Rane N, Choudhary S, Rane J (2023) Education 4.0 and 5.0: integrating artificial intelligence (AI) for personalized and adaptive learning. Available at SSRN 4638365

Rosé CP, Wang Y-C, Cui Y, Arguello J, Stegmann K, Weinberger A, Fischer F (2008) Analyzing collaborative learning processes automatically: exploiting the advances of computational linguistics in computer-supported collaborative learning. Int J Comput-Support Collab Learn 3(3):237–271

Sanusi IT, Olaleye SA, Oyelere SS, Dixon RA (2022) Investigating learners' competencies for artificial intelligence education in an African K-12 setting. Comput Educ Open 3:100083

Schwartz R, Dodge J, Smith NA, Etzioni O (2020) Green AI. Commun ACM 63(12):54–63

Shah P (2023) AI and the future of education: teaching in the age of artificial intelligence. Wiley

Shi L (2023) Exploring the emerging roles of artificial intelligence in teaching practices and its relationships with teachers [Doctoral dissertation, University of Georgia]. University of Georgia Open Scholar. https://openscholar.uga.edu/record/2921/files/Lehong%20Dissertation%20Final_Teacher%20and%20AI%20relationship.pdf

Siemens G, Gasevic D (2012) Guest editorial: learning and knowledge of analytics. Educ Technol Soc 15(3):1–2

Soller A, Martinez A, Jermann P, Muehlenbrock M (2005) From mirroring to guiding: a review of state-of-the-art technology for supporting collaborative learning. Int J Artif Intell Educ 15(4):261–290

van Wynsberghe A (2021) Sustainable AI: AI for sustainability and the sustainability of AI. AI Ethics 1:213–218. https://doi.org/10.1007/s43681-021-00043-6

Vygotsky LS (1978) Mind in society: the development of higher psychological processes. Harvard University Press

Woolf BP, Burleson W, Arroyo I, Dragon T, Cooper DG, Picard RW (2009) Affect-aware tutors: recognizing and responding to student affect. Int J Learn Technol 4(3–4):129–164

Woolf BP, Lane HC, Chaudhri VK, Kolodner JL (2013) AI Grand Challenges for Education. AI Magazine, 34(4):66–84. https://doi.org/10.1609/aimag.v34i4.2484

Zhai X, Chu X, Chai CS, Jong MSY, Istenic A, Spector M et al (2021) A review of artificial intelligence (AI) in education from 2010 to 2020. Complexity 2021(1):8812542

Zhang N, Liu Y (2024) Design and implementation of virtual laboratories for higher education sustainability: a case study of Nankai University. Front Educ 8:1322263. https://doi.org/10.3389/feduc.2023.1322263

Advancing Sustainable Agriculture Through Artificial Intelligence: Harvesting Greener Future

2

Ananya Pandey, Jipson Joseph, and Achyutananda Mishra

Abstract

The growing population of the globe is expected to reach 10 billion by 2050. Its demands have led to widespread food insecurities and hunger, which conventional farming failed to address. To meet such a grim scenario, the United Nations drafted the Sustainable Development Goals (SDGs) to end hunger (SDG 2), promote sustainable consumption and production (SDG 12), and promote life on land (SDG 15). To achieve these goals, the global farming system needs to make a substantial transition by adopting alternative methods and approaches. The concept of sustainable agriculture is an integrated system of animal and plant production practices to satisfy human food and fiber requirements. It enhances environmental quality, efficient use of energy resources, economic viability of farm operation, and quality of life of farmers and society. It has emerged as a viable alternative to meet such goals. Sustainable agriculture practice incorporates the true spirit of the abovementioned SDGs by combining the aspirations of the present and future generations.

Smart technologies, which include the use of robotics, intelligent sprayers, satellite drones, Internet of Things (IoT) devices, and climate sensors, appear as supplementary tools for sustainable agriculture to mitigate the accelerated demands of food across the globe. The advancement of artificial intelligence (AI)/IoT technologies has the potential to monitor the agricultural environment to ensure high-quality products. According to the World Economic Forum (WEF), the integration of AI into the farming sector can provide an adequate food supply to the people. Such integration of smart technologies with sustain-

A. Pandey · J. Joseph · A. Mishra (✉)
School of Law, Christ University, Bangalore, Karnataka, India
e-mail: ananya.pandey@res.christuniversity.in; jipson.joseph@res.christuniversity.in; achyuta.mishra@christuniversity.in

B. Thokchom et al. (eds.), *Artificial Intelligence Enabled Real Time Environmental Monitoring*, https://doi.org/10.1007/978-981-95-3767-9_2

able agriculture has significant advantages for the world in achieving sustainable development goals.

However, smart technologies are not free from challenges due to their novelty, complications related to the control and operation of IoT/AI machines, data sharing and management, interoperability and large amounts of data ownership, analysis, and storage.

Keeping in view the above, this chapter explores the integration and interaction of AI-enabled smart technologies and smart agriculture analysing key challenges involved in this regard.

Keywords

Artificial intelligence (AI) · Conventional farming · Internet of Things (IoT) · Sustainable agriculture, Smart technologies, Technological advancements

2.1 Introduction

The environmental effects of traditional agricultural operations have been a source of increasing concern in recent years. The overuse of water supplies, chemical fertilisers, and pesticides has done significant damage to soil quality, biodiversity, and the whole ecosystem. Accordingly, there emerged an increasing need for sustainable agricultural methods that guarantee effective crop production without compromising environmental quality. Artificial intelligence (AI) is gaining popularity in the agriculture sector as it is deemed to support sustainable farming methods. AI is the creation of computer systems, capable of activities like learning, problem-solving, and decision-making that normally require human intellect. AI in agriculture refers to the application of sophisticated algorithms and data analysis methods that enhance farm operations and facilitate well-informed decision-making.

An increasingly vital element of sustainable agriculture is smart farming. It is the use of techniques that satisfy the need for food production guaranteeing long-term ecological balance (Wu et al. 2010). Although farming has been associated with manual work for generations, it is physically demanding and takes a lot of time. Thanks to technological breakthroughs and AI, farmers are now capable of increasing their productivity while reducing its environmental impact. AI in sustainable agriculture has several advantages from reducing resource use to raising yield potential. Farmers become capable of precisely monitoring their 'crop health' by utilising the data-driven insights of AI-powered equipment and sensors. This allows them to avoid the blanket-spraying of chemicals into the whole field by modifying their irrigation schedules or applying targeted treatments just when it is necessary.

Sustainability includes all those methods and approaches that are meant to preserve and protect our planet. Sustainable development is the developmental paradigm that echoes the principle of sustainability. This strategy is intended to address the needs of both current and future generations. Sustainable development is formally defined as a 'development that meets the needs of the present without compromising the ability of future generations to meet their own needs' (UN World Commission on Environment and Development 1987). The developmental paradigm was particularly focused on the satisfaction of the needs of the present generation without any consideration toward environmental sustainability. However, the sustainable development approach taking into account the exploitation of the environment insisted on transforming the developmental policies with environmental concerns.

'Sustainable agriculture' refers to an integrated system of plant and animal production techniques with a site-specific application that will, over time, fulfill human demands for food and fibre, improve the state of the environment and the foundation of natural resources that support the agricultural sector, and utilise non-renewable resources and agricultural resources as efficiently as possible. Besides, it includes natural biological cycles and controls, maintains the farm operations' financial sustainability and improves the standard of living for farmers and the community at large (USDA, Sustainable Agriculture n.d.).

Compared to sustainable agriculture, there is no widely accepted definition of AI, but it can be defined as any software technology that possesses at least one of the following capabilities: perception (including audio, visual, textual, and tactile—e.g. face recognition); decision-making (such as medical diagnosis systems); prediction (like weather forecasts); automatic knowledge extraction and pattern recognition from data (for instance, the identification of fake news circles in social media); interactive communication (e.g. social robots or chatbots); and logical reasoning (such as theory development from premises) (Vinuesa et al. 2020).

The unparalleled progress in AI has induced the agricultural sector to make use of its benefits by incorporating its various applications. Integrating AI into sustainable farming has several benefits. The ability of farmers to make accurate judgements about water use, fertiliser application, and pest control is made possible by artificial intelligence algorithms that evaluate massive quantities of data including weather patterns, soil conditions, and crop health. Reduced resource use, less chemical use, and more overall efficiency are the results of this improvement.

This chapter therefore analyses the use of artificial intelligence in achieving and advancing the goal of sustainable agriculture by examining its historical background, application, advantages and current trends. Besides, the intersection between SDGs, AI and agriculture and the role of government in supporting and promoting the use of artificial intelligence in advancing sustainable agriculture will also be discussed.

2.2 Evolution of AI in Agriculture

In 1985, McKinion and Lemmon made the first attempt to use AI in agriculture when they developed GOSSYM (a dynamic cotton (Gossypium hirsutum L.)), a model for simulating cotton crops by using an expert system to improve cotton output taking into account climate, irrigation, fertiliser, weed control, and cultivation parameters (Zha 2020). The evolution of the use of AI in agriculture can be classified into four phases where the technological advancements take a new and upgraded shape at every phase:

(a) Phase I (between 1950 and 1980s)
(b) Phase II (between 1990 and 2000s)
(c) Phase III (between 2000 and 2017)
(d) Phase IV (2017 onwards)

Firstly, Phase I is the early stage, i.e. between 1950 and 1980s. This period recognised the first attempts to use AI in farming. Researchers investigated the fundamental uses of technological advances in the 1950s and 1960s, including rule-based systems for agricultural production and managerial decision-making. Expert systems were employed later in the 1970s and 1980s to help farmers with crop selection, pest management, and soil management choices. The highly advanced computers may be fed with data interpreted by professionals in agriculture and agronomists (Rani et al. 2011). The data collection and analysis helped the farmers to improve their farming outputs and the farming machinery reduced their physical labour. This advancement led to what was widely called 'precision agriculture/precision farming' proposed by Pierre Robert in the early 1980s (Nelson 2023).

Secondly, Phase II is the stage of remote sensing and precision agriculture, i.e. between 1990s and 2000s. Drones and imagery from satellites are examples of remote sensing technologies that were developed in the 1990s to monitor crop health, soil moisture, and yield output. Precision farming has become a prominent field in agriculture maximising farmers' inputs like planting, weeding, spraying, and so forth. Algorithms and machine learning have been used to analyse massive datasets produced by sensor equipment and drones. Accurate weather and agricultural yield predictions were made possible by these systems (Karunathilake et al. 2023).

Thirdly, Phase III relates to the emergence of robotics and automated machines, i.e. from 2000 to 2017. The use of AI-driven robotics for field activities such as planting, harvesting and other tasks is closely linked to this stage. AI methods were employed by agricultural robots to identify crops such as computer vision and to enhance performance such as machine learning. Drone use and autonomous cars have started to be used for spraying, mapping, and agricultural monitoring, but they became popular to farmers in the next phase only (Wakchaure et al. 2023).

Lastly, Phase IV deals with the evolution of smart farming and the Internet of Things (IoT) in the contemporary scenario. This phase is also called agriculture 4.0 as it incorporates and propagates the technological advancements of Industry 4.0 into the agricultural sector (Liu et al. 2021). Using this method, real-time data in agriculture can be collected and exchanged for the analysis of insect outbreaks, weather patterns, crop conditions, and other relevant aspects that influence production. Predictive analysis is being used to assist farmers in scheduling planting and irrigation and managing pest control methods. Furthermore, in the present scenario, sustainable farming methods are being promoted by AI technology. For instance, by maximising resource efficiency, AI systems may limit chemical consumption, improve crop rotations, and reduce their negative effects on the environment (Ksolves 2023).

All these four phases show the gradual and systematic growth of the incorporation and use of AI in agriculture affirming that the industrial and technological advancements have contributed to the growth of agriculture by empowering it with smartness and productivity.

2.3 Application of AI in Sustainable Agriculture

Artificial intelligence technology provides creative ways to maximise resource use and minimise the negative effects of farming on the environment by reducing waste. With the help of artificial intelligence integration, sustainable methods like smart irrigation systems and precision agriculture can be employed to improve agricultural output and safeguard the environment. There are several potential applications of AI in sustainable agriculture that can be classified into the following:

2.3.1 Monitoring Crop and Soil Health

This application helps to build agricultural metrics over hundreds of acres of cultivable land in an efficient manner; remote sensing (RS) methods, hyperspectral photography, and 3D laser scanning are essential. From the standpoints of time and effort, it might bring about a dramatic change in the way farmers monitor their farmlands (Dharmaraj and Vijayanand 2018). Using satellite imaging and remote sensing is one of the main ways of using artificial intelligence in crop monitoring. Satellite imaging is the collection of the earth's image making use of the imaging satellites. It collects the data in the form of electromagnetic radiation and it is transmitted to the ground stations through radio waves. Remote sensing is the process of collecting the data without physical contact or direct involvement. Remote sensing functions in the agriculture sector by making use of satellites, aeroplanes, or drones to gather data about crops including the details on their health, growth, and yield. Artificial intelligence (AI) systems can offer important insights regarding crop status by evaluating the spectral information acquired by these references.

Large agricultural areas can be seen in a wider perspective through satellite imaging that helps the experts to make a thorough analysis of the field and the crops enabling farmers to keep an overall check on their productivity. By using AI algorithms to evaluate the abovementioned data, the scientists equip the farmers with adequate knowledge to discover areas that need attention such as insect infestations or nutrient deficits and to realise the current trends and anomalies in the field. Furthermore, it also helps them to predict yields and monitor crop health (Mana et al. 2024). In short, remote sensing and satellite imageries are utilised by the experts in the field to empower farmers with adequate information converting their conventional farming methods into smart farming techniques for increasing their productivity and ensuring food security.

2.3.2 Weather Prediction

Farmers fail to make the best decisions about soil preparation, seeding, and harvesting due to unpredictable climate change. Many times, unrealistic weather forecasts and unpredictable climate changes have caused severe damage to the farmers. But with the use of AI-supported weather forecasting that provides appropriate information and analysis on weather, farmers will be able to prepare the sort of crop to produce, the seeds to sow, and the process to harvest the crop. Besides, price forecasting can help farmers maximise their profits by giving them a clearer estimate of the price of crops over the following several weeks (Patel and Bala 2022). Large volumes of unstructured and heterogeneous data shall be analysed using AI approaches that can then be utilised to identify and take advantage of the whole agricultural production. These AI approaches are necessary to make sense of the ever-increasing volume of data and to meet the demanding new requirements in weather forecast (WF), climate monitoring (CM), and decadal prediction (DP). Using AI approaches can result in less human development effort, more effective use of computational resources, and higher-quality forecasts all at once (Dewitte et al. 2021). Real-time weather and climate data shall be gathered with the use of smart sensors that are linked to the Internet of Things. A comprehensive prediction helps farmers assess their crop needs more effectively. There needs to exist certain systems that can also provide notifications to farmers to enable them to protect their crops in the case of extreme weather (AlZubi and Galyna 2023).

2.3.3 Intelligent Spraying

Using sprayers, agricultural spraying applies insecticides, fungicides, and herbicides to field crops for crop protection. Using blanket spraying, conventional sprayers evenly cover the whole field. However, variables like weather, soil conditions, and variations in pest pressure within a field are not taken into consideration by uniform treatment. With the use of sensors, data, and variable rate systems, smart spraying technology seeks to optimise spraying to maximise efficiency and minimise environmental effects (Saiwa 2024). Precision spraying, another name for smart spraying, uses real-time data to more precisely target spraying. Sensors keep an eye on things like soil moisture, crop canopy height, wind speed, air temperature, and the presence of pests or diseases. Control systems that adjust nozzle outputs and spray patterns use this data as input. Global positioning system (GPS) mapping makes it possible to document the locations of sprayed areas. Generally, as opposed to uniform blanket spraying, smart sprayers only administer chemicals in a required place at a stipulated time and quantity (Javaid et al. 2022).

The use of smart spraying technology in agriculture is being fuelled by several causes. Demand is being driven by issues about environmental pollution from agricultural chemical runoff and the increased use of precision agriculture. Stricter laws governing the use of chemicals and the safety of operators encourage application optimisation. In addition, smart sprayers aid in combating weeds' growing tolerance to herbicides. The cost of parts like variable rate pumps, sensors, and controllers has also dropped, increasing farmers' return on investment (Saiwa 2024).

2.3.4 Pest Management

AI detection systems use cameras and sensors that record information about movement, heat, and sound to automatically identify and monitor pests. By identifying pests and comparing this data to large databases using machine learning algorithms, AI can suggest treatment strategies (Field Routes n.d.). AI's use in pest control is in line with the growing need for environment-friendly agricultural methods. One of the severe threats to this method is the uncontrollable use of pesticides. AI contributes to the development of environment-friendly agriculture by reducing reliance on chemical pesticides. This shift towards sustainable pest avoidance benefits the long-term profitability of farming operations in addition to the advantageous improvements to the environment. AI-powered insect avoidance techniques can also assist farmers in meeting the regulatory requirements related to sustainable agriculture and obtain necessary certifications.

The adoption of integrated pest management (IPM) techniques, which stress the use of several pest control strategies in a coordinated and sustainable way, can be made possible by AI technology. Through the integration of AI into IPM programmes, farmers may optimise the efficacy of non-chemical control strategies such as cultural practices or biological control agents (Deguine et al. 2021). By reducing the need for pesticides and maintaining controllable pest populations, this all-encompassing approach promotes ecological balance and resilience in agricultural settings.

2.3.5 Yield Prediction and Enhancement

Yield mapping, a vital component of the historically unprecedented advancements in precision farming frameworks, gives farmers the ability to see spatial variation throughout the field. This technique is useful for planning future actions and managing the results of previous mappings. The majority of its references are related to the process of obtaining georeferenced data on harvest yield and quality (Talaviya et al. 2020). It contributes to an accurate crop production prediction model with the help of which farmers can make decisions concerning the choice of crop, time of its planting and growing, etc. (Van Klompenburg et al. 2020).

Crop yield prediction is the technique of estimating how much agricultural output, such as crops, will be collected in a particular region for a set amount of time using data and technology. Development of informed predictions about the upcoming crop depends on a variety of variables including past data, weather patterns, soil quality, and agricultural techniques. It facilitates the making of well-informed decisions of the farmers, legislators, and other stakeholders in the food supply chain about planting, resource allocation, and food security. It is similar to projecting crop output in the future to guarantee effective and sustainable agriculture. Predicting crop yield is an intricate task that aims at predicting how much food will be gathered in a given area during a certain growing season. Technological developments, data-driven analysis, and the integration of several crop growth and productivity-influencing elements work together to create this process. It is the art and science of foreseeing nature's abundance to improve food security, allocate resources as efficiently as possible, and make educated judgements (Mundhe et al. 2023).

2.3.6 Aerial Survey and Imaging Using Drones

In a mechanical context, remotely operated unmanned aircraft are known as automatons, unmanned aeronautical vehicles (UAVs), or unmanned aircraft systems (UAS). They collaborate with the GPS and other installed sensors. In the field of agriculture, drones are being used for disaster management, weed identification, crop health monitoring, livestock and animal monitoring, and irrigation equipment monitoring. Agriculture is being greatly impacted by remote sensing, which uses UAVs to capture, process, and analyse images (Talaviya et al. 2020). Drones enable tech-savvy farmers to make well-informed judgements about a sustainable and profitable agricultural environment by collecting copious amounts of data. Drones fill huge field blocks without compaction of the soil or structural damage in contrast to typical manned power sprayers. This characteristic is essential for maintaining the integrity of the soil. Furthermore, drones are incredibly versatile when it comes to traversing intricate and small field plots. To put things in perspective, over 80% of Indian farmers farm less than five acres (Vislavath 2024). Hence, the incorporation and usage of drone technology in agriculture would be beneficial to India's varied and rolling terrain and also to small-scale farmers.

2.3.7 Produce Grading and Sorting

After the crops are harvested, farmers can still benefit from AI computer vision. Imaging algorithms may be used not only to identify pests, diseases, and faults in plants throughout their growth but also help to separate the 'good' products from the bad ones. Computer vision can automate the sorting and grading process with an accuracy rate and speed far greater than even a trained expert in the field by examining the size, shape, colour, and volume of fruits and vegetables (Rizzoli 2021). It enhances the production and supply chain helping the farmers to reach their customers without much labour and delay.

2.3.8 Robotics and Sustainable Agriculture

In agriculture, robotics is mostly used to expedite monotonous and repetitive processes such as spraying, mowing, sowing, harvesting, controlling weeds, picking, and ultimately sporting goods and packing. Fresh produce supply chains have found a variety of usages for the automation of advanced robots, cloud computing, blockchain, and big data. Field robots are utilised in precision farming to manage weeds helping to reduce the need for hazardous and overuse of chemical sprays on crops and soil. This progressive approach contributes to sustainable agricultural practices (Bhagat et al. 2022).

2.4 Advantages of AI in Sustainable Agriculture

Traditionally, physical labour has been a key component of farming that also helped to ensure long-term ecological stability. With the advancement of science and technology, many new machines, pesticides, and fertilisers including harmful chemicals like endosulfan and ammonium nitrate started to occupy the agrarian sector leading to more production. However, it also had an adversarial impact on environmental stability. However, AI has brought about a paradigm shift leading to more efficiency and less environmental effect. AI has a wide range of applications in sustainable agriculture, including improved production and resource efficiency. AI-powered solutions minimise the need for indiscriminate pesticide use by providing accurate crop health monitoring and focused actions. Some of the significant advantages of the use of AI for sustainable farming can be classified as follows:

2.4.1 Enhancement of Yield Efficiency

The World Economic Forum reports that AI-based sustainable farming led to an 8% rise in sales prices. During an 18-month trial experiment period that included three consecutive planting cycles, the application of AI techniques in farming produced outstanding results. Plant growth per acre increased by 21%, while farmers also saw a 9% and 5% decrease in pesticide and fertiliser use, respectively (Janakiram 2024).

2.4.2 Minimisation of Environmental Impact

AI-powered systems assist in better decision-making, pinpoint areas of improvement, and suggest routes of transportation with the help of precise and real-time accessible data. Consequently, there shall be a large reduction in expenses, a productivity management, and a decrease in carbon emissions (Hasani 2023). The three SDGs (SDGs 13, 14, and 15) that are concerned with combating climate change, life below the ocean, and life on land will benefit from this sustainable farming. AI will help low-carbon energy systems that integrate renewable energy sources and energy efficiency highly, all of which are necessary to combat climate change. AI has the potential to enhance ecological health as well. AI can help reach SDG target 14.1, which calls for the prevention and drastic reduction of all forms of maritime pollution, by using algorithms to automatically identify potential oil spills. SDG target 15.3, which asks for halting desertification and repairing degraded land and soil, is another advantageous illustration (Vinuesa et al. 2020).

2.4.2.1 Reduction of Carbon Footprint

AI makes it possible to analyse energy flows in real time and forecast supply and demand, which helps to increase the energy system's overall efficiency and lower the power mix's carbon intensity by reducing the carbon footprint (Climate Seed 2024). Rising atmospheric concentrations of carbon dioxide (CO2) pose a threat to the world's climate, biodiversity, and resource availability. It is imperative to keep an eye on the variables of various AI techniques since they have an impact on the effectiveness of the system. Rising atmospheric concentrations of carbon dioxide (CO2) pose a threat to the world's climate, biodiversity, and resource availability (Nunes 2023).

2.4.3 Minimisation of Efforts

The robotic application of AI is replacing human labour in the agricultural sector by boosting its use and dislodging occasions. For example, in certain higher-tech countries such as Japan, robots are routinely used in agribusinesses to replace human resources. However, this is not always the case because technology also generates new employment prospects for people by substituting them to boost productivity and decrease the need for human labour (Duggal 2024).

2.4.4 Time Efficiency

Time is money in the fast-paced environment of today. Comprehending the significance of efficiency and cost-effectiveness is essential for the present generation of developmental policies. AI is a potent tool that can assist you in achieving both of these targets. Through process optimisation, strategy optimisation, and eventually time and money savings, AI has completely transformed several sectors, including marketing and agriculture. Up to 40% of working hours are predicted to be enhanced or automated by generative AI. Agriculture will benefit from this time reduction, among other sectors (McKendrick 2023).

2.4.5 Profitability

In the agri-food industry, AI has the potential to boost production, reduce waste, and promote sustainability, thereby leading to profitability (Ahuja and Mehra 2023). By

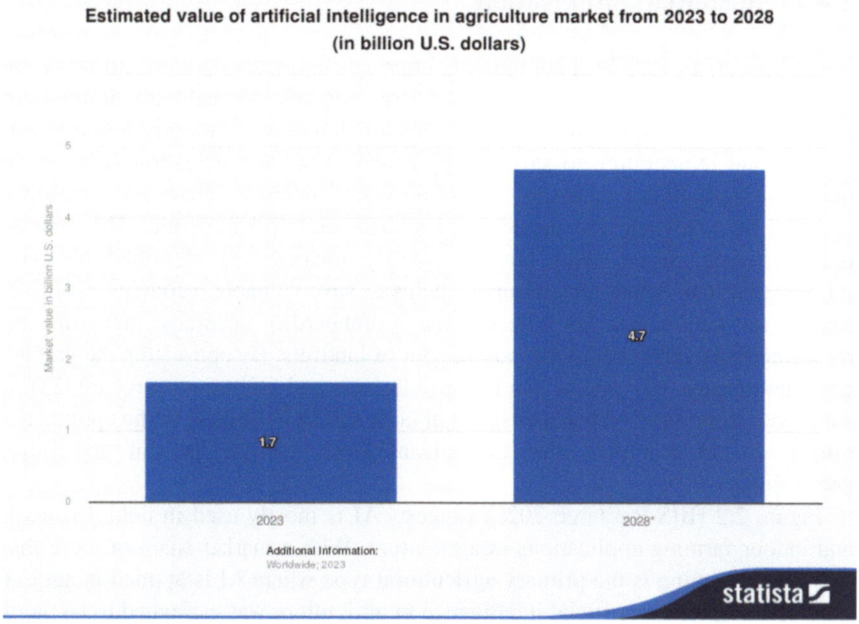

Fig. 2.1 Estimated value of artificial intelligence in agriculture market from 2023 to 2028. (Source: Statista; Future Farming 2023)

using AI, manufacturers like agricultural and industrial sectors may improve resource allocation, simplify operations, and reduce waste by projecting future market trends, consumer behaviour, and operational demands. It helps matching goods and services to consumer needs and reduce expenses while simultaneously increasing profitability and improving sustainability (CloudIQ n.d.).

The *x*-axis in Fig. 2.1 represents the years from 2022 to 2028 and the *y*-axis represents the value of US dollars ($) in billion. According to the estimates, artificial intelligence will be worth around 1.7 billion US dollars in the global agriculture sector in 2023, and by 2028, that amount is expected to rise to 4.7 billion.

2.4.6 Sustainability

AI is emerging as a disruptive factor in environmental biotechnology at a time when environmental concerns are growing. The innovative fusion of ecology and technology is driving the development of environmentally sound management practices (Medium 2024). AI in agriculture has a lot to offer to the environment, including fewer carbon emissions and resource use through precision farming. Additionally, it helps maintain the health of the soil, which is essential for profitable agriculture (Amini n.d.).

2.4.7 Resource Optimisation

Smart networks with AI capabilities can improve energy distribution, cut waste and encourage the integration of renewable energy sources. AI-enabled building management systems can identify trends in tenant behaviour and modify lighting, ventilation, and temperature to save energy usage. AI-driven irrigation systems can maximise agricultural yields by delivering water precisely where and when it is required by analysing the meteorological data and soil moisture. This process reduces waste. Water leaks may be swiftly located and identified by using AI-powered leak detection systems. It helps to save valuable resources. AI-driven robots can sort recyclables with previously unheard-of accuracy, increasing the recycling rates and keeping the garbage out of landfills. By optimising the garbage collection routes, AI systems can lower pollutants and fuel usage (Sroka 2023). In short, the adoption of AI based systems in farming can contribute to the optimal and proper use of resources that are advantageous for the present and future generations.

Figure 2.2 (BIS Research 2020) suggests AI is mostly used in field, livestock, and indoor farming applications in agriculture. With a market share of more than 60%, field farming is the primary agricultural type where AI is applied in agriculture. The market for artificial intelligence in agriculture was estimated to be worth almost 1.1 billion dollars in 2019 and is expected to reach a value of over 3.8 billion dollars by 2024 (Statista 2023).

2.5 SDGs, Sustainable Agriculture, and AI

An ever-widening array of industries are being shaped by artificial intelligence (AI). For example, both short- and long-term effects of AI are anticipated in the fields of global productivity, equality and inclusion, environmental results, and several other areas (Vinuesa et al. 2020). A thorough integration of food and agriculture-related issues is seen in all 17 SDGs and their 169 objectives. SDG 1 aims to end poverty, while SDG 2 is committed to eradicating hunger, enhancing food security and nutrition, and advancing sustainable agriculture and contains objectives about social protection, land rights, and resilience. Various SDGs refer to the relationship between food security and natural sources like SDG 14 on oceans and marine resources; SDG 15 on ecosystems, biodiversity, forests, and land; and other SDGs on water, energy, gender, climate change, and consumption and production (FAO n.d.). SDGs do not merely stand for the protection of natural resources, but they also want to ensure food security and other concerns of the present generation. However, they try to secure the aspirations of the present generation without harming the natural resources for the future generation (Fig. 2.3).

The United Nations set the basic goals for sustainable development (SDGs) in 2015, which can be achieved with the help of AI. The SDGs seek to safeguard everyone's prosperity, eradicate poverty, and save the environment. AI can assist in achieving these objectives by fostering innovation across several industries, cutting

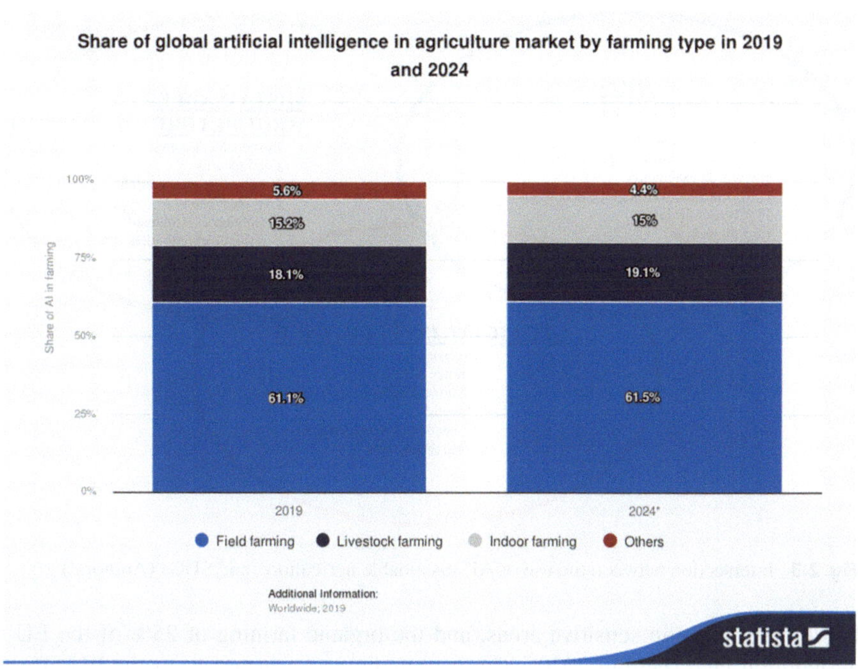

Fig. 2.2 Share of global artificial intelligence in agriculture market by farming type in 2019 and 2024. (Source: BIS Research 2020)

waste, and increasing efficiency. For instance, AI can aid global and national initiatives to lower poverty, foster economic growth, and enhance access to healthcare and education. It can also help achieve environmental objectives including biodiversity preservation, greenhouse gas emission reduction, and the promotion of sustainable resource use. The efficient use of AI can contribute to the effective realisation of SDGs. Organisations must use caution, nevertheless, while implementing AI, so as not to adversely affect sustainable development policies and programmes. This entails safeguarding security and privacy, expanding advantages, and assuring the moral and responsible use of AI (Ghengini 2023). A few goals that are at least impacted by agriculture are goal 6 (clean water and sanitation), which is due to the sustainable management of water for irrigation and water conservation; goal number 12 (responsible consumption and production), which is to guarantee sustainable production patterns on arable land; and goal number 13 (climate action), which is impacted by agriculture because it has the potential to be a contributor to the solution rather than a victim. Goal 2 (Zero Hunger) is also particularly relevant in this arena as it directly supports sustainable agriculture, while Goal 15 (Life on Land) combats land degradation and biodiversity loss (Zürner et al. 2023). Lately, policies and declarations have been put out or imposed to promote the sustainable growth of agriculture. One such attempt is the European Union's 2020-approved European Green Deal, which outlines objectives for 2030. These include a 55% decrease in net greenhouse gas emissions from 1990 levels, a 50% reduction in pesticide use, a

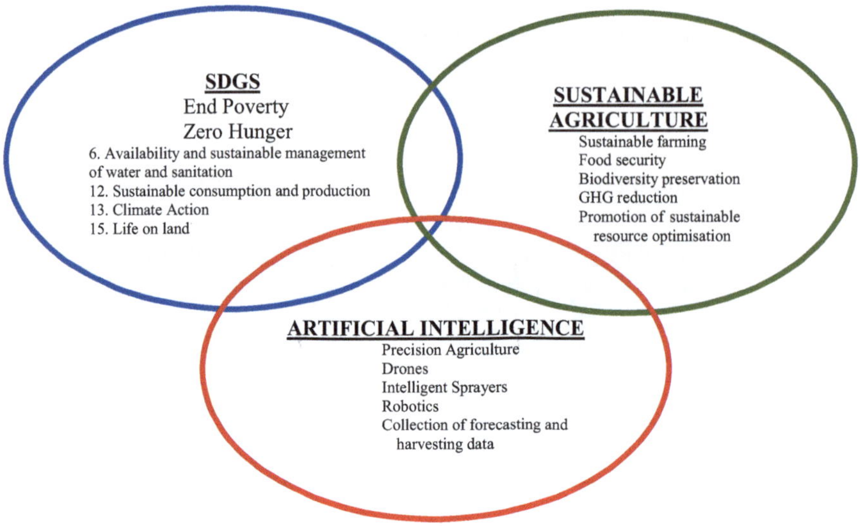

Fig. 2.3 Intersection between the use of AI, sustainable agriculture, and SDGs (Authors')

ban on pesticides in sensitive areas, and the organic farming of 25% of the EU's agricultural land (Fetting 2020). The various proactive measures of the EU affirm that there needs to be meaningful global, regional, and national policy design for the realisation of SDGs and AI-integrated sustainable farming.

As per the figures of the rising population, the demand for food will be increasing heavily in the coming years. Sustainable intensification (SI) is a crucial strategy for meeting food demands while reducing its adverse environmental consequences. SI does this crucial task by increasing the production of crops per area while maintaining or even reducing its adverse environmental effects (Zürner et al. 2023). SI does not address how to achieve that goal, but several alternative agricultural strategies do.

2.6 Challenges Involved

For many farmers, implementing AI in agriculture is a major barrier albeit having many advantages. The main cause is that it necessitates huge expenditures for infrastructure, technology, and training. Although AI has numerous advantages for agriculture, farmers find it challenging to completely adopt these new technologies for a variety of reasons.

2.6.1 Collection of Data

Although precision agriculture has grown over the last three decades, two major issues have prevented machine learning predictions from being widely used in the field. *Firstly*, a sizable percentage of farmers and producers have not embraced precision-Agriculture's utilities (there are several studies and explanations for this). *Secondly*, utilising precision-Agriculture technology does not include gathering all relevant data via IoT or other channels to make predictions (Mirali 2024). Since the collection of data is inevitable for smart farming, the shade of these issues acts as a barrier to its proper implementation.

2.6.2 Ownership of Data

Who is the Ag-sector data owner? This question has a more complicated answer than one could think. It is true that producers—farmers, greenhouse managers, breeders, agronomists, etc.—own and have access to data, but this is not the full story. Owners have access to certain vital data, but unless they pay for access and get training, they don't have instant access to many other sources of pertinent data, such as satellite imaging or historical environmental data of the region. Data scientists and engineers, who are qualified to collect and combine data from many sources, might effectively fill this gap, especially concerning farmers' lack of access to environmental and satellite data (Mirali 2024). These scientists and technical experts are mostly workers of multinational or transnational companies (MNCs/TNCs), and hence, this data is ultimately owned by them and the farmers will have to depend on them for the availability of data. The governments have to ensure that the necessary data is available to the farmers with less cost or no cost for the effectiveness of smart farming. Exploiting the uneducated farmers taking advantage of ownership of data should not be supported.

2.6.3 Data Privacy

Another conflating problem with the sharing of farming data is data privacy. Large-scale AI applications in agriculture can handle intellectual property concerns via aggregating, encrypting, and anonymising data. When it comes to the intellectual property of farming data, there are suitable non-disclosure agreements (NDAs) and AI approaches that may be used to analyse the data and derive valuable insights without including disputes over data granularity or geographical aspects (Mirali 2024). Agriculture-related data sources can be called big data on account of their volume, velocity, and variety. Accordingly, the data privacy concerns of the farmers are high as a clear and detailed exposition of the responsible use of the collected data is not given to them. Data security and data privacy are sometimes interchangeably used. But data security stands for the protection of data from adversary attacks, whereas data privacy calls for the responsible use of data respecting the desire of the users and protecting data from any unauthorised access. Data privacy governs various elements such as the collection, analysis, storage, and access of data. Both data security and data privacy concerns are very important for the effectiveness of smart farming (Amiri-Zarandi et al. 2022).

2.6.4 Significant Upfront Expenses

Even if AI outcomes can be reasonably priced in the medium to long run, the initial investment may need to be quite valuable. Adopting AI into farming may be questionable for the time being given the financial struggles of many agribusinesses and farms, particularly small-scale producers and those in developing nations. However, as technology advances, the cost of implementing AI in agriculture could decrease. Companies might also look into financing options including grants from the government or private investment (MATR 2023). Increasing productivity with less impact on the environment should be the objective of governmental policies, and to achieve this goal, they should support the farmers to incorporate the latest technological advancements into their farming practices.

2.6.5 Insufficient Hands-On Experience with Emerging Technologies

Globally, there are differences in the level of technical improvement in several aspects of agriculture. While certain areas might fully benefit from AI, there are several obstacles in nations where modern agriculture equipment is rare. Technology businesses may need to adopt a visionary strategy if they hope to conduct business in areas where agriculture is becoming more and more prevalent. Apart from providing their products, companies also need to provide farmers and agribusiness owners who are prepared to take on novel outcomes with training and continuous assistance (MATR 2023).

2.6.6 Restricted Access for Farmers on a Small Scale

Although the incorporation of AI into the field of agriculture has the potential to completely transform the sector, there are some disadvantages as well. Small-scale farmers have restricted access to such kinds of technological advancements, which is one of its main drawbacks. The cost of implementing and maintaining AI technology might make it unaffordable for many small-scale farmers who lack the capital to invest in it. Small-scale farmers are left behind, while large-scale farmers profit from AI, creating a digital divide between the two groups of farmers. This has the potential to worsen already-existing disparities in the agriculture industry and impede the development of sustainable agriculture. Finding solutions to increase small-scale farmers' access to and affordability of AI technology is crucial (Ali 2023).

2.6.7 Excessive Dependence on AI Predictions

One of the possible drawbacks of artificial intelligence in agriculture is an overreliance on its forecasts. Although AI may provide farmers with insightful analysis and helpful suggestions, it is crucial to keep in mind that technology is flawed and subject to error. Farmers who depend too much on AI forecasts can miss crucial elements like soil conditions or weather trends that could have an impact on their crops. Furthermore, biases or inaccuracies might be introduced into AI systems, resulting in suggestions or forecasts that are not correct. Consequently, rather than depending only on AI, farmers should utilise it as a tool to augment their expertise and understanding. Farmers can make better decisions by integrating AI information with their own experience (Ali 2023). AI should not be considered a substitute for human intellect, skills, and expertise, but it should be utilised only as a tool.

2.7 Potential Future Developments

Collaboration and information exchange between farmers, academics, and technology developers are vital due to the fast advancement of developing technologies in artificial intelligence farming. The integration of these technologies and their potential to address the world's food problem, which concerns sustainability, can be hastened and optimised by establishing partnerships and organising forums for communication. A design of collaboration and partnership is an effective way to promote smart agriculture among farmers. The appropriate governments should coordinate this collaborative design for the true welfare of the country and its economic growth and food security.

2.7.1 Drones and Robots

Among the cutting-edge newly developed technology in AI farming are drones and robotics. They simplify operations, save labour costs, and increase efficiency by performing precise activities including aerial imaging, planting, spraying, and harvesting. They also have the advantage of covering large agricultural areas within less time, augmenting their productivity. With unparalleled accuracy, farmers can now monitor crops, evaluate plant health, and carry out targeted treatments thanks to these technologies (Javaid et al. 2022).

2.7.2 Blockchain Technology

A further advancement in artificial intelligence farming is blockchain technology. Blockchain improves food safety, supply chain management, and traceability by generating an unchangeable, visible record. The provenance, quality, and travel of agricultural goods may be seen in real time by farmers and customers, fostering sustainability and confidence along the whole value chain (Rejeb et al. 2020).

2.7.3 Use of Biotechnology

Biotechnology and genetic editing are expanding the possibilities for disease resistance and agricultural enhancement. Scientists can improve plant characteristics, boost yields, and create robust crops by using precise gene editing techniques like clustered regularly interspaced short palindromic repeats (CRISPR). It can help to identify and characterise the genes that are responsible for diseases in the crops. By enabling the creation of disease-resistant cultivars, biotechnology lessens the need for chemical pesticides and promotes environmentally friendly agricultural methods (Zaidi et al. 2020). All these can also contribute to increasing the quality of crops.

2.7.4 Precision Agriculture

Data-driven decision-making in agriculture is being driven by algorithms from machine learning and AI. Through the analysis of large datasets gathered from sensors, satellites, and other sources, these technologies can offer important insights for improving pest control, fertiliser application, and irrigation. Real-time data allows farmers to make well-informed decisions, which boosts output and optimises resource use (Tantalaki et al. 2019).

2.7.5 Vertical Farming

The ideas of traditional agriculture are being redefined by hydroponics and vertical farming. It is an advanced technique of farming in which crops are cultivated in vertically stacked layers designed for sustainable food production. These techniques boost food security, lower transportation costs, and allow for year-round cultivation of crops in regulated conditions with little water and land requirements. AI is integrated into vertical farming systems to monitor and adjust environmental variables for maximum growth and output (Oh and Lu 2023).

2.8 Role of Government in Advancing Sustainable Agriculture

Sustainable initiatives are a crucial component of corporations' current strategy for attaining sustainable development, since they attempt to promote environmental, social, and economic development (Barbosa 2024). The integrity of agri-food systems is adversely affected by unsustainable agricultural practices such as burning residue, deforestation, the use of particular fertilisers or pesticides, utilising a basic crop structure, unsustainable water management, and the use of fossil fuels. Their adversarial impacts include greenhouse gas emissions, soil erosion, soil deterioration and desertification, water scarcity, loss of ecosystem services that agricultural systems rely on, and a decrease in the amount of land that can be farmed (Ibrahim and Johansson 2021). Governmental organisations should possess the comprehensive vision necessary to identify the requirements and strategies for putting sustainable agriculture techniques into effect. Government assistance has been crucial in addressing environmental issues since it is believed that governments should strive to achieve economic growth without harming social and environmental sustainability (Zhou et al. 2021). Both in developed and developing economies, the government has imposed varying degrees and goals of support for sustainable agriculture. The Common Agricultural Policy (CAP) in Europe aims to lessen the negative impacts of unsustainable farming practices. Subsidies from the European Union (EU) have been granted with consideration for social and sustainable issues such as the preservation of agricultural jobs (Blasi et al. 2023). Over the past few decades, China has instituted regulations and subsidies aimed at safeguarding the financial interests of farmers while simultaneously increasing the output and calibre of agri-food goods. Their current objective is to improve agricultural environmental resource protection. However, developing nations have been critical of rich nations' agricultural subsidies, claiming that these subsidies have contributed to low prices and continued poverty in developing nations. Large emerging economies are currently seeing government support for agricultural production catch up to that of established nations (Borsellino et al. 2020). The policy initiatives of the governments towards sustainable farming, irrespective of their economic status on the global level, are very much essential for its growth. The absence or lack of governmental support is one of the stringent barriers to its application and promotion.

2.9 Conclusion and Suggestions

The use of AI in agriculture is revolutionising sustainable agricultural practices. Its capacity to handle enormous datasets reduces environmental impact by enabling educated decisions to be made about insect management, fertilisation, and irrigation. AI plays a critical role in promoting sustainable agricultural methods through its contributions to real-time crop monitoring and precision agriculture. Accepting AI is a critical step towards a sustainable agricultural future as well as technical progress. The possibilities for integrating AI into sustainable farming are staggering. AI algorithms will grow more accurate and productive as technology develops, allowing the present generation to address the challenging environmental issues. AI integration empowers the farmers to properly manage their land, safeguard natural resources, and create a more resilient and sustainable agricultural future. The rapid advancement of AI must be accompanied by the regulatory understanding and supervision that AI-based technologies require to facilitate sustainable growth. If this isn't done, there may be gaps in ethical norms, safety, and openness.

Using AI in agriculture could potentially solve some of the most serious food-related issues that the globe is facing in the present times. AI can help optimise land usage, enhance crop resilience, and raise food production in light of the world's expanding population and a finite amount of arable land. Farmers can better forecast and manage the effects of pests, illnesses, and climate change by utilising predictive models and sophisticated analytics. This will ensure a more sustainable and safe food supply for coming generations. Interdisciplinary cooperation is also essential for the future of AI farming. To share best practices, create novel solutions, and exchange knowledge, there needs to be effective collaboration between scientists, engineers, farmers, technologists, and policymakers. Collaboration and the use of a range of skills might help the farmers to hasten the adoption and integration of AI technology, resulting in more effective, durable, and environmentally friendly agricultural systems. Establishing a suitable legal and legislative framework is the first step in this approach. These strategies will help focus AI's enormous potential on achieving the SDGs and providing the most possible value to both the people and the environment, besides addressing the needs of the present and future generations.

Regulatory insight, in which decision-makers have enough knowledge about AI's problems to be able to create sensible policy, ought to come before regulatory oversight. Gaining such knowledge is even more crucial than monitoring as poorly thought-out policy will probably be at best ineffectual and at worst harmful.

References

Ahuja S, Mehra P (2023) Sustainable artificial intelligence solutions for agricultural efficiency and carbon footprint reduction in India. In: Agricultural economics and agri-food business. Intechopen

Ali A (2023) Advantages and disadvantages of artificial intelligence in agriculture » Hubvela. https://hubvela.com/hub/technology/advantages-disadvantages-artificial-intelligence/agriculture/#google_vignette. Accessed 15 Apr 2024

AlZubi AA, Galyna K (2023) Artificial intelligence and internet of things for sustainable farming and smart agriculture. IEEE Access 11:78686

Amini (n.d.) The role of AI in sustainable farming practices: promoting environmental conservation. https://www.amini.ai/articles/the-role-of-ai-in-sustainable-farming-practices-promoting-environmental-conservation#:~:text=AI%20in%20farming%20offers%20significant,critical%20aspect%20of%20productive%20agriculture. Accessed 11 Apr 2024

Amiri-Zarandi M, Dara RA, Duncan E, Fraser EDG (2022) Big data privacy in smart farming: a review. Sustainability 14(15):9120

Barbosa MW (2024) Government support mechanisms for sustainable agriculture: a systematic literature review and future research agenda. Sustainability 16(5):2185

Bhagat PR, Naz F, Magda R (2022) Artificial intelligence solutions enabling sustainable agriculture: a bibliometric analysis. PLoS One 17(6):e0268989

BIS Research (2020) Share of global artificial intelligence in agriculture market by farming type in 2019 and 2024 [Graph]. In Statista. Retrieved August 05, 2025, from https://www.statista.com/statistics/1174399/global-ai-in-agriculture-market-by-farming-type/

Blasi E, Rossi ES, Zabala JÁ, Fosci L, Sorrentino A (2023) Are citizens willing to pay for the ecosystem services supported by common agricultural policy? A non-market valuation by choice experiment. Sci Total Environ 893:164783

Borsellino V, Schimmenti E, El Bilali H (2020) Agri-food markets towards sustainable patterns. Sustainability 12(6):2193

Climate Seed (2024) AI is an alley in reducing carbon footprint. https://climateseed.com/blog/reducing-carbon-footprint-with-artificial-intelligence#:~:text=Reducing%20emissions%20through%20intelligent%20trade%2Doffs&text=AI%20enables%20real%2Dtime%20analysis,intensity%20of%20the%20electricity%20mix. Accessed 12 Apr 2024

CloudIQ (n.d.) How can companies take advantage of AI to reduce costs, increase profits and enhance sustainability? https://www.cloudiqtech.com/how-can-companies-take-advantage-of-ai-to-reduce-costs-and-increase-profits-and-sustainability/#:~:text=By%20predicting%20future%20market%20trends,and%20services%20with%20market%20demands. Accessed 12 Apr 2024

Deguine JP, Aubertot JN, Flor RJ, Lescourret F, Wyckhuys KA, Ratnadass A (2021) Integrated pest management: good intentions, hard realities. A review. Agron Sustain Dev 41(3):38

Dewitte S, Cornelis JP, Müller R, Munteanu A (2021) Artificial intelligence revolutionises weather forecast, climate monitoring and decadal prediction. Remote Sens 13(16):3209

Dharmaraj V, Vijayanand C (2018) Artificial intelligence (AI) in agriculture. Int J Curr Microbiol App Sci 7(12):2122–2128

Duggal N (2024) Advantages and disadvantages of artificial intelligence. Simplilearn. https://www.simplilearn.com/advantages-and-disadvantages-of-artificial-intelligence-article. Accessed 12 Apr 2024

FAO (n.d.) Food and agriculture organization of the United Nations (no date). https://www.fao.org/evaluation/list/sdgs/en. Accessed 15 Apr 2024

Fetting C (2020) The European green deal. ESDN Report 2:53

Field Routes (n.d.) AI in pest control: a game changer for the industry. https://www.fieldroutes.com/blog/pest-control-ai#:~:text=AI%20detection%20systems%20automate%20pest,pests%20and%20recommend%20treatment%20plans. Accessed 11 Apr 2024

Future Farming (2023) Estimated value of artificial intelligence in agriculture market from 2023 to 2028 (in billion U.S. dollars) [Graph]. In Statista. Retrieved August 05, 2025, from https://www.statista.com/statistics/1326924/ai-in-agriculture-marketvalue-worldwide/

Gebresenbet G, Bosona T, Patterson D, Persson H, Fischer B, Mandaluniz N, Chirici G, Zacepins A, Komasilovs V, Pitulac T, Nasirahmadi A (2023) A concept for application of integrated digital technologies to enhance future smart agricultural systems. Smart Agric Technol 5:100255

Ghengini L (2023) 8 ways artificial intelligence can contribute to environmental conservation. 2030 builders. https://2030.builders/8-ways-ai-can-contribute-to-environmental-conservation/. Accessed 11 Apr 2024

Hasani V (2023) 6 Ways AI can help reduce carbon emissions. Sustamize https://www.sustamize.com/blog/6-ways-ai-can-help-reduce-carbon-emissions#:~:text=By%20having%20access%20to%20real,efficiency%2C%20and%20reduced%20carbon%20emissions. Accessed 11 Apr 2024

Ibrahim MA, Johansson M (2021) Attitudes to climate change adaptation in agriculture–a case study of Öland, Sweden. J Rural Stud 86:1–15

Janakiram MSV (2024) How Indian farmers are using AI to increase crop yield. Forbes. https://www.forbes.com/sites/janakirammsv/?sh=805f665b2872. Accessed 11 Apr 2024

Javaid M, Haleem A, Singh RP, Suman R (2022) Enhancing smart farming through the applications of agriculture 4.0 technologies. Int J Intell Netw 3:150–164

Karunathilake EMBM, Le AT, Heo S, Chung YS, Mansoor S (2023) The path to smart farming: innovations and opportunities in precision agriculture. Agriculture 13(8):1593

Ksolves (2023) Harvesting success: empowering agriculture through big data implementation. https://www.ksolves.com/blog/big-data/harvesting-success-empowering-agriculture-through-big-data-implementation. Accessed 10 Apr 2024

Liu Y, Xiaoyuan M., Shu L, Hancke GP, Abu-Mahfouz AM (2021) From industry 4.0 to agriculture 4.0: current status, enabling technologies, and research challenges. IEEE Trans Ind Inform 17(6):4322–4334

Mana AA, Allouhi A, Hamrani A, Rahman S, El Jamaoui I, Jayachandran K (2024) Sustainable AI-based production agriculture: exploring AI applications and implications in agricultural practices. Smart Agric Technol 7:100416

MATR (More agricultural technology resources) (2023). https://www.agritechtomorrow.com/news/2023/11/27/ai-in-agriculture-challenges-benefits-and-use-cases/15093/. Accessed 15 Apr 2024

McKendrick J (2023) How generative AI could give us Back at least 40% of our time. Forbes. https://www.forbes.com/sites/joemckendrick/2023/05/24/how-generative-ai-could-give-us-back-at-least-40-of-our-time/?sh=65f408ce2efc. Accessed 12 Apr 2024

Medium (2024) Revolutionizing green technology: the pivotal role of AI in environmental sustainability. https://medium.com/ai-ink/revolutionizing-green-technology-the-pivotal-role-of-ai-in-environmental-sustainability-26f99db4561d. Accessed 11 Apr 2024

Mirali M (2024) AI in agriculture: opportunities and challenges. https://graindatasolutions.com/data-science-artificial-intelligence-applications-agricultural-practices-opportunities-challenges/

Mundhe S, Sanap J, Jadhav P, Kalsadkar V, Das C (2023) Forecasting crop yield for sustainable agriculture. Int J Adv Res Sci Commun Technol 3(2):29–34

Nelson B (2023) The evolution of precision agriculture and policy implications. Market Intel. https://www.fb.org/market-intel/the-evolution-of-precision-agriculture-and-policy-implications. Accessed 15 Apr 2024

Nunes LJ (2023) The rising threat of atmospheric CO2: a review on the causes, impacts, and mitigation strategies. Environments 10(4):66

Oh S, Lu C (2023) Vertical farming-smart urban agriculture for enhancing resilience and sustainability in food security. J Hortic Sci Biotechnol 98(2):133–140

Patel J, Bala J (2022) Artificial intelligence-a multidimensional approach to smart farming. In: Futuristic trends in agriculture engineering & food sciences. ISBN: 978-93-95632-65-2, IIP Proceedings, vol 2, Book 9, Part 1, Chapter-26

Rani PMN, Rajesh T, Saravanan R (2011) Expert systems in agriculture: a review. J Comput Sci Appl 3(1):59–71

Rejeb A, Keogh JG, Zailani S, Treiblmaier H, Rejeb K (2020) Blockchain technology in the food industry: a review of potentials, challenges and future research directions. Logistics 4(4):27

Rizzoli A (2021) 8 Practical applications of AI in agriculture. V7. https://www.v7labs.com/blog/ai-in-agriculture. Accessed 11 Apr 2024

Saiwa (2024) The rise of smart spraying technology in agriculture. https://saiwa.ai/blog/smart-spraying-technology-in-agriculture/. Accessed 10 Apr 2024

Sroka N (2023) How AI is revolutionising solid waste management. Swana. https://swana.org/news/blog/swana-post/swana-blog/2023/12/11/how-ai-is-revolutionizing-solid-waste-management#:~:text=AI%2Dpowered%20robots%20and%20machines,reduces%20the%20risk%20of%20contamination. Accessed 15 Apr 2024

Talaviya T, Shah D, Patel N, Yagnik H, Shah M (2020) Implementation of artificial intelligence in agriculture for optimisation of irrigation and application of pesticides and herbicides. Artif Intell Agric 4:58–73

Tantalaki N, Souravlas S, Roumeliotis M (2019) Data-driven decision making in precision agriculture: the rise of big data in agricultural systems. J Agric Food Inf 20(4):344–380

U.S. Department of Agriculture (USDA) Sustainable Agriculture, National Agricultural Library (n.d.). https://www.nal.usda.gov/farms-and-agricultural-production-systems/sustainable-agriculture. Accessed 15 Apr 2024

UN World Commission on Environment and Development (1987) UN World Commission on Environment and Development, ed., Report of the World Commission on Environment and Development: our common future. Environment and Society Portal. https://www.environmentandsociety.org/mml/un-world-commission-environment-and-development-ed-report-world-commission-environment-and. Accessed 14 June 2024

Van Klompenburg T, Kassahun A, Catal C (2020) Crop yield prediction using machine learning: a systematic literature review. Comput Electron Agric 177:105709

Vinuesa R, Azizpour H, Leite I, Balaam M, Dignum V, Domisch S, Felländer A, Langhans SD, Tegmark M, Fuso Nerini F (2020) The role of artificial intelligence in achieving the sustainable development goals. Nat Commun 11(1):1–10

Vislavath PK (2024) Drones steer agriculture towards a sustainable future. Outlook Planet. https://planet.outlookindia.com/opinions/drones-steer-agriculture-towards-a-sustainable-future-news-416827#:~:text=With%20their%20capacity%20for%20comprehensive,damage%20to%20the%20physical%20structure. Accessed 11 Apr 2024

Wakchaure M, Patle BK, Mahindrakar AK (2023) Application of AI techniques and robotics in agriculture: a review. Artif Intell Life Sci 3:100057

Wu J, Ping L, Ge X, Wang Y, Fu J (2010) Cloud storage as the infrastructure of cloud computing. In: 2010 International conference on intelligent computing and cognitive informatics. IEEE, pp 380–383

Zaidi SSEA, Mahas A, Vanderschuren H, Mahfouz MM (2020) Engineering crops of the future: CRISPR approaches to develop climate-resilient and disease-resistant plants. Genome Biol 21(1):289

Zha J (2020) Artificial intelligence in agriculture. J Phys Conf Ser 1693(1):012058. IOP Publishing

Zhou Y, He Z, Zhao S (2021) How do government subsidies affect the strategic choices of enterprises and individuals in agricultural waste recycling? Sustain Prod Consum 28:1687–1698

Zürner S, Deutschländer LP, Schieck M, Franczyk B (2023) Sustainable development of AI applications in agriculture: a review. Procedia Comput Sci 225:3546–3553

Artificial Intelligence for Sustainable Water Resource Management

3

Madhukar Singh, Kriti Mishra, Sujatro Ray Chowdhuri, Keisham Radhapyari, Mayuri Pandey, and Shashi Kant Singh

Abstract

Water is fundamental to life, yet its sustainable management is increasingly under threat due to rapid population growth, urbanization, climate variability, and rising pollution levels. Traditional water management strategies, often fragmented, reactive, and sector-specific, are struggling to cope with the growing complexity, uncertainty, and scale of modern water challenges. In this context, artificial intelligence (AI) emerges as a transformative enabler, capable of enhancing efficiency, foresight, and resilience in Sustainable Water Resource Management (SWRM). This chapter explores the integration of AI, particularly machine learning (ML) and deep learning (DL) techniques—into the framework of Sustainable Water Resource Management (SWRM), aligned with the targets of Sustainable Development Goal 6 (clean water and sanitation for all).

This chapter first provides foundational overview of the global water cycle and the stark reality of freshwater availability, emphasizing the growing pressure on groundwater and surface water systems. The chapter outlines the limitations of conventional water management and positions Integrated Water Resource Management (IWRM) as a holistic but incomplete framework unless supported by advanced data-driven tools. AI, with its capacity to analyze complex datasets,

M. Singh (✉) · K. Mishra
Central Ground Water Board, SUO, New Delhi, India

S. R. Chowdhuri · S. K. Singh
Central Ground Water Board, CHQ, Faridabad, India

K. Radhapyari
Central Ground Water Board, NER, Guwahati, India

M. Pandey
Banaras Hindu University, Varanasi, India

B. Thokchom et al. (eds.), *Artificial Intelligence Enabled Real Time Environmental Monitoring*, https://doi.org/10.1007/978-981-95-3767-9_3

identify patterns, and support real-time decisions, offers new pathways for adaptive and inclusive water governance.

The chapter presents a structured taxonomy of AI applications across key domains of agriculture, industry, domestic supply, wastewater treatment, groundwater management, and climate resilience. From smart irrigation and leak detection to flood forecasting, aquifer health monitoring, and policy simulation, AI enables more precise and proactive water management.

It also addresses critical challenges such as data quality, algorithm transparency, ethical considerations, and the need for institutional readiness, highlighting the need for responsible, inclusive adaption and implementation. While AI is not a panacea, its thoughtful integration can usher in a new era of water governance, one that is responsive, equitable, and climate-resilient. As the world faces escalating water stress, AI offers a timely and essential tool to support sustainability, justice, and resilience in water resource management.

Keywords

Artificial intelligence (AI) · Sustainable Water Resource Management (SWRM) · Machine learning (ML) · Climate resilience · Groundwater management · Smart irrigation · Water governance

Abbreviations

AI	artificial intelligence
CGWB	Central Ground Water Board
CNN	convolutional neural network
DL	deep learning
GWP	Global Water Partnership
IEA	International Energy Agency
IFPRI	International Food Policy Research Institute
IPCC	Intergovernmental Panel on Climate Change
IWRM	Integrated Water Resource Management
ML	machine learning
NN	neural network
RNN	recurrent neural network
SDG	Sustainable Development Goals
SWRM	Sustainable Water Resource Management
UN DESA	United Nations Department of Economic and Social Affairs
UNEP	United Nations Environment Programme
UNGA	United Nations General Assembly
WSSCC	Water Supply and Sanitation Collaborative Council
WWAP	World Water Assessment Programme

3.1 Introduction

Next to air, water is the most fundamental requirement for life. Although water covers about 71% of the planet, only about 2.5% is fresh (Durack 2015). Nearly 3.6 billion people or 47% of the global population are living in areas that experience water scarcity for at least 1 month each year (WWAP 2018) with half a billion people in the world facing severe water scarcity all year round (Mekonnen and Hoekstra 2012). The current water utilization is already near its critical threshold of sustainability. By 2050, global freshwater stress owing to increasing population and associated developmental needs, water use will increase significantly, and nearly 5 billion people will be affected by freshwater scarcity (Boretti and Rosa 2019). The ever-growing pollution of already stressed water resources is further complicating the situation by making them unfit for utilization (IFPRI 2015). Moreover, climate change characterized by rising temperatures and more frequent extreme events such as heatwaves, heavy rains, floods, sudden droughts, and persistent droughts have become critical concerns for water security (IPCC 2014). Against this backdrop, the United Nations General Assembly adopted the Sustainable Development Goals (SDG) in 2015. Goal no. 6 of SDG is clean water and sanitation (UNGA 2017). It aims to ensure the availability and sustainability of water and sanitation for all.

Management of water resources has been going on since time immemorial. Most of the ancient civilizations developed at places where water for human needs and agriculture was readily available (Yannopoulos et al. 2015). In the Indus Valley civilization (~3000–1500 BCE), extensive network of reservoirs, wells, canals, and rainwater-harvesting techniques were developed (Nair 2004). In medieval times rapid technological progress happened in society, but was focused on resource exploitation. In the twentieth century, the same trend continued. Even in the Water Conference in 1977 where the necessity of coordination within the water sector was emphasized, the challenges due to high water demand and negative environmental impacts of irrigation were not approached sufficiently (Snellen and Schrevel 2004). It was in the 1990s when shortcomings in traditional water management, like quality issues, overexploitation, ecosystem degradation, and social concerns, were given attention. Till then water-related issues had become multidimensional, multisectoral, and multiregional (Biswas 2004). To overcome these issues, at the International Conference on Environment and Water in Dublin in the year 1992, four important guiding principles were determined (Xie 2006). These principles became the backbone of a holistic approach known as integrated water resource management (IWRM). The IWRM is based on core concepts of integration, decentralization, participation, and sustainability (Xie 2006).

Considering the importance and magnitude of the challenges in water resources management, the deployment of the most advanced technologies becomes necessary. Artificial intelligence (AI), which is a branch of the computer science discipline, is involved in developing computer programs that can complete tasks that would typically require human intelligence (Saleh 2019). In the past few decades, AI has emerged as a solution to overcome the challenges presented by the complexity, dynamics, and nonlinearity in hydrological systems (Nourani et al. 2014). AI techniques offer a promising alternative or supplement to conventional water resource management techniques (Chang et al. 2016). Within the domain of water management, AI can take center stage in addressing the nuanced challenges arising from escalating water scarcity, climate fluctuations, quality, resource management, and burgeoning demand (Mohseni et al. 2023). Machine learning, a subset of AI, systems excel at predicting, forecasting, and recognizing patterns and anomalies, further enhancing their efficacy in the realm of water resource optimization (Razavi et al. 2022). In this chapter, we shall discuss the status of water resources, sustainable water resource management, and areas of AI application as it holds the potential to revolutionize future water resource management practices by playing a key role in fostering informed decision-making and promoting the sustainable utilization of this critical resource.

3.2 Water Cycle and Water Distribution

Understanding the water cycle is essential as it sets the foundation for understanding the complexities involved in water resources. The water cycle describes where water is found on Earth and how it moves (Braga et al. 2014). The water cycle is the continuous, natural movement of water among its reservoirs of the ocean, atmosphere, cryosphere, and land in liquid, solid, and gaseous forms (Stocker et al. 2013). Water can be stored in the atmosphere, on Earth's surface, below the ground, or in biological resources. It can be in a liquid, solid, or gaseous state. Water moves between the places it is stored at large scales to very small scales. Natural and anthropogenic agents affect where water is stored, how it moves, and how clean it is. The water cycle encompasses hydrological processes at all scales within the hydrosphere. The main driving force for the water cycle is solar radiation. The hydrological cycle is evident in land-water-atmosphere interactions through the exchange of energy and water (Kleidon and Renner 2013).

As it moves, water can transform into a liquid, a solid, or a gas. The different ways in which water moves between pools are known as fluxes (Brutsaert 2005). Circulation mixes water in the oceans and evaporation transports water vapor in the atmosphere. Water moves between the atmosphere and the Earth's surface through evaporation, evapotranspiration, and precipitation. Water moves across the land surface through snowmelt, runoff, and streamflow. Through infiltration and percolation, water seeps into the ground. When underground, groundwater flows within aquifers and can return to the surface through springs or from natural groundwater discharge into rivers and oceans.

Humans alter the water cycle (Wang et al. 2010). We redirect rivers, build dams and canals, and pump water from aquifers. We use water from rivers, lakes, reservoirs, and groundwater aquifers. We use that water to supply our homes and communities, for agriculture and livestock, and in industrial activities like power generation, mining, and manufacturing. Greenhouse gas emissions activities (Piao et al. 2007), as well as water conservation and consumption actions, influence the water cycle (Haddeland et al. 2014). The amount of available water for use in particular geographic areas depends on how much water is available in the pool (water quantity). Water availability also depends on when and how fast water moves from one pool to another (water timing), how much water is used (water use), and how clean the water is (water quality). Human activities affect water quality that directly reduces the availability of utilizable water resources. Climate change is also affecting the water cycle. It affects water quality, quantity, timing, and use.

Globally total water reserve is 1385.9 million km^3. But nearly 97.5% of this is saline, meaning we cannot utilize it directly. Nearly 96.6% water is in the oceans and seas remaining 0.9% is in the form of saline groundwater or lakes. For this chapter, our discussion will be limited to the freshwater resources that are essential for human survival.

Freshwater constitutes only about 2.5% of the total volume of water on Earth, and two-thirds of this freshwater is locked in glaciers and ice caps (Durack 2015; Abbott et al. 2019). Just 0.77% of all freshwater is held in aquifers, lakes, rivers, etc.

Let's understand this with a different crude analogy, suppose we have 1000 bottles of water 1 L each. A total of 975 bottles are discarded because the water in them is too saline to use even for agriculture. We are left with 25 bottles of fresh water. Nearly 17 bottles of the remaining 25 are also not available for use because it is locked in the polar ice caps and mountain glaciers. Of the remaining 8 bottles, 7.5 are in as groundwater and the remaining half-liter represents all the other available fresh water in lakes, rivers, swamps, the atmosphere, and even all the biological water. Percentage values or analogy do not tell exactly how much water is available. So, here we will discuss the volume in absolute terms. A total of 35.02 million km^3 of fresh water is available (Abbott et al. 2019), of which 24.06 million km^3 is frozen in the form of glaciers and permanent ice covers. Moreover, 10.53 million km^3 is groundwater. Only, 91,000 km^3 is in lakes and 2120 km^3 is in river flow.

The above quantification of water is based on the type of reservoir or pool. Generally, all these reservoirs form an open system and are connected. The flux of water can move from one reservoir to the other as we discussed previously. Although the total water reserves on the Earth are constant, fluxes can change their form as well as their location.

3.3 Challenges to Water Resource Management

Over the course of the past century, there has been a substantial sixfold increase in global water consumption (Wada et al. 2017). This trend is persistent and shows no signs of abating. The surge in water usage is attributed to various factors, like population growth, economic development, and evolving consumption patterns. Projections indicate a continuous rise in global water demand, particularly in line with the anticipated increase in the world population from nearly 8 billion in 2022 to an estimated 9.7 billion by 2050 (UN DESA 2017) (United Nations Department of Economic and Social Affairs). This population growth is expected to be concentrated in urban areas, with around two-thirds of the global population residing in cities. Similar to the demographic changes, the global gross domestic product

(GDP) is poised for significant expansion. Projections suggest that between 2022 and 2050, the global GDP will increase substantially, approximately 2.5 times its current value. Anticipated global trends present significant challenges to water security, particularly in agriculture and energy production, both of which are known for their substantial water requirements.

In the contemporary context, the global water demand is estimated to be approximately 4600 km^3 per year, with projections indicating a notable increase of 20%–30% to a range between 5500 and 6000 km^3 per year by the year 2050 (Burek et al. 2016). Notably, the agricultural sector dominates global water uses, constituting about 70% of the total, with the majority allocated for irrigation purposes, and projections show that by 2050, food demand will increase by nearly 60%.

Industrial sector consumes roughly one-fifth of total water use (WWAP 2018). Future projections (Burek et al. 2016) anticipate a surge in overall water demand from the industrial sector across most regions globally, with exceptions of Northern America and Western and Southern Europe. The Organisation for Economic Co-operation and Development (OECD) predicts a remarkable 400% increase in water demand for manufacturing from 2000 to 2050. The growing production of biofuels, as highlighted by the International Energy Agency (IEA) in 2012, also contributes significantly to the surge in water consumption, posing challenges to sustainable water management.

The utilization of water for domestic purposes, constituting approximately 10% of global water withdrawals, is poised for a substantial increase across most regions of the world. In relative terms, the most significant upsurges in domestic water demand are anticipated in the subregions of Africa and Asia, where the demand could potentially more than triple (Burek et al. 2016). This projected growth in domestic water demand is primarily attributed to the anticipated expansion of water supply services in urban settlements. The rapid urbanization trend, coupled with population growth, is expected to drive an increased reliance on domestic water services, necessitating strategic planning and sustainable management practices to ensure equitable access to safe and sufficient water resources in the face of growing global demands.

Groundwater use globally, mainly for agriculture, amounted to nearly 800 km^3 per year in the 2010s, with India, the United States, China, Iran, and Pakistan (in descending order) accounting for 67% of total abstractions worldwide (Burek et al. 2016). Water withdrawals for irrigation have been identified as the primary driver of groundwater depletion worldwide. In India, 87% of groundwater withdrawal is for irrigation purposes (CGWB 2023). A particularly alarming revelation is that a third of the world's largest groundwater systems are already under significant stress (Richey et al. 2015). This indicates a critical state of depletion and overuse. Furthermore, the trends in groundwater extraction raise concerns about the increasing reliance on nonrenewable groundwater, which is unequivocally an unsustainable trajectory. These trends underscore the urgency for concerted efforts in sustainable water management to address the escalating risks of depleting water resources, especially at the regional and local levels where the impacts are more pronounced and immediate.

Since the 1990s, the state of water pollution has exhibited a concerning deterioration across nearly all rivers in Africa, Asia, and Latin America, as highlighted by the United Nations Environment Programme (UNEP) in 2016. This downward trend in water quality is anticipated to intensify in the coming decades, posing heightened threats to human health, environmental well-being, and the overarching goal of sustainable development, as indicated in a joint report by Veolia and the International Food Policy Research Institute (IFPRI) in 2015. A substantial contributing factor to this escalating challenge is the discharge of approximately 80% of all industrial and municipal wastewater directly into the environment without undergoing any form of prior treatment. According to the World Water Assessment Programme (WWAP) in 2014, this widespread lack of treatment results in a progressively deteriorating overall water quality, with adverse implications for both human health and ecosystems. Globally, the predominant water quality issue revolves around nutrient loading, which is frequently intertwined with pathogen loading, as outlined by UNEP in 2016. This multifaceted problem not only underscores the urgent need for improved wastewater management practices but also emphasizes the critical importance of addressing water pollution as an integral part of broader sustainability initiatives.

The intensification of agriculture has led to a notable surge in global chemical usage (De et al. 2014). Recognizing the urgency of the situation, a report from the Special Rapporteur on the right to food (UN General Assembly 2017) emphasizes the critical need for improved policies governing pesticide use. Compounding this challenge are contaminants of emerging concern, which continue to evolve and proliferate, often detected at concentrations surpassing anticipated levels (Sauvé and Desrosiers 2014). These contaminants include pharmaceuticals, hormones, industrial chemicals, personal care products, flame retardants, detergents, perfluorinated compounds, caffeine, fragrances, cyanotoxins, nanomaterials, and antimicrobial cleaning agents. Solid and liquid wastes emerging from industries and households contaminate freshwater sources and make them unfit for utilization. When pollution loading crosses a certain threshold, aquatic life gets adversely affected.

Adding to the complexity, climate change, impacting the land-water-atmospheric system has accelerated temporal and spatial changes in the water cycle (Taylor et al. 2013; Milly and Dunne 2016) as well as amplified regional and local water shortages (Schewe et al. 2014). According to findings presented by the Intergovernmental Panel on Climate Change (IPCC) in 2014, climatic shifts result in wetter regions experiencing heightened precipitation levels, while drier regions undergo more pronounced aridity. Higher water temperatures accelerate the depletion of dissolved oxygen, and extreme rainfall events are expected to introduce higher concentrations of pollutants into water bodies. Shifts in the spatial and temporal patterns of precipitation influence surface water flows, impacting dilution effects. Simultaneously, rising temperatures contribute to increased evaporation from open surfaces and soils, while augmented transpiration by vegetation potentially reduces overall water availability (Hipsey and Arheimer 2013).

These changes underscore the importance of proactive planning and the expeditious implementation of strategic, rational, and efficacious management approaches.

Swift and well-devised countermeasures are essential to address the escalating challenges to water security as a result of these multidimensional global changes (Burek et al. 2016). The delicate interplay of increasing demand and changing climatic patterns underscores the urgency for comprehensive and adaptive water resource management strategies at every scale.

Understanding the gravity of current water availability challenges requires a comprehensive evaluation by comparing current water uses to their maximum sustainable levels. The global water uses, currently standing at approximately 4600 km^3 per year, are already close to the maximum sustainable thresholds (Mekonnen and Hoekstra 2012). Notably, these global figures do not represent more severe challenges at regional and local levels, underscoring the need for nuanced assessments.

Another important and complex challenge is water governance. Effective governance is essential for evolving or improving water management (Pahl-Wostl 2009). The International Conference on Freshwater in Bonn identified the priority action area in which water governance was the first of three (WSSCC 2001). The first UN World Water Development Report stated that the "world's water crisis is one of water governance, essentially caused by the ways we mismanage water" (WWAP 2018). Formulation of appropriate water policy, legislation, institutional arrangement, ensuring meaningful stakeholder participation, and implementation are all part of water governance. The governance model needs to be ever-evolving based on the data and feedback.

So, the challenges in water resource management are a spectrum of complexities, ranging from issues of quantity and quality to governance and sustainability. With escalating global demand driven by population growth, urbanization, and industrial expansion, ensuring equitable access to clean water while mitigating scarcity and pollution remains paramount. Climate change catalyzes these challenges, altering precipitation patterns and intensifying extreme weather events, further stressing water supplies. Balancing competing demands for water across sectors, coupled with outdated infrastructure and inefficient allocation practices, underscores the urgent need for adaptive and integrated management approaches. In an era marked by escalating global water scarcity, effective water management becomes necessary for addressing these challenges and ensuring sustainable development.

3.4 Sustainable Water Resource Management

As water demand continues to increase in tandem with these multidirectional developments, there arises an urgent need for comprehensive water management strategies to ensure the sustainable utilization of this critical resource in the face of growing global challenges. Population surge and accelerating urbanization necessitate greater food production and industrial output, increasing water requirements. While increasing pollution of water bodies reduces freshwater availability. Water demand is increasing over time and supply is getting limited, placing unprecedented pressure on water resources.

Traditional water management approaches are typically tailored to specific water pools, issues, or locations, addressing challenges in a localized manner. However, these conventional methods may prove inadequate in larger contexts and changing climatic conditions. As we have discussed previously, water resources operate as open systems, it becomes challenging for traditional practices to manage them in their entirety.

Managing water resources comprehensively requires an examination of diverse environmental, social, and economic factors. On September 25, 2015, the UN General Assembly adopted the 2030 Development Agenda titled "Transforming our world: the 2030 Agenda for Sustainable Development." This agenda has 17 Sustainable Development Goals (SDG) and the associated 169 targets. Goal no. 6 of these SDGs is to "ensure availability and sustainable management of water and sanitation for all by 2030." Goal 6 has associated targets like implementing Integrated Water Resource Management (IWRM) at all levels and the protection and restoration of water-related ecosystems. From these targets, it can be understood that Sustainable Water Resource Management (SWRM) is a broader concept encompassing IWRM and ensuring the sustainability of water-related ecosystems.

IWRM represents a systemic approach that promotes integrated development and management of water, land, and other related resource. The aim is to maximize economic and social welfare without degrading vital ecosystems. IWRM seeks to address various dimensions, acknowledging the intricate nature of water systems. Recognizing the finite nature of water resources and the potential adverse impacts, IWRM aligns with the broader context of sustainable development. It aims to facilitate the coordinated development and management of water and related resources to foster equitable social and economic development without compromising the sustainability of vital ecosystems, as articulated by the Global Water Partnership in 2000.

SWRM is a broader approach that seeks to balance the utilization of water resources with the preservation of ecological integrity and social equity over the long term. It involves optimizing water use efficiency, protecting water quality, and promoting equitable access to water resources for current and future generations. SWRM recognizes the interconnectedness of environmental, economic, and social dimensions, emphasizing the need for integrated and adaptive strategies. By addressing the limitations of conventional water management practices and considering the finite nature of water resources, SWRM aims to achieve a harmonious balance between human needs, ecosystem health, and the preservation of water resources for sustained well-being and development. The bottom line is we need sufficient water resources of reasonably good quality for all our existing and future needs without damaging the environment.

3.5 Introduction of Artificial Intelligence (AI)

In the dynamic landscape of water management, the integration of artificial intelligence (AI) emerges as a transformative force, offering innovative solutions to address the complex challenges associated with water scarcity and resource optimization (Alsaiari et al. 2023).

As conventional approaches face limitations in handling the intricacies of modern water management, AI presents itself as a technological enabler, providing unprecedented opportunities for efficiency, precision, and adaptability.

Before delving into the concept of AI in SWRM, it's imperative to go through the fundamental concepts of AI, its constituents, and its application in water resource management. So, the initial inquiry arises: what exactly is AI? AI is the science and engineering of making intelligent machines, especially intelligent computer programs. It is related to the similar task of using computers to understand human intelligence, but AI does not have to confine itself to biologically observable methods. In simple words, AI refers to the development of computer programs that can perform tasks that typically require human intelligence (McCarthy 2004).

3.5.1 Artificial Intelligence (AI)

AI is a broad field within computer science that aims to create intelligent machines capable of mimicking human-like cognitive functions such as learning, problem-solving, perception, language understanding, and decision-making. It can be categorized into narrow AI (focused on specific tasks) and general AI (possessing human-like intelligence across various domains). All the existing AI systems are considered narrow AI.

3.5.2 Components of AI

AI encompasses a range of techniques and approaches (Fig. 3.1), including computer vision, robotics, natural language processing, automation, expert systems, knowledge-based systems, and machine learning (Kumar et al. 2019).

An interesting and core component of AI is machine learning (M.L.), which is a subset of AI that focuses on developing algorithms and statistical models to enable machines to perform tasks without being explicitly programmed (Kearnes et al. 1994). It involves learning from data, identifying patterns, and making predictions or decisions. ML algorithms include supervised learning, unsupervised learning, semi-supervised, and reinforcement learning.

Neural networks (NN), a subfield of ML, are powerful models inspired by the structure and function of the human brain. These networks consist of interconnected nodes arranged in layers, including an input layer, one or more hidden layers, and an output layer (Bishop 2006). Nodes, also known as neurons, process information by applying weights to input features and activation functions. During training, the network adjusts these weights based on the error between predicted and actual outputs, a process known as backpropagation. Neural networks can take various forms, such as feedforward neural networks (FNN) for sequential information flow, recurrent neural networks (RNN) for capturing temporal dependencies, convolutional neural networks (CNN) for image processing, and generative adversarial networks (GAN) for generating realistic data. Neural networks have revolutionized machine

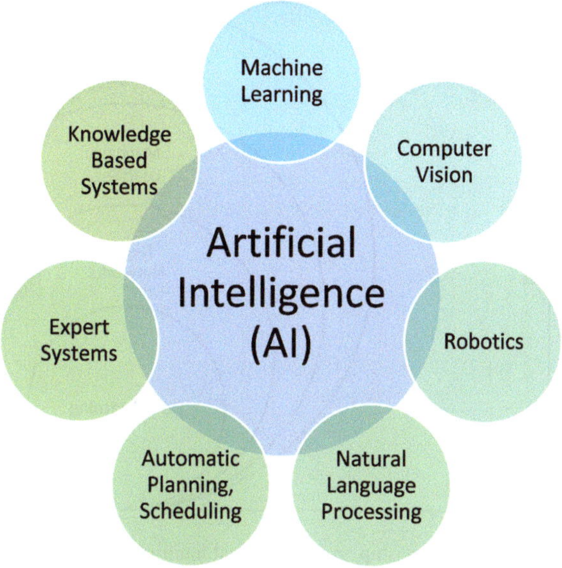

Fig. 3.1 Main components of artificial intelligence

learning, demonstrating remarkable success in tasks like image and speech recognition, natural language processing, healthcare diagnostics, and more. The deep learning paradigm, utilizing neural networks with multiple hidden layers, has particularly fuelled advancements, making neural networks a central component in the contemporary machine learning landscape.

Deep learning (DL) is again a subfield of machine learning, which involves the use of artificial neural networks with multiple layers (deep neural networks) to learn and make decisions (Bishop 2006). DL has been particularly successful in tasks such as image and speech recognition, natural language processing, and playing strategic games. It is characterized by the ability to automatically learn hierarchical representations of data.

In summary, AI is the overarching concept of creating intelligent machines, ML is a subset of AI focused on learning from data, and DL is a specialized form of ML using deep neural networks for complex pattern recognition and decision-making. While AI covers a wide range of approaches, ML and DL specifically address the learning aspect, with DL representing a more sophisticated and layered approach to feature learning (Fig. 3.2).

Based on the learning approach of the computer program, ML can broadly be classified into the following categories (Mohammed et al. 2016):

- **Supervised learning**: The ML model is trained on a labeled dataset, where each input is associated with the correct output. The model learns to map inputs to outputs by generalizing patterns from the training data. The most general supervised learning tasks are classification and regression.

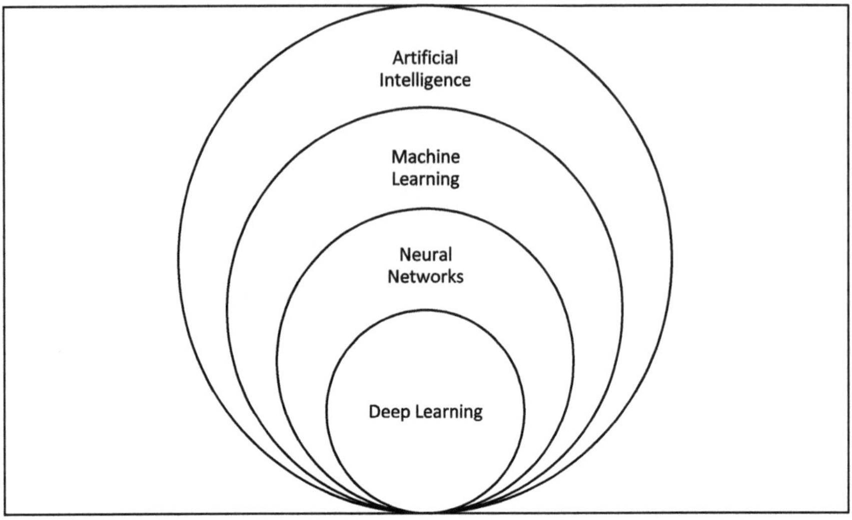

Fig. 3.2 Relationship among artificial intelligence, machine learning, neural networks, and deep learning

- **Unsupervised learning**: The ML model is given data without explicit instructions on what to do with it. The system must find patterns, relationships, or structures in the data without labeled guidance. Unsupervised learning analyzes unlabeled datasets without the need for human interference.
- **Semi-supervised learning**: This is a mixed model or a combination of supervised and unsupervised learning, where the model is trained on a dataset that contains both labeled and unlabeled examples. The goal of a semi-supervised learning model is to achieve a better outcome than that produced using the labeled data alone from the model.
- **Reinforcement learning**: The ML agent learns by interacting with an environment and receiving feedback in the form of rewards or penalties. The agent aims to maximize cumulative rewards over time. It is a powerful tool that can help increase automation or optimize the operational efficiency of systems.

The summarized ML type, the fundamental principle for algorithms or models, and the type of output that can be achieved (Sarker 2021) are given in Table 3.1.

3.6 Machine Learning Algorithms or Models

Machine learning algorithms are computational procedures or mathematical models that enable computers to learn from data and make predictions or decisions without being explicitly programmed. These algorithms form the core components of machine learning systems, which aim to identify patterns, relationships, and insights within data.

Table 3.1 Type of machine learning, its model fundamental, and output type

Sl no	Learning type	Model	Output
1.	Supervised	Algorithms or models learn from labeled data	Classification, regression
2.	Unsupervised	Algorithms or models learn from unlabeled data	Clustering, association
3.	Semi-supervised	Algorithms or models learn from labeled as well as unlabelled data	Classification, clustering
4.	Reinforcement	Algorithms or models are based on reward or penalty	Classification, control

Several factors influence the selection of machine learning algorithms or models for a particular task. The choice often depends on the characteristics of the data and the goals of the specific problem. Apart from these two, many other factors influence the appropriate algorithm selection such as computational resources, scalability, model complexity, etc. Table 3.2 provides a list of some common algorithms used in ML based on the type of desired outcome (Sarker 2021). These algorithms can be applied standalone or in combination depending on the data type, desired output, and accuracy (Ghobadi and Kang 2023). The model that is created by the combination of two or more algorithms is called a hybrid model.

3.7 Role of Artificial Intelligence and Machine Learning in Sustainable Water Resource Management

The role of AI in water management revolves around its capacity to process large amounts of data, identify patterns, and make real-time predictions or recommendations, thereby empowering decision-makers with valuable insights. AI and its subdivision of machine learning have the potential to solve problems across water systems. AI can act as a technological catalyst, enabling water management strategies to evolve beyond traditional methods (Ghobadi and Kang 2023). Because of its ability to analyze data at scale, adapt to changing conditions, and inform decision-making positions, AI can be an efficient tool in the pursuit of efficient, equitable, and sustainable water resource management practices.

Traditional water management practices face inherent complexities and uncertainties that may challenge their effectiveness in the face of changing environmental and societal conditions. One significant complexity lies in the intricate interplay of natural systems, where surface water and groundwater dynamics, influenced by geological, climatic, and hydrological factors, create intricate and often unpredictable patterns. Traditional approaches may struggle to account for these complexities, leading to challenges in accurately estimating water availability and distribution. Moreover, the complexities due to changes in land-use patterns, population growth, and urbanization further compound the uncertainties.

Table 3.2 List of common problems and associated algorithms or models in machine learning

Type of problem	Algorithm or model	When to apply	Conditions
Classification	Logistic regression	Appropriate for binary or multiclass classification problems	Assumes a linear relationship between features and log-odds of the output
	Decision trees	Effective for both binary and multiclass classification tasks	Useful when interpretability is important; captures nonlinear relationships
	Random forest	Ensembles of decision trees for binary and multiclass problems	Suitable when seeking improved accuracy and variable importance
	Support vector machines (SVM)	Initially designed for binary classification; extends to multiclass	Effective when finding a hyperplane that maximally separates classes is crucial
	k-nearest neighbors (k-NN)	Effective for both binary and multiclass classification	Useful when local patterns are important, but maybe computationally expensive
	Naive Bayes	Particularly useful for text classification problems	Assumes independence between features, given the class; works well with sparse data
	Neural networks	Powerful for complex, nonlinear relationships in both binary and multiclass tasks	Suitable for large datasets and problems with intricate patterns; computationally intensive
	Linear discriminant analysis (LDA)	Effective for binary or multiclass classification	Assumes features are normally distributed with the same covariance matrix
Regression	Linear regression	Suitable for predicting a continuous target variable	Assumes a linear relationship between input features and the output
	Decision trees (regression trees)	Effective for predicting continuous values	Useful when interpretability is important; captures nonlinear relationships
	Neural network regression	When the relationship between input features and the output is highly nonlinear and complex	Suitable for situations with large amounts of data
	Polynomial regression	Suitable when the true relationship between variables is not linear	Useful when higher-degree polynomial terms are expected to improve the model
	Support vector machines (SVR)	Suitable for capturing nonlinear relationships	Effective when dealing with high-dimensional feature spaces

(continued)

Table 3.2 (continued)

Type of problem	Algorithm or model	When to apply	Conditions
Clustering	K-means clustering	Suitable for partitioning data into k clusters	Assumes spherical clusters, work well when cluster sizes are relatively equal
	Hierarchical clustering	Useful for exploring hierarchical relationships within the data	Can be agglomerative (bottom-up) or divisive (top-down), depending on the desired approach
	DBSCAN (density-based spatial clustering of applications with noise)	Effective for identifying clusters with varying shapes and densities	Defines clusters as dense regions separated by less dense regions
	Mean shift	Suitable for finding modes or centroids of data density.	Works well for non-parametric density estimation
	Gaussian mixture models (GMM)	Appropriate for probabilistic modeling of data points	Assumes data points are generated from a mixture of Gaussian distributions
	BIRCH (balanced iterative reducing and clustering using hierarchies)	Suitable for large datasets and incrementally updating clusters	Builds a tree structure to represent clusters hierarchically
	Fuzzy C-means	Appropriate when data points can belong to multiple clusters with varying degrees	Each data point has a membership value for each cluster
Association	Apriori algorithm	Suitable for discovering frequent itemsets and generating association rules	Requires multiple passes through the data and is effective for moderate-sized datasets
	Eclat algorithm	Appropriate for finding frequent itemsets without candidate generation	Generally faster than Apriori, particularly when the dataset is sparse
	FP-growth (frequent pattern growth) algorithm	Efficient for mining frequent itemsets using a divide-and-conquer strategy	Particularly useful for large datasets with many transactions

(continued)

Table 3.2 (continued)

Type of problem	Algorithm or model	When to apply	Conditions
Control	PID (proportional-integral-derivative) control with machine learning enhancements	Combining traditional PID control with machine learning enhancements is useful for systems with complex dynamics	Enhancements can include adaptive tuning of PID parameters based on the system's behavior
	Model predictive control (MPC) with machine learning models	MPC is enhanced by incorporating machine learning models for predicting system behavior	Useful when there is uncertainty or nonlinearity in the system dynamics
	Deep reinforcement learning	Deep neural networks to represent complex policies or value functions	Effective for problems with high-dimensional state and action spaces, such as robotics and autonomous vehicles
	Neurocontrol	Combines neural network models with control systems for adaptive and learning-based control	Useful for systems with nonlinear and time-varying dynamics
	Fuzzy logic control	Incorporates fuzzy logic reasoning to handle uncertainty in control decisions	Effective when dealing with imprecise or ambiguous information

In the face of competing water demands from various sectors, including agriculture, industry, and households. Traditional practices, typically designed to meet singular objectives, may struggle to be effective. Furthermore, the temporal and spatial variability of water availability adds another layer of complexity. Traditional methods may not adequately capture the dynamic nature of water resources, leading to suboptimal allocation strategies that may not align with current needs.

The approach for AI applications in water management is driven by a multiple set of goals aimed at addressing the challenges of water scarcity and sustainable resource utilization. By leveraging AI, the intention is to advance the development of sustainable water management practices that align with global sustainability goals. Equitable access to water is a critical focus, with AI-driven strategies. AI applications in water management seek to integrate cutting-edge technologies to build resilience, promote efficiency, and contribute to a more sustainable and equitable future for water resource management worldwide.

3.8 AI Applications in Various Fields of Water Resource Management

The dynamic nature of water-related issues requires innovative solutions in water management. Traditional approaches, while valuable, often struggle to adapt swiftly to the complexities presented by changing conditions. It becomes essential to explore innovation in the pursuit of sustainable and resilient water resource management. Application of AI and its various components can be used in different aspects of water resource management. Standalone algorithms as well as multiple algorithm-based models can be programmed depending upon the earlier discussed conditions. In this section, we shall briefly discuss the potential AI applications for water resource management in specific aspects of agriculture, industrial, domestic, wastewater, climate change, water storage, groundwater, and water governance.

A. **Agriculture Sector**.
Current water demand for agriculture is the highest (nearly 70% of the total water requirement). Studies suggest that global crop irrigation water requirement can increase by 23% to 42% by 2050 above the 2010 levels (Burek et al. 2016). Artificial intelligence (AI) can play a significant role in transforming agriculture water management by providing advanced technologies and solutions to address various challenges. Here are some key applications of AI in agriculture water management:

(i) *Soil Moisture Management.*
AI-based sensors and devices monitor soil moisture levels in real time. Machine learning algorithms analyze (Karandish and Šimůnek 2016) this data to provide insights into the soil's water-holding capacity, enabling farmers to optimize irrigation and prevent waterlogging or drought conditions. Kalman filters, support vector machines (SVM), and random forest models can be developed for real-time soil moisture prediction.

(ii) *Precision Irrigation.*
AI is used to optimize irrigation practices by incorporating data from sensors, satellites, and weather forecasts. Machine learning algorithms analyze this data to determine the precise amount and timing of irrigation needed for different crops, reducing water wastage and enhancing water use efficiency (Seyedzadeh et al. 2020). Decision trees, random forests, support vector machines (SVM), and neural network algorithms are used for irrigation scheduling.

(iii) *Crop Monitoring and Management.*
Drones and satellite imagery of agriculture fields are analyzed by AI models for crop health, moisture levels, and stress conditions. This information assists farmers in making data-driven decisions regarding irrigation schedules, identifying areas that require more attention, and optimizing resource allocation (Goap et al. 2018). Algorithms like convolutional neural networks (CNN), image segmentation, and object detection can be used for analyzing and interpreting satellite and drone imagery.

(iv) *Smart Water Distribution.*
Data from soil moisture sensors and weather forecasts are analyzed to manage the distribution of water resources efficiently (Shi et al. 2019). This helps in preventing over-irrigation or under-irrigation, leading to improved water conservation and crop yield. Reinforcement learning for optimizing water distribution and genetic algorithms for resource allocation are best suited.

(v) *Predictive Modeling for Water Availability.*
Machine learning models analyze historical climate data, soil conditions, and water usage patterns to predict future water availability (Sidhu et al. 2020). This information aids farmers and water resource managers in planning for potential water shortages and adapting their strategies accordingly. Long short-term memory (LSTM) networks and time series analysis can be applied to predict future water availability.

(vi) *Automated Irrigation Systems.*
AI-powered automated irrigation systems utilize real-time data and machine learning algorithms to control irrigation equipment. These systems adjust water delivery based on crop needs, weather conditions, and soil moisture levels, promoting water efficiency and reducing operational costs (Blasi et al. 2021). Fuzzy logic control, proportional-integral-derivative (PID) controllers, and model predictive control models are used for developing automated irrigation systems.

(vii) *Crop Selection and Planning.*
Environmental factors, historical data, and market trends are analyzed by the ML models to suggest suitable crops for specific regions. This helps farmers make informed decisions about crop selection, considering water availability and maximizing yields (Kumar et al. 2015). Models like machine learning-based recommender systems and decision trees for crop recommendation can be helpful.

(viii) *Water Quality Monitoring.*
AI is used to analyze water quality parameters in irrigation sources. Sensors and AI algorithms detect contaminants, salinity levels, and other factors affecting water quality (Maroju et al. 2023). This information guides farmers in selecting appropriate water sources for irrigation.

(ix) *Decision Support Systems.*
Decision support systems integrate data from various sources to provide actionable insights. These systems assist farmers in making informed decisions related to irrigation, crop management, and resource allocation based on real-time information (Navarro-Hellín et al. 2016). Expert systems, rule-based systems, and multi-criteria decision analysis (MCDA) are a few to name for providing decision support.

(x) *Supply Chain Optimization.*
AI can be applied to optimize the supply chain, ensuring efficient water use from the source to the field (Omidzade et al. 2020). Predictive analytics help in planning and managing water resources along the entire agri-

cultural value chain, reducing losses and promoting sustainability. Genetic algorithms, linear programming, and ant colony optimization models can help in optimizing water use in the agricultural supply chain.

B. **Industrial Water Sector**.

Currently industrial sector consumes nearly 20% of global water utilization. With the expansion of the global economy, the demand is expected to increase. The integration of AI in industrial water management can bring a transformative shift in optimizing water usage, enhancing efficiency, and ensuring sustainable practices within industrial operations.

(i) *Water Usage Monitoring and Data Analytics.*
ML models analyze real-time and historical water usage data, identifying patterns, trends, and anomalies. This insight helps industries understand their water consumption patterns and potential areas for improvement (Mamo et al. 2013). Time series analysis, data clustering, and regression analysis models can be used for this application.

(ii) *Process Optimization for Water Efficiency.*
Industrial processes can be optimized using AI in such a way that water usage can be minimized while maintaining production efficiency. Genetic algorithms (Lavric et al. 2005) can be applied to find optimal process parameters, while reinforcement learning adapts strategies based on real-time feedback. Genetic algorithms and reinforcement learning are best suited for process optimization ensuring maximum water use efficiency. This may help in reducing the water footprint in the industrial sector.

(iii) *Water Recycling and Treatment Optimization.*
Many industrial complexes are mandated to have effluent treatment plants mandatorily so that the wastewater can be treated at the source. AI assists in optimizing water treatment processes, determining the optimal conditions for recycling, and predicting water quality parameters (Nayak et al. 2018; Van Fan et al. 2020). This ensures that recycled water meets the required standards for industrial use. Optimization algorithms and neural networks for water quality prediction are used for this purpose.

(iv) *Smart Water Metering and Control Systems.*
Smart water metering systems with AI capabilities enable real-time monitoring and control of water usage (Günther et al. 2015). Control theory for adaptive control algorithms adjusts water flow and pressure based on demand fluctuations, ensuring efficient utilization. This helps in reducing the overuse and wastage of water resources.

(v) *Water Price Forecasting and Budgeting.*
Like energy price forecasting and budgeting, AI algorithms like time series analysis and machine learning predictive models analyze historical water pricing data and consumption patterns to forecast future water prices (Lee and Ko 2011). This assists both consumers and utility companies supplying water in budgeting and financial planning.

(vi) *Water Risk Assessment and Scenario Planning.*
AI models assess potential water-related risks and simulate different scenarios. The Monte Carlo simulation algorithm (Manno 1999) can help industries prepare for water scarcity events, regulatory changes, and other uncertainties.

These AI applications empower industries to optimize water demand management, reduce water consumption, and enhance overall sustainability in water resource usage. The specific algorithms chosen depend on the complexity of the industrial processes, available data, and the desired outcomes for water efficiency.

C. **Domestic Water Sector**.
Currently, the water requirement of the domestic sector globally is nearly 10% of the total water supply. Considering the large population still does not have access to safe, reliable, and sufficient water supply, expected population growth will put further pressure on available water resources. AI can play a pivotal role in optimizing both domestic water demand and supply management, ensuring efficient use of water resources. Here are key applications and associated algorithms in the context of domestic water demand and supply management:

(i) *Smart Water Metering and Leak Detection.*
AI models can analyze water consumption patterns in real time, identifying anomalies that may indicate leaks or inefficient water use. Smart meters equipped with AI capabilities (Salloom et al. 2022) provide valuable data for both consumers and utility companies. Anomaly detection and time series analysis models are used for leak detection water consumption patterns.

(ii) *Optimization of Water Distribution Networks.*
AI can help in designing efficient water distribution networks (Parvaze et al. 2023), considering factors such as pipe diameter, pump operation, water demand, and reservoir levels. Genetic algorithms find optimal solutions for network design and operation.

(iii) *Predictive Maintenance for Water Infrastructure.*
Failure and leaks in water infrastructure can cause serious losses and interrupt the water supply. AI models can predict potential failures or leaks in water infrastructure by analyzing data from sensors and historical maintenance records (Beh et al. 2017). Predictive maintenance reduces downtime and prevents water losses in the distribution system.

(iv) *Water Quality Monitoring of Domestic Water Supply.*
AI applications monitor water quality parameters in domestic water supplies, ensuring that tap water meets safety standards (Maroju et al. 2023). Sensor fusion enhances the accuracy of water quality assessments. Sensor fusion and neural network-like models are helpful for real-time and predictive water quality monitoring.

(v) *Behavioral Analytics for Water Conservation.*
Based on household-level water consumption patterns and user behaviors, AI can be used to provide personalized recommendations for water conservation (Arsene et al. 2023). This helps in nudging individual behavior toward sustainable water use. Clustering analysis ML models can be used for this purpose.

(vi) *Emergency Response Systems.*
AI can predict and recognize patterns associated with water emergencies, enabling rapid response to leaks, contamination events, or supply disruptions. This enhances the resilience of the water supply system.

The application of AI in domestic water management holds immense promise for revolutionizing the way households manage and conserve water resources. By leveraging AI-driven technologies such as smart meters, predictive analytics, and automated leak detection systems, households can optimize water usage, detect inefficiencies, and mitigate wastage in real-time.

D. **AI in Domestic and Industrial Wastewater Management**.
Freshwater resources are under serious threat from harmful pollutants caused by human activities (Ray et al. 2023). A vast amount of wastewater is produced every day from domestic and industrial sources; most of it contains toxic pollutants and is directly released into the environment without being treated (Boretti and Rosa 2019). AI models are becoming more and more popular in waste water-related fields. They have been employed for the prediction and optimization of the wastewater treatment process. Here are key applications and associated algorithms in the context of domestic and industrial wastewater management:

(i) ***Real-Time Monitoring and Control.***
AI-driven monitoring systems collect real-time data from sensors in wastewater treatment plants (Zhang et al. 2020) and process them to ensure the smooth functioning of the treatment process. Control algorithms (Lee et al. 2008) optimize treatment processes, adjusting parameters like chemical dosing and aeration in response to changing conditions.

(ii) ***Predictive Maintenance for Treatment Facilities.***
Potential issues in wastewater treatment infrastructure can be predicted by AI systems analyzing historical data and patterns (Ribalta et al. 2021). Machine learning predictive models, anomaly detection helps prevent equipment failures, reducing downtime and improving overall system reliability.

(iii) ***Effluent Quality Monitoring.***
AI models monitor effluent quality based on various parameters, to maintain it with environmental standards. Neural networks (Xie et al. 2022) analyze complex relationships within wastewater composition to improve monitoring accuracy and alert or adjust the treatment process accordingly.

(iv) ***Optimization of Treatment Processes.***
The wastewater treatment process involves complex design and operating conditions. AI algorithms can optimize treatment processes. AI models can establish effective complex relationships between the various input and output variables, thereby finding optimal configurations and operational parameters (Nayak et al. 2018). Genetic algorithms iteratively search for solutions that maximize treatment efficiency.

(v) ***Resource Recovery and Reuse.***
AI can facilitate the identification and process of opportunities for resource recovery from wastewater (Abidli et al. 2022), such as extracting energy, nutrients, and metals. Reinforcement learning can optimize strategies for safe resource recovery from wastewater.

(vi) ***Energy Efficiency in Wastewater Treatment.***
The wastewater treatment process is highly energy-intensive. AI models can regulate energy consumption in wastewater treatment processes by adjusting equipment operation based on the changes in the variables (Oulebsire et al. 2019). Machine learning models like optimization algorithms forecast energy requirements for efficient management.

(vii) ***Smart Sewer Systems.***
AI-enabled smart sewer systems monitor sewer networks, detecting anomalies, blockages, or overflows (Huber et al. 2006). Real-time analytics and data fusion enhance the overall performance of sewer infrastructure.

(viii) ***Regulatory Compliance and Reporting.***
Regulatory compliances are an essential part of wastewater treatment facilities. AI applications assist in regulatory compliance by interpreting environmental regulations (Amor and Dimyadi 2021) through natural language processing (NLP). Rule-based systems automate compliance checks and reporting processes.

These AI applications contribute to sustainable wastewater management practices, promoting efficient treatment, reducing environmental impact, and ensuring compliance with regulatory standards. The specific algorithms selected depend on the complexity of the wastewater treatment processes and the desired outcomes for optimization and environmental protection.

E. **Climate Variability and Change**.
Climate change and artificial intelligence are among the most powerful trends of this century. The ongoing shifts in climate patterns, including alterations in precipitation, temperature, and the frequency of extreme weather events, underscore the urgency of innovative solutions (IPCC 2014). Rising global temperatures and unpredictable climatic phenomena demand adaptive strategies that can dynamically respond to emerging challenges, ensuring the continued availability and reliability of water resources. AI plays a crucial role in particularly three important and relevant areas: mitigation, adaption, and resilience.

Here are key applications and associated algorithms in the context of using AI for assessing and addressing the impacts of climate change on water resources:

(i) ***Climate Modeling and Prediction.***
Climate modeling at any geographic scale is a very complex process because of the number of variables involved. Sophisticated climate models and accurate predictions of temperature, precipitation, and other climate variables can be done using advanced ML models (Schneider et al. 2017). Ensemble learning combines multiple models to provide more accurate and robust climate predictions.

(ii) ***Extreme Event Prediction.***
Deep learning models based on neural networks (Liu et al. 2016) analyze historical climate data to predict extreme weather events such as floods, droughts, heat waves, and cyclones. This information can aid in preparedness and response planning and minimizing the loss of life and property.

(iii) ***Hydrological Modeling and Flood Forecasting.***
ML algorithms enhance hydrological models by assimilating, analyzing, and interpreting real-time data from various sources (Ghiggi et al. 2019). Support vector machines and other machine learning techniques improve flood forecasting accuracy, providing early warnings to vulnerable regions.

(iv) ***Water Availability and Scarcity Assessment.***
Data from remote sensing satellites can be processed by AI systems to assess water availability and identify regions at risk of water scarcity (Yang et al. 2016). Data analytics algorithms analyze historical and real-time information to predict future water availability trends.

(v) ***Vulnerability Assessment of Water Infrastructure.***
Risk assessment models can be created for the evaluation of the vulnerability of water infrastructure to climate change impacts (Catani 2021). These models can help in the mitigation of the hazards associated with to failure of this infrastructure. GIS integrates spatial data to assess potential threats and AI models plan adaptive measures.

(vi) ***Ecosystem Monitoring and Conservation.***
Ecosystem degradation due to human actions and the ill effects of climate change is rapidly increasing. AI applications use random forest classifiers (Rodriguez-Galiano et al. 2012), computer vision, and image processing to monitor ecosystems affected by climate change. This includes assessing changes in vegetation, land use, and water bodies for conservation efforts.

(vii) ***Drought Prediction and Management.***
AI models analyze historical climate and precipitation data and combine them with the real-time scenario to predict drought conditions. Machine learning predictive models (Deo and Sahin 2015) provide early warnings, allowing for proactive drought management strategies.

(viii) ***Climate-Resilient Water Resource Planning***.
Decision support systems aid in climate-resilient water resource planning. AI can revolutionize these systems with its ability to synthesize large multivariable data and comprehend them in various scenarios. Multi-criteria decision analysis (MCDA) models consider multiple criteria to prioritize and adapt water management strategies in response to climate change.

(ix) ***Adaptive Governance and Policy Development***.
AI, through natural language processing (NLP) and sentiment analysis, assesses public opinion and stakeholder sentiments related to climate change policies (Surminski et al. 2018). This information supports adaptive governance and policy development for sustainable water resource management.

These AI applications contribute to a comprehensive understanding of climate change impacts on water resources and empower decision-makers to implement adaptive strategies for sustainable water management. The choice of models depends on the specific objectives of the applications and the type of data available for analysis.

F. **AI in Surface Water Storage Management**.
When a dam or similar obstruction is placed across the natural or man-made flow of the water, reservoirs are formed upstream. Reservoirs together with dams are designed to control the flow that allows the management of water resources, permitting storage of water for consumption, electricity production, protection against floods, and many other applications (Castelletti et al. 2008). AI has immense potential in surface water storage management to optimize reservoir operations, enhance water storage efficiency, and improve decision-making processes. Here are key applications and associated algorithms in the context of AI for surface water storage management:

(i) ***Reservoir Operation Optimization***.
Reinforcement learning and genetic algorithms can be utilized to optimize reservoir operations. These algorithms learn from historical data and real-time information to develop strategies that maximize water storage for summers and release based on changing conditions (Karunanithi et al. 1994).

(ii) ***Real-Time Water Level Prediction***.
Real-time water level information is essential for reservoir operation and management. AI models analyze historical water level data and meteorological information to predict real-time water levels in reservoirs (Hipni et al. 2013). Time series analysis and machine learning predictive models aid in proactive decision-making for storage and release operations.

(iii) ***Hydrological Modeling and Forecasting***.
Random forest (Zarei et al. 2021), support vector machines, and other machine learning algorithms improve the accuracy of precipitation forecast, streamflow, and sediment inflow predictions. These accurate predictions help update hydrological models created for reservoir management.

(iv) ***Climate Change Impact Assessment.***
The impact of climate change on surface water storage is uncertain and depends upon many parameters. Ensemble learning combines multiple models for climate predictions, while Bayesian networks help evaluate the uncertainty (Sperotto et al. 2017) associated with climate change scenarios. These AI models can help in forecasting the impact of climate change on storage management.

(v) ***Optimization of Water Release Policies.***
Dams and reservoirs are created for multipurpose needs like water for irrigation, power generation, flood control, environmental flow, fisheries, etc. Multi-objective optimization models and fuzzy logic can assist in developing water release policies that balance competing needs (Wang et al. 2009).

(vi) ***Sedimentation Prediction and Management.***
Sedimentation in the reservoir not only reduces the capacity but also has the potential to damage the dam infrastructure. AI models, including neural networks and ensemble models, predict sedimentation levels in reservoirs (Nourani et al. 2021). This information guides reservoir management strategies to mitigate the impact of sedimentation on water storage capacity.

(vii) ***Dam Infrastructure Safety Monitoring.***
Dam failures due to various reasons especially because of aging are becoming common, putting the downstream life and property in danger. AI-enabled IoT analytics and structural health monitoring models assess the condition of dam infrastructure (Hariri-Ardebili et al. 2023). This ensures early detection of potential issues, contributing to the overall safety and reliability of surface water storage facilities.

(viii) ***Water Quality Monitoring and Management.***
AI applications monitor and predict water quality parameters in surface water storage (Batur et al. 2018), ensuring that stored water meets safety standards. Sensor fusion enhances the accuracy of water quality assessments.

(ix) ***Adaptive Water Storage Planning.***
AI-driven decision support systems aid in adaptive water storage planning (Hossain et al. 2014). Reinforcement learning algorithms adapt storage plans based on real-time data, climate conditions, and changing water demand.

These AI applications in surface water storage management contribute to optimizing reservoir operations, improving water storage efficiency, and ensuring the sustainable utilization of surface water resources. The selection of algorithms depends on the specific objectives, data availability, and characteristics of the surface water storage system.

G. **Groundwater Resource Management.**
Groundwater, stored in aquifers, acts as insurance against the water crisis. However, increased global demand for water has led to the indiscriminate

exploitation of this partially renewable resource. Since groundwater is a complex system and direct groundwater monitoring for large areas is difficult, the use of artificial intelligence algorithms can be very helpful in its management (Basahi et al. 2018). Artificial intelligence is increasingly employed in groundwater resource management to optimize extraction, monitor aquifer health, and improve decision-making processes. Here are key applications and associated algorithms in the context of AI for groundwater resource management:

(i) ***Groundwater Extraction Optimization.***
Unsustainable groundwater extraction is the biggest management challenge today. AI algorithms optimize groundwater extraction strategies by learning from historical data and real-time information. Reinforcement learning and genetic algorithms (Perea et al. 2020) contribute to developing sustainable extraction plans that balance water demand with aquifer health.

(ii) ***Aquifer Health Monitoring.***
Aquifer health can be determined from broadly two parameters. First is the water quantity and second is water quality. If the aquifer is desaturated beyond a limit, the recovery may not be possible. The same is true with groundwater contamination; if the aquifer is contaminated, then it becomes very difficult to mitigate it. AI models analyze historical groundwater levels and quality data to monitor the health of aquifers. Predictive models and time series analysis help identify trends, anomalies, and potential issues related to groundwater storage and quality.

(iii) ***Real-Time Groundwater Level Prediction.***
Groundwater is an invisible resource; the water level acts in a pulse of the aquifer. AI models utilize historical groundwater level data and meteorological information to predict real-time groundwater levels. Support vector machines and neural networks (Mohanty et al. 2015) contribute to accurate and timely predictions for effective groundwater resource management.

(iv) ***Groundwater Recharge Estimation.***
Aquifers get recharged from rainfall or due to lateral and vertical flows within the system. Till the extraction is balanced by the recharge, especially from rainfall, the exploitation may be considered sustainable. AI applications process data from remote sensing satellites and machine learning models to estimate groundwater recharge (Samani et al. 2022). These tools aid in understanding the replenishment of aquifers and optimizing groundwater resource availability.

(v) ***Climate Change Impact Assessment.***
Climate change can impact groundwater in many ways. AI can be applied to assess the impact of climate change on groundwater resources (Coppola et al. 2003). Ensemble learning combines multiple climate models based on ANN, and Bayesian networks help evaluate the uncertainty associated with climate change scenarios affecting groundwater availability.

(vi) ***Salinity and Contaminant Monitoring.***
Severe contamination or saline ingress may adversely affect the aquifers. AI applications monitor and predict groundwater salinity and contaminant levels (Singha et al. 2021) so that mitigation measures can be taken in advance. Sensor fusion enhances the accuracy of water quality assessments, ensuring sustainable groundwater resource management practices.

(vii) ***Smart Well Infrastructure Monitoring.***
AI-enabled IoT analytics and structural health monitoring assess the condition of well infrastructure (Chen et al. 2017). This ensures early detection of potential issues, contributing to the overall safety and reliability of groundwater resource management.

(viii) ***Decision Support Systems for Groundwater Management.***
Decision support systems (Ataei et al. 2024) aid in groundwater management decisions. Reinforcement learning algorithms adapt extraction plans based on real-time data, changing aquifer conditions, and evolving water demand.

These AI applications in groundwater resource management contribute to optimizing extraction, maintaining aquifer health, and ensuring sustainable utilization of groundwater resources.

H. **AI in Water Resource Policy Formulation.**
Water governance can be defined as sociopolitical, economic, and administrative systems that regulate water use and resource management. It determines who gets what water, when and how, and who has the right to water. Generally, there a policies and laws related to water, and then rules are framed with institutional mechanisms to enforce them. AI can play a significant role in shaping water resource policy formulation, enforcement, monitoring, and impact assessment by providing advanced tools and methodologies for data analysis, decision support, and scenario planning. Here are key applications and associated aspects of AI in water resource policy formulation:

(i) ***Data-Driven Policy Insights.***
AI-based systems can process large datasets, including historical records, satellite imagery, and sensor data, to extract valuable insights. These insights inform policymakers about water availability, usage patterns, and environmental conditions, supporting evidence-based policy formulation (Reis et al. 2019a). The policy so formulated will be more practical and evidence-based.

(ii) ***Optimization of Water Allocation.***
Domestic-agriculture-industrial sectors have competing water demands. AI-driven optimization algorithms assist in allocating water resources efficiently (Desouza et al. 2020). These algorithms consider various factors, including demand, environmental constraints, and infrastructure limitations, to optimize water distribution and usage, guiding policy decisions.

(iii) ***Regulatory Compliance Monitoring.***
Regulatory institutions get overwhelmed in compliance monitoring as per law. AI, particularly NLP, processes regulatory documents and compliance

reports to ensure adherence to water governance policies (Amor and Dimyadi 2021). Machine learning models can assess and monitor compliance, facilitating policy enforcement and adjustments.

(iv) ***Scenario Planning and Risk Assessment***.
Modeling techniques such as Bayesian networks facilitate scenario planning and risk assessment as per changing supply and demand variables (Reis et al. 2019b). Decision-makers can use these tools to evaluate the potential impacts of different policy decisions on water resources and ecosystems

(v) ***Policy Impact Simulation***.
Impact assessment is an important aspect of any governance model. It provides valuable insights into the areas for improvement. AI-driven simulation models and decision support systems allow policymakers to simulate the impact of proposed or enforced policies (Valle-Cruz et al. 2019). This aids in understanding how different policy scenarios may affect water availability, quality, and usage, helping refine policy formulations and implementation.

(vi) ***Dynamic Policy Adaptation***.
AI can support dynamic policy adaptation based on changing conditions (Dwivedi et al. 2019). These systems particularly reinforcement learning and adaptive systems continuously learn from real-time data, allowing policies to evolve in response to emerging challenges or opportunities.

(vii) ***Policy Communication and Transparency***.
Clear, top-down, and bottom-up communication is key for regulation. AI aids in generating clear and accessible policy communication (Androutsopoulou et al. 2019) through natural language generation (NLG). Additionally, data visualization tools help present complex water resource information to policymakers and the public, enhancing transparency in policy formulation.

These applications showcase how AI technologies can assist policymakers in crafting more informed, adaptive, and effective water resource policies. The integration of AI contributes to evidence-based decision-making and the development of policies that address the dynamic challenges of water resource management.

3.9 Challenges and Risks

We have discussed numerous advantages of AI applications in water resource management. They range from the global scale to the household level. However, applications of AI are not free from the challenges and risks. If not anticipated and mitigated, may give rise to serious and unexpected problems (Richards et al. 2023). Hence identification of challenges and risks and their mitigation must be responsibly and preemptively done to ensure effectiveness and ethical use. Striking a balance between the benefits of AI-driven innovation and the protection of privacy,

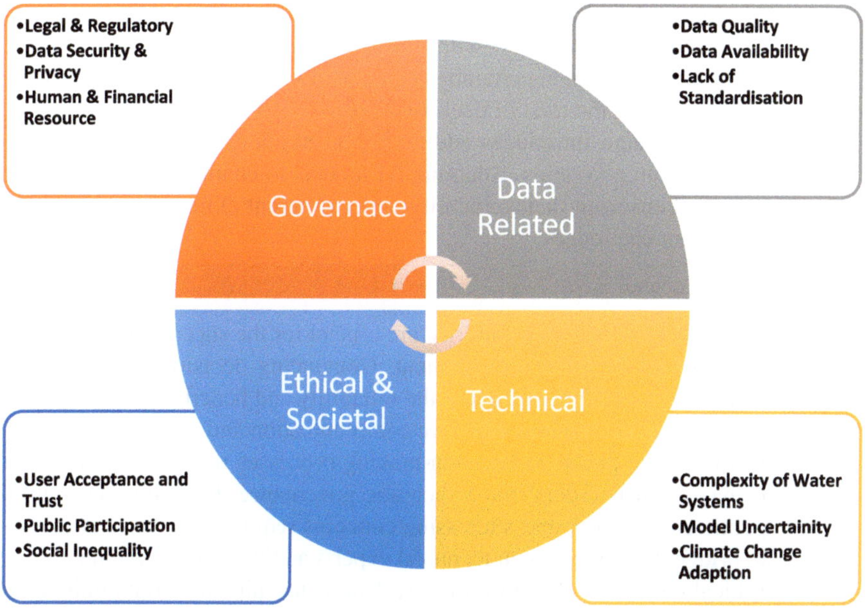

Fig. 3.3 Challenges associated with the application of AI in Sustainable Water Resource Management

equity, and ethical principles remains a formidable challenge in sustainable water resource management. Challenges related to AI application in SWRM can be divided into four groups: data-related, technical, ethical and social, and governance-related as given in Fig. 3.3.

(a) Data related.

(i) AI models rely on high-quality data, ensuring the integrity of the data is a big challenge and if the data used for training the models is biased or of poor quality, it can lead to biased predictions and inaccurate decision-making.

(ii) AI algorithms require abundant data. In many cases, water-related data may be sparse, inconsistent, or of varying quality, which can affect the accuracy and reliability of AI models.

(iii) There is a lack of standardized formats and protocols for water data collection and sharing. Inconsistent data formats and communication protocols hinder the seamless integration of AI applications across different water management systems.

(b) Technical.

(i) Water resource systems are inherently complex, with interconnected variables and dynamic interactions. Developing AI models that can accurately represent and predict the behavior of such complex systems is challenging.

(ii) AI models, especially complex ones like deep learning algorithms, can be perceived as "black boxes" with limited explainability. Understanding and communicating the uncertainties associated with AI predictions is essential for gaining the trust of decision-makers and stakeholders.
(iii) Climate change introduces additional uncertainties and complexities to water resource systems. Adapting AI models to changing climate conditions and ensuring their reliability under different climate scenarios pose significant challenges.

(c) Ethical and societal.

(i) Stakeholder acceptance and trust are crucial for the successful implementation of AI in water management. Convincing decision-makers, water managers, and the public about the reliability and benefits of AI models is a challenge that requires effective communication and education.
(ii) Involving the public in decision-making processes related to AI in water management is crucial for democratic governance. Ensuring meaningful public engagement and addressing concerns require innovative strategies to bridge the gap between technical experts and the general public.
(iii) Ethical concerns related to privacy, data ownership, and community consent need to be carefully addressed when implementing AI in water management. Additionally, ensuring that AI technologies benefit all members of society and do not exacerbate existing disparities is crucial.

(d) Governance.

(i) Ensure compliance with existing regulations and develop new regulations where necessary to govern the use of AI in water management. Establish clear guidelines for ethical conduct, data protection, and accountability within the regulatory framework.
(ii) Defining ownership and governance structures for water-related data poses a challenge. Legal frameworks must address issues related to data ownership, sharing, and access rights among stakeholders in the water management ecosystem.
(iii) The successful implementation of AI in water management requires a workforce with the necessary skills in both water science and AI technologies. Building the capacity of professionals to understand, operate, and interpret AI results is a challenge that requires targeted training programs. High upfront costs, including infrastructure, training, and maintenance, must be justified by the long-term benefits in terms of improved water management efficiency and sustainability.

Risks are intertwined with the challenges associated with AI applications. Each challenge if not addressed properly will lead to an associated risk. Implementation skills alone are insufficient; without robust mechanisms to identify and remediate errors, AI can do more harm than good. There are serious risks associated with the AI models from catchment to the household level. Water security is widely regarded as a vital aspect of national security frameworks in numerous countries. Given this significance, any potential risks stemming from the use of AI in water resources management could result in severe negative consequences.

3.10 Conclusion

Challenges in water resource management are complex, encompassing a range of interconnected variables. One of the primary challenges is water scarcity, exacerbated by factors such as population growth, urbanization, and climate change, which intensify competition for limited freshwater supplies. Additionally, water pollution from industrial, agricultural, and urban sources threatens the quality of already scarce resources. Inequitable access to water resources can lead to conflicts over water rights and usage. Insufficient regulatory frameworks and fragmented governance structures hinder coordinated action and long-term planning. Addressing these challenges necessitates integrated approaches that prioritize sustainability, resilience, and equity while fostering collaboration among diverse stakeholders.

Leveraging innovative technologies in management strategies can be a game changer in water resource management. Throughout this discourse, we have explored how AI tools and methodologies can revolutionize the way we perceive, utilize, and conserve water resources. AI presents a powerful means of optimizing water allocation and distribution. AI enables more precise forecasting of water availability and facilitates real-time decision-making to ensure efficient utilization of limited resources across sectors. By leveraging advanced data analytics and machine learning algorithms, water managers and stakeholders can gain invaluable insights related to resource management. In climate change scenarios, with the help of various simulations and assessing their potential impacts, decision-makers can proactively identify and mitigate risks such as droughts, floods, and water quality degradation. Furthermore, AI enables the development of intelligent monitoring and control systems that can respond dynamically to changing conditions.

However, while the promise of AI in water management is undeniable, it is essential to recognize and address potential challenges and limitations. These include concerns related to data privacy, algorithm bias, and the ethical implications of automated decision-making. Additionally, the adoption of AI may require significant investments in infrastructure, capacity building, and stakeholder engagement to ensure inclusive and equitable outcomes.

In conclusion, the application of AI in sustainable water resource management represents a paradigm shift toward a more integrated, adaptive, and resilient approach to water governance. By harnessing the power of AI, we can unlock new opportunities for innovation, collaboration, and collective action toward achieving water security for present and future generations. However, realizing this vision requires a concerted effort to harness the full potential of AI while addressing its associated challenges and ensuring ethical and responsible deployment. Through strategic investment, collaboration, and commitment, we can harness the transformative potential of AI to safeguard our most precious resource—water.

References

Abbott BW, Bishop K, Zarnetske JP et al (2019) Human domination of the global water cycle is absent from depictions and perceptions. Nat Geosci 12:533–540. https://doi.org/10.1038/s41561-019-0374-y

Abidli A, Huang Y, Ben Rejeb Z, Zaoui A, Park CB (2022) Sustainable and efficient technologies for removal and recovery of toxic and valuable metals from wastewater: recent progress, challenges, and future perspectives. Chemosphere 292:133102

Alsaiari AO, Moustafa EB, Alhumade H, Abulkhair H, Elsheikh A (2023) Coupled artificial neural network with artificial rabbits optimizer for predicting water. Adv Eng Softw 175:103315., ISSN 0965-9978. https://doi.org/10.1016/j.advengsoft.2022.103315

Amor R, Dimyadi J (2021) The promise of automated compliance checking. Dev Built Environ 5:100039., ISSN 2666-1659. https://doi.org/10.1016/j.dibe.2020.100039

Androutsopoulou A, Karacapilidis N, Loukis E, Charalabidis Y (2019) Transforming the communication between citizens and government through AI-guided chatbots. Gov Inf Q 36(2):358–367., ISSN 0740-624X,. https://doi.org/10.1016/j.giq.2018.10.001

Arsene D, Predescu A, Truică C-O, Apostol E-S, Mocanu M (2023) Decision support strategies for household water consumption behaviors based on advanced recommender systems. Water 15:2550

Ataei P, Takhtravan A, Gheibi M, Chahkandi B, Faramarz MG, Wacławek S, Fathollahi-Fard AM, Behzadian K (2024) An intelligent decision support system for groundwater supply management and electromechanical infrastructure controls. Heliyon:e25036., ISSN 2405-8440. https://doi.org/10.1016/j.heliyon.2024.e25036

Basahi JM, Masoud MHZ, Rajmohan N (2018) Effect of a flash flood on trace metal pollution in the groundwater-WadiBaysh Basin, western Saudi Arabia. J Afr Earth Sci 147:338–351

Beh EHY, Zheng F, Dandy GC, Maier HR, Kapelan Z (2017) Robust optimization of water infrastructure planning under deep uncertainty using metamodels. Environ Model Softw 93:92–105

Bishop CM (2006) Pattern recognition and machine learning. Springer, New York. ISBN 978-0-387-31073-2

Biswas AK (2004) Integrated water resources management: a reassessment – a water forum contribution. Water Int 29(2):248–256

Blasi AH, Abbadi MA, Al-Huweimel R (2021) Machine learning approach for an automatic irrigation system in Southern Jordan Valley. Eng Technol Appl Sci Res 11(1):6609–6613. https://doi.org/10.48084/etasr.3944

Boretti A, Rosa L (2019) Reassessing the projections of the world water development report. NPJ Clean Water 2:15. https://doi.org/10.1038/s41545-019-0039-9

Braga, B., Chartres, C., Cosgrove, W.J., da Cunha, L.V., Gleick, P.H., Kabat, P., Kadi, M.A., Loucks, D.P., Lundqvist, J., Narain, S. and Xia, J., 2014. Water and the future of humanity: revisiting water security. Calouste Gulbenkian Foundation

Brutsaert W (2005) Hydrology: an introduction. Cambridge University Press, Cambridge

Burek P, Satoh Y, Fischer GW, Kahil MT, Scherzer A, Tramberend S, Nava LF, Wada Y, Eisner S, Flörke M, Hanasaki N, Magnuszewski P, Cosgrove B, Wiberg D (2016) Water futures and solution – fast track initiative (final report)

Castelletti A, Pianosi F, Soncini-Sessa R (2008) Water reservoir control under economic, social and environmental constraints. Automatica 44:1595–1607. https://doi.org/10.1016/j.automatica.2008.03.003

Catani F (2021) Landslide detection by deep learning of non-nadiral and crowdsourced optical images. Landslides 18:1025–1044

CGWB (2023) National Compilation on Dynamic Ground Water Resource Assessment of India

Chang FJ, Wang YC, Tsai WP (2016) Modelling intelligent water resources allocation for multi-users. Water Resour Manag 30:1395–1413

Chen G, Long T, Xiong J, Bai Y (2017) Multiple random forests modeling for urban water consumption forecasting. Water Resour Manag 31(15):4715–4729

Coppola E Jr, Szidarovszky F, Poulton M, Charles E (2003) Artificial neural network approach for predicting transient water levels in a multilayered groundwater system under variable state, pumping, and climate conditions. J Hydrol Eng 8(6):348–360
De A, Bose R, Kumar A, Mozumdar S (2014) Targeted delivery of pesticides using biodegradable polymeric nanoparticles. Springer Briefs in Molecular Science
Deo RC, Sahin M (2015) Application of the extreme learning machine algorithm for the prediction of monthly Effective Drought Index in Eastern Australia. Atmos Res 153:512–525
Desouza KC, Dawson GS, Chenok D (2020) Designing, developing, and deploying artificial intelligence systems: lessons from and for the public sector. Bus Horiz 63(2):205–213. https://doi.org/10.1016/j.bushor.2019.11.004
Durack PJ (2015) Ocean salinity and the global water cycle. Oceanography 28(1):20–31. https://doi.org/10.5670/oceanog.2015.03
Dwivedi YK, Rana NP, Jeyaraj A et al (2019) Re-examining the unified theory of acceptance and use of technology (UTAUT): towards a revised theoretical model. InfSyst Front 21:719–734. https://doi.org/10.1007/s10796-017-9774-y
Ghiggi G, Humphrey V, Seneviratne SI, Gudmundsson L (2019) GRUN: an observations-based global gridded runoff dataset from 1902 to 2014. Earth Syst Sci Data Discuss 2019:1–32
Ghobadi F, Kang D (2023) Application of machine learning in water resources management: a systematic literature review. Water 15:620. https://doi.org/10.3390/w15040620
Goap A, Sharma D, Shukla AK, Rama Krishna C (2018) An IoT-based smart irrigation management system using Machine learning and open source technologies. Comput Electron Agric 155:41–49. https://doi.org/10.1016/j.compag.2018.09.040
Günther M, Camhy D, Steffelbauer D, Neumayer M, Fuchs-Hanusch D (2015) Showcasing a smart water network based on an experimental water distribution system. Proc Eng 119:450–457
Haddeland I, Heinke J, Biemans H, Eisner S, Flörke M, Hanasaki N, Ludwig F, Masaki Y, Schewe J, Stacke T (2014) Global water resources are affected by human interventions and climate change. Proc Natl Acad Sci U S A 111(9):3251–3256
Hariri-Ardebili MA, Mahdavi G, Nuss LK, Lall U (2023) The role of artificial intelligence and digital technologies in dam engineering: narrative review and outlook. Eng Appl Artif Intell 126(Part A):106813., ISSN 0952-1976. https://doi.org/10.1016/j.engappai.2023.106813
Hipni A, El-shafie A, Najah A, Karim OA, Hussain A, Mukhlisin M (2013) Daily forecasting of dam water levels: comparing a support vector machine (SVM) model with adaptive neuro-fuzzy inference system (ANFIS). Water Resour Manag 27(10):3803–3823. https://doi.org/10.1007/s11269-013-0382-4
Hipsey M, Arheimer B (2013) Challenges for water-quality research in the new IAHS decade on: hydrology under societal and environmental change. IAHS-AISH Proc Rep 361:17–29
Huber WC, Rossman LA, Dickinson RE, Singh VP, Fervert DK (eds) (2006) EPA Storm Water Management Model, SWMM5 Chapter 14 in Watershed models. CRC Press, Boca Raton. ISBN 13: 978-0-8493-3609-6, ISBN 10: 0-8493- 3609-0
IEA (2012) World energy outlook 2012. IEA, Paris. https://www.iea.org/reports/world-energy-outlook-2012. Licence: CC BY 4.0
International Energy Agency (IEA) (2012) Chapter 17: Water for energy: is energy becoming a thirstier resource? In: World energy outlook 2012. IEA, Paris. www.iea.org/publications/freepublications/publication/WEO2012_free.pdf
IPCC (2013) Climate change 2013: the physical science basis. In: Stocker TF et al (eds) Contribution of Working Group I to the Fifth Assessment Report of the Intergovernmental Panel on Climate Change. Cambridge University Press, Cambridge/New YorkUSA, p 1535. https://doi.org/10.1017/cbo9781107415324
IPCC (2014) Climate change 2014: synthesis report. In: Core Writing Team, Pachauri RK, Meyer LA (eds) Contribution of Working Groups I, II, and III to the Fifth Assessment Report of the Intergovernmental Panel on Climate Change. IPCC, Geneva, 151 pp
Karandish F, Šimůnek J (2016) A comparison of numerical and machine-learning modeling of soil water content with limited input data. J Hydrol 543:892–909. https://doi.org/10.1016/j.jhydrol.2016.11.007

Karunanithi N, Grenney WJ, Whitley D, Bovee K (1994) Neural networks for river flow prediction. J Comput Civ Eng 8:201–220

Kearns MJ, Vazirani UV, Vazirani U (1994) An introduction to computational learning theory. MIT Press

Kleidon A, Renner M (2013) Thermodynamic limits of hydrologic cycling within the Earth system: concepts, estimates and implications. Hydrol Earth Syst Sci 17:2873–2892. https://doi.org/10.5194/hess-17-2873-2013

Kumar R, Singh M, Kumar P, Singh J (2015) Crop selection method to maximize crop yield rate using machine learning technique. https://doi.org/10.1109/ICSTM.2015.7225403

Kumar S, Vijay, Sheshadri KN (2019) Applications of Artificial Intelligence in Academic Libraries. Int J Comput Sci Eng 7:136–140. https://doi.org/10.26438/ijcse/v7si16.136140

Lavric V, Iancu P, Pleşu V (2005) Genetic algorithm optimisation of water consumption and wastewater network topology. J Clean Prod 13(15):1405–1415. ISSN 0959-6526

Lee C-M, Ko C-N (2011) Short-term load forecasting using lifting scheme and Arima models. Expert Syst Appl 38(5):5902–5911. https://doi.org/10.1016/j.eswa.2010.11.033

Lee MW, Hong SH, Choi H, Kim J-H, Lee DS, Park JM (2008) Real-time remote monitoring of small-scaled biological wastewater treatment plants by a multivariate statistical process control and neural network-based software sensors. Process Biochem 43:1107–1113

Mamo TG, Juran I, Shahrour I (2013) Urban water demand forecasting using the stochastic nature of short-term historical water demand and supply pattern. J Water Resour Hydraul Eng 2:92–103

Manno I., Introduction to the Monte Carlo Method, AkadémiaiKiadó, Budapest, 1999

Maroju RG, Choudhari SG, Shaikh MK, Borkar SK, Mendhe H (2023) Application of artificial intelligence in the management of drinking water: a narrative review. Cureus 15(11):e49344. https://doi.org/10.7759/cureus.49344. PMID: 38146561; PMCID: PMC10749683

McCarthy J (2004) What is artificial intelligence?

Mekonnen M, Hoekstra A (2012) A global assessment of the water footprint of farm animal products. Ecosystems 15. https://doi.org/10.1007/s10021-011-9517-8

Milly PCD, Dunne KA (2016) Potential evapotranspiration and continental drying. Nat Clim Chang 6(10):946–949. https://doi.org/10.1038/nclimate3046

Mohammed M, Khan M.B., Bashier Mohammed B.E., Machine learning: algorithms and applications. CRC Press, 2016

Mohanty S, Jha MK, Raul SK, Panda RK, Sudheer KP (2015) Using artificial neural network approach for simultaneous forecasting of weekly groundwater levels at multiple sites. Water Resour Manag 29(15):5521–5532

Mohseni U, Agnihotri PG, Pande CB, Durin B (2023) Understanding the climate change and land use impact on streamflow in the present and future under CMIP6 climate scenarios for the Parvara Mula Basin, India. Water 15:1753

Nair KS (2004) Role of water in the development of civilization in India – a review of ancient literature, traditional practices and beliefs. Int Assoc Hydrol Sci 286:160–166

Navarro-Hellín H, Martínez-del-Rincon J, Domingo-Miguel R, Soto-Valles F, Torres-Sánchez R (2016) A decision support system for managing irrigation in agriculture. Comput Electron Agric 124:121–131

Nayak M, Dhanarajan G, Dineshkumar R, Sen R (2018) Artificial intelligence-driven process optimization for cleaner production of biomass with co-valorization of wastewater and fue gas in an algal biorefinery. J Claen Prod 201:1092–1100. https://doi.org/10.1016/j.jclepro.2018.08.048

Nourani V, Baghanam AH, Adamowski J, Kisi O (2014) Application of hybrid wavelet-artificial intelligence models in hydrology: a review. J Hydrol 514.358-377

Nourani V, Gokcekus H, Gelete G (2021) Estimation of suspended sediment load using artificial intelligence-based ensemble model. Complexity 2021(6633760):19. https://doi.org/10.1155/2021/6633760

Omidzade F, Ghodousi H, Shahverdi K (2020) Comparing Fuzzy SARSA learning and ant colony optimization algorithms in water delivery scheduling under water shortage conditions. J Irrig Drain Eng 146(9):04020028

Oulebsir R, Lefkir A, Safri A, Bermad A (2019) Optimization of the energy consumption in activated sludge process using deep learning selective modeling. Biomass Bioenergy 132:105420
Pahl-Wostl C (2009) A conceptual framework for analyzing adaptive capacity and multi-level learning processes in resource governance regimes. Glob Environ Chang 19:354–365
Parvaze S, Kumar R, Khan JN et al (2023) Optimization of water distribution systems using genetic algorithms: a review. Arch Comput Methods Eng 30:4209–4244
Perea RG, Moreno MÁ, da Silva Baptista VB, Córcoles JI (2020) Decision support system based on genetic algorithms to optimize the daily management of water abstraction from multiple groundwater supply sources. Water Resour Manag 34(15):4739–4755
Piao S, Friedlingstein P, Ciais P, de Noblet-Ducoudré N, Labat D, Zaehle S (2007) Changes in climate and land use have a larger direct impact than rising CO_2 on global river runoff trends. Proc Natl Acad Sci USA 104:15242–15247
Ray SS, Verma RK, Singh A, Ganesapillai M, Kwon Y-N (2023) A holistic review of how artificial intelligence has redefined water treatment and seawater desalination processes. Desalination 546:116221
Razavi S, Hannah DM, Elshorbagy A, Kumar S, Marshall L, Solomatine DP, Dezfuli A, Sadegh M, Famiglietti J (2022) Coevolution of machine learning and process-based modelling to revolutionize earth and environmental Sciences a perspective. Hydrol Process 36:e14596. https://doi.org/10.1002/hyp.14596
Reis J, Santo PE, Melão N (2019a) Impacts of artificial intelligence on public administration: a systematic literature review. In: 2019 14th Iberian conference on information systems and technologies (CISTI), pp 1–7. https://doi.org/10.23919/CISTI.2019.8760893
Reis J, Santo PMF d E, Melão NFR (2019b) Artificial intelligence in government services: a systematic literature review. In: Rocha Á, Adeli H, Reis L, Costanzo S (eds) New knowledge in information systems and technologies, vol 930. Springer. https://doi.org/10.1007/978-3-030-16181-1_23
Ribalta M, Mateu C, Bejar R, Rubión E, Echeverria L, Alegre FJV, Corominas L (2021) Sediment level prediction of a combined sewer system using spatial features. Sustainability 13:4013
Richards CE, Tzachor A, Avin S et al (2023) Rewards, risks, and responsible deployment of artificial intelligence in water systems. Nat Water 1:422–432. https://doi.org/10.1038/s44221-023-00069-6
Richey S, Thomas B, Lo M-H, Reager JT, Famiglietti J, Voss K, Swenson S, Rodell M (2015) Quantifying renewable groundwater stress with GRACE. Water Resour Res 51. https://doi.org/10.1002/2015WR017349
Rodriguez-Galiano VF, Ghimire B, Rogan J, Chica-Olmo M, Rigol-Sanchez JP (2012) An assessment of the effectiveness of a random forest classifier for land-cover classification. ISPRS J Photogramm Remote Sens 67:93–104
Saleh Z (2019) Artificial intelligence definition, ethics and standards
Salloom T, Kaynak O, Yu X, He W (2022) Proportional integral derivative booster for neural networks-based time-series prediction: case of water demand prediction. Eng Appl Artif Intell 108:104570. ISSN 0952-1976
Samani S, Vadiati M, Delkash M, Bonakdari H (2022) A hybrid wavelet–machine learning model for qanat water flow prediction. Acta Geophys 1:1–19
Sarker IH (2021) Machine learning: algorithms, real-world applications and research directions. SN Comput Sci 2:160. https://doi.org/10.1007/s42979-021-00592-x
Sauvé S, Desrosiers M (2014) A review of what is an emerging contaminant. Chem Cent J 8:15
Schewe J et al (2014) Multimodel assessment of water scarcity under climate change. Proc Natl Acad Sci 111(9):3245–3250. https://doi.org/10.1073/pnas.1222460110
Schneider T, Lan SW, Stuart A, Teixeira J (2017) Earth system modeling 2.0: a blueprint for models that learn from observations and targeted high-resolution simulations. Geophys Res Lett 44:12396–12417
Seyedzadeh A, Maroufpoor S, Maroufpoor E, Shiri J, Bozorg-Haddad O, Gavazi F (2020) Artificial intelligence approach to estimate discharge of drip tape irrigation based on temperature and pressure. Agric Water Manag 228:105905. https://doi.org/10.1016/j.agwat.2019.105905

Shi X, Han W, Zhao T, Tang J (2019) Decision support system for variable rate irrigation based on UAV multispectral remote sensing. Sensors (Basel) 19(13):2880. https://doi.org/10.3390/s19132880

Sidhu RK, Kumar R, Rana PS (2020) Long short-term memory neural network-based multi-level model for smart irrigation. Mod Phys Lett B 34(36):2050418. https://doi.org/10.1142/S0217984920504187

Singha S, Pasupuleti S, Singha SS, Singh R, Kumar S (2021) Prediction of groundwater quality using efficient machine learning technique. Chemosphere 276:130265

Snellen WB, Schrevel A (2004) IWRMforsustainable use of water: 50 years of international experience with the concept of integrated water management. In: Proceedings of The Netherlands conference on water for food and ecosystems, vol 31

Sperotto A, Molina J-L, Torresan S, Critto A, Marcomini A (2017) Reviewing Bayesian networks potentials for climate change impacts assessment and management: a multi-risk perspective. J Environ Manag 202(Part 1):320–331., ISSN 0301-4797,. https://doi.org/10.1016/j.jenvman.2017.07.044

Stocker TF, Qin D, Plattner GK, Tignor M, Allen SK, Boschung J, Nauels A, Xia Y, Bex V, Midgley PM 2013 Cambridge University Press, Cambridge/New York 33–115, https://doi.org/10.1017/cbo9781107415324.005

Surminski S, Manuela Di Mauro J, Baglee AR, Connell RK, Hankinson J, Haworth AR, Ingirige B, Proverbs D (2018) Assessing climate risks across different business sectors and industries: an investigation of methodological challenges at national scale for the UK. Philos Trans R Soc A Math Phys Eng Sci 376:20170307

Taylor CM et al (2013) Modeling soil moisture–precipitation feedback in the Sahel: importance of spatial scale versus convective parameterization. Geophys Res Lett 40(23):6213–6218. https://doi.org/10.1002/2013gl058511

Tindall JA, Campbell AA, U.S. Geological Surv., Water Security: National & Global Issues (2010). https://pubs.usgs.gov/fs/2010/3106/pdf/FS10-3106.pdf

United Nations (2015) Transforming our world: the 2030 Agenda for Sustainable Development

United Nations Department of Economic and Social Affairs (UNDESA) (2017) World population prospects: key findings and advance tables – the 2017 revision. Working Paper No. ESA/P/WP/248 (UNDESA, Population Division, New York

United Nations Environment Programme (UNEP) (2016) A snapshot of the world's water quality: towards a global assessment. UNEP, Nairobi

United Nations General Assembly (UNGA) (2017) Report on pesticides and human rights 2017

United Nations World Water Assessment Programme (WWAP) (2003) Water for people, water for life: the United Nations World Water Development Report. United Nations Educational, Scientific and Cultural Organization (UNESCO), Paris

United Nations World Water Assessment Programme (WWAP). The United Nations

United Nations World Water Assessment Programme (WWAP) (2014) The United Nations World Water Development Report, Water and Energy. UNESCO, Paris

Valle-Cruz D, Alejandro Ruvalcaba-Gomez E, Sandoval-Almazan R, Ignacio Criado J (2019) A review of artificial intelligence in government and its potential from a public policy perspective. In: Proceedings of the 20th annual international conference on digital government research, pp 91–99. https://doi.org/10.1145/3325112.3325242

Van FY, Chin HH, Klemeš JJ et al (2020) Optimisation and process design tools for cleaner production. J Clean Prod 247:119181

Veolia/IFPRI (International Food Policy Research Institute) (2016) The murky future of global water quality: new global study projects rapid deterioration in water quality. IFPRI/Veolia, Washington, DC/Chicago

Wada Y, Bierkens MF, Roo AD, Dirmeyer PA, Famiglietti JS, Hanasaki N, Konar M, Liu J, Müller Schmied H, Oki T, Pokhrel Y (2017) Human–water interface in hydrological modeling: current status and future directions. Hydrol Earth Syst Sci 21:4169–4193

Wang WC, Chau KW, Cheng CT, Qiu L (2009) A comparison of the performance of several artificial intelligence methods for forecasting monthly discharge time series. J Hydrol 374:294–306. https://doi.org/10.1016/j.jhydrol.2009.06.019

Wang J, Hong Y, Gourley J, Adhikari P, Li L, Su F (2010) Quantitative assessment of climate change and human impacts on long-term hydrologic response: a case study in a sub-basin of the Yellow River, China. Int J Climatol 30:2130–2137

Water Supply and Sanitation Collaborative Council (WSSCC) (2001) Ministerial conference on Freshwater; Water Supply and Sanitation Collaborative Council (WSSCC). Declaration of the Bonn International Germany, Bonn

World Bank Group (2016) High and dry: climate change, water, and the economy 1. https://openknowledge.worldbank.org/bitstream/handle/10986/23665/K8517%20Executive%20Summary.pdf

World Econ. Forum (2018) The global risks report 2018. http://www3.weforum.org/docs/WEF_GRR18_Report.pdf

World Water Assessment Programme (Nations Unies) (2018) The United Nations World Water Development Report. United Nations Educational, Scientific and Cultural Organization, New York

World Water Development Report (2014) Water and energy. UNESCO, Paris

Xie M (2006) Integrated water resources management (IWRM)- introduction to principles and practices. In: Africa regional workshop on IWRM. World Bank Institute, Nairobi

Xie Y, Chen Y, Lian Q, Yin H, Peng J, Sheng M, Wang Y (2022) Enhancing real-time prediction of effluent water quality of wastewater treatment plant based on improved feedforward neural network coupled with optimization algorithm. Water 14:1053

Yang Y, Guan H, Batelaan O et al (2016) Contrasting responses of water use efficiency to drought across global terrestrial ecosystems. Sci Rep 6:23284. https://doi.org/10.1038/srep23284

Yannopoulos SI, Lyberatos G, Theodossiou N, Li W, Valipour M, Tamburrino A, Angelakis AN (2015) evolution of water lifting devices (pumps) over the centuries worldwide. Water 7:5031–5060. https://doi.org/10.3390/w7095031

Zarei M, Najarchi M, Mastouri R (2021) Bias correction of global ensemble precipitation forecasts by random forest method. Earth Sci Inform 14:677–689. https://doi.org/10.1007/s12145-021-00577-7

Zhang W, Tooker NB, Mueller AV (2020) Enabling wastewater treatment process automation: leveraging innovations in real-time sensing, data analysis, and online controls. Environ Sci Water Res Technol 6:2973–2992

Artificial Intelligence for Monitoring and Forecasting Air Quality

4

Monika Yadav, Pragya Pathak, and Rituraj Kashyap

Abstract

Air quality degradation due to pollution is a pressing global issue impacting public health and environmental sustainability. Traditional monitoring methods often struggle to provide timely and precise data necessary for effective intervention. In recent years, artificial intelligence (AI) has emerged as a transformative tool in the field of air quality management. AI models leverage advanced algorithms and large-scale data analytics to enhance the accuracy, speed, and predictive capabilities of air quality monitoring systems. This abstract explores the role of AI in monitoring air quality, highlighting its ability to identify pollution sources, predict pollutant levels, and facilitate data-driven decision-making.

Keywords

Air quality · Artificial intelligence · Pollutants · AQI

4.1 Introduction

Governments and the general public have recently given air pollution in emerging nations more attention as it has become a critical issue due to the rapid development of industry and traffic (Zheng et al. 2016). The most significant element that directly affects disease incidence and lowers life quality in cities and metropolitan areas is

M. Yadav (✉)
Department of Applied Sciences & Humanities, Faculty of Engineering & Technology, Rama University, Kanpur, Uttar Pradesh, India

P. Pathak · R. Kashyap
Department of Computer Science & Engineering, Faculty of Engineering & Technology, Rama University, Kanpur, Uttar Pradesh, India

B. Thokchom et al. (eds.), *Artificial Intelligence Enabled Real Time Environmental Monitoring*, https://doi.org/10.1007/978-981-95-3767-9_4

the quality of the air (Cheng et al. 2014). Many nations, mostly in the industrialized world, have installed sizable air quality monitoring systems in urban areas as part of air pollution management programs to gather data in real time and notify authorities and citizens about their present state of the air (Refer to waqi.info.) (Montalvo et al. 2022). The fundamental role of environmental protection work is environmental monitoring and early warning, both in terms of long-term development and scientific decision-making (Yang and Wang 2017). Particulate matter (PM), nitrogen dioxide (NO_2), sulfur dioxide (SO_2), ozone (O_3), carbon monoxide (CO), and volatile organic compounds (VOCs) are some of the pollutants that cause this problem (Montalvo et al. 2022).

Particulate matter: Particulate matter (PM) is a complicated mixture of airborne solid and liquid particles that is released into the atmosphere after the burning of coal and fuels such as petrol, diesel, and wood. Nitrogen oxides and organic molecules found in the environment can also react chemically to form it. PM can also originate from vegetation and animals. PM emissions in large cities are ascribed to coal-fired power plants, vehicles, and trucks (Sierra-Vargas and Teran 2012). The components of air pollution known as particulate matter (PM) are carbonaceous particles along with reactive metals and organic compounds that have been adsorbed onto them. Sulfates, nitrates, polycyclic aromatic hydrocarbons, endotoxins, and metals like copper, iron, zinc, nickel, and vanadium are often found as components of PM (Hamanaka and Mutlu 2018). Nitrogen dioxide: Nitrogen dioxide (NO_2) is an oxidizing free radical that can start several harmful processes in biological systems. It is believed that both endogenously and exogenously generated NO_2 can cause several disorders (Faustini et al. 2014). Sulfur dioxide: When fuels containing sulfur, like coal and oil, are burned and refined petroleum is produced, sulfur dioxide (SO_2) is created. Acid is created when SO_2 dissolves in water vapor. Sulfates and other respirable particulates are created when SO_2 reacts with other gases and airborne particles (Chen et al. 2007). Ozone: Ozone or O_3 is a significant atmospheric trace gas that shields people and the environment from UV light absorption in the stratosphere (Wei et al. 2022). Climate changes are expected to influence future levels of surface ozone (O_3), a strong oxidant that has adverse impacts on health and ecosystems (Doherty et al. 2013). Carbon monoxide: The most prevalent and pervasive air contaminant in the environment is carbon monoxide. One of the gases that is always present because of human technology is carbon monoxide. It is mostly formed by the incomplete combustion or oxidation of carbonaceous materials or chemicals that are utilized as fuel for internal combustion engine transportation, industrial heating and processing, and residential building heating (Jaffe 1968). Volatile organic compounds (VOCs): The creation of ozone and photochemical oxidants linked to urban pollution is significantly influenced by VOCs. Determining the volatile organic compounds (VOCs) in the atmosphere has garnered more attention in recent decades. Every day, a significant number of VOCs are emitted from both fixed and mobile sources (Hester and Harrison 1995).

4.2 Air Pollution

An urgent worldwide problem, air pollution is made worse by urbanization, industrialization, and population expansion. It originates from multiple sources, such as car exhaust, industrial pollutants, and agricultural practices (Zhou et al. 2019). There are several sources of air pollution, but the main one is the burning of fossil fuel products. The source, chemical makeup, size, and method of release of air pollutants into interior or outdoor environments can all be used to categorize them (Bernstein et al. 2004). Air pollution refers to the presence of harmful substances in the atmosphere, typically caused by human activities. The primary source of air pollution is the combustion of fossil fuels in vehicles, aircraft, and power plants. When these fuels are burned, they release pollutants such as carbon dioxide, carbon monoxide, sulfur dioxide, and particulate matter into the air. These pollutants can have adverse effects on human health and the environment. Carbon dioxide, for example, contributes to climate change, while carbon monoxide can impair lung function. Sulfur dioxide can lead to respiratory issues, and particulate matter can cause respiratory and cardiovascular problems. Overall, air pollution poses significant risks to both human well-being and the natural environment (Sharma et al. 2013).

4.2.1 Source of Atmospheric Pollution

There are many different sources of air pollution as shown in Fig. 4.1. Primary pollutants are those that enter the atmosphere directly from sources. Others, referred to as secondary pollutants, are created in the atmosphere as a consequence of chemical reactions with existing pollutants and atmospheric gases (Holman 1999). A range of pollution sources, including fuel burning, resuspended dust, and geogenic sources, are the source of heavy metals found in air particles. Air pollution is a result of these sources' various heavy metal emissions into the atmosphere. For instance, the burning of coal contributes significantly to the emissions of heavy metals such as manganese (Mn), arsenic (As), chromium (Cr), nickel (Ni), and lead (Pb) (Liu et al. 2019).

Several contaminants, including inorganic ions, organic pollutants, organometallic compounds, radioactive isotopes, gaseous pollutants, and nanoparticles, have gravely contaminated the environment. They are classified as heavy metals because of their high density or large atomic weight. The term "heavy metal" is now used to refer to metallic chemical elements and metalloids that are hazardous to both humans and the environment (Briffa et al. 2020). These metals are naturally occurring in the crust of the planet and can be found in fluids, rocks, sediments, soils, and microorganisms at background quantities that are normal. Increased metal concentrations in the environment may result from anthropogenic releases of these elements. Heavy metals linger in the environment because they are incapable of being broken down or eliminated. In addition to being naturally occurring components of the earth's crust, heavy metals are also released by human activity. Then, because they are incapable of being broken down or eliminated, they become persistent

Fig. 4.1 Various prominent sources of air pollution

environmental pollutants. Certain elevated levels of heavy metals in soils may have a natural genesis due to the weathering of the underlying bedrock (Mohammed et al. 2011).

The combustion of fossil fuels (such as oil, coal, petroleum, and diesel fuel) from vehicle, industrial, and power sources as well as from vegetative sources (such as wood, plants, and materials) is the main source of the complex mixtures of gases and particles with condensed organic matter that plague urban areas with air pollution. Numerous physical, chemical, and/or photochemical processes that take place in the atmosphere throughout the harmful air pollutants' residence period in this environmental compartment decide their fate. Apart from the direct impact of air pollution on human health, it is thought that atmospheric input is the main source of some harmful chemicals found in lakes and streams (Fishbein 1990). Open burning can take place at a collecting point, like a dump, or at the location where the garbage is produced, like a home. Open burning of residential or domestic rubbish is a widespread practice that occurs in both developed and developing nations. There are several air pollutants produced by the open burning of waste, including greenhouse gases (GHGs), reactive trace gases, particle matter (PM), and hazardous chemicals, both in residential settings and at dump sites (Wiedinmyer et al. 2014).

4.2.2 Impact of Pollution on the Environment

Successful implementation of air quality management has been shown in numerous developed nations across the globe. In general, there has been a decrease in pollutant concentrations along with improvements in health measures (Ayala et al. 2012). Metal mining has the potential to significantly contribute to pollution in the environment. When mining occurs, pollutants are discharged into the environment, which can affect ecosystems and endanger human and wildlife health, particularly in the long run (Ghosh et al. 2021). For all living things, the environment is their fundamental source of life. Maintaining life and fostering well-being requires a clean, healthy atmosphere. There can be major health hazards when the environment is damaged, whether by toxins from industrial processes, mining, or other sources. In a damaged ecosystem, exposure to dangerous substances can cause a variety of health problems, from neurological diseases to cancer. As a result, preserving a pure and clean atmosphere is essential to guaranteeing the well-being and vitality of every living thing (Ghosh et al. 2016). Environmental contamination can result in preconceptional mutagenicity (both maternal and paternal) or postconceptional teratogenicity (maternal) that leads to congenital abnormalities. Chromosome abnormalities and disorders brought on by novel mutations are possible outcomes of preconceptional mutagenic effects (Dolk and Vrijheid 2003).

4.2.3 Impact of Pollution on Humans

Human health can be affected in a variety of ways by the various air pollution compositions, exposure times, doses, and the fact that people are typically exposed to mixes of pollutants rather than single substances. Effects on human health might vary from nausea and dyspnea to skin irritation and cancer. They also include immune system suppression, which results in a variety of illnesses, birth abnormalities, and severe developmental delays in children (Kampa and Castanas 2008). Both the scientific and political domains are now debating air pollution and its impact on human health. On the other hand, natural sources of air pollution have always existed (Pénard-Morand and Annesi-Maesano 2004). The primary purpose of studies on air pollution epidemiology is to determine how ambient air pollution affects a population's health individually. Acute or chronic health impacts are also possible. Acute effects result from exposures over time and are temporary. Although cumulative effects of exposure are more likely to cause chronic consequences, lifetime exposure may also have more complex roles. If the exposure has enough variance, that is an important design factor for statistical analysis. The temporal and/or spatial distributions of pollution are the origins of variation in exposure to ambient air pollution. The spatial distribution is expanded to encompass individual microenvironments and personal behaviors when taking into account personal exposure to pollution. Due to the difficulty of accurately estimating relevant exposures for each participant, ambient air pollution measurements, often obtained from central site monitors, have been employed in the majority of air pollution research (Dominici et al. 2003).

Effects of air pollution and health hazards to those living near and on roads may arise from chronic traffic like rush hour traffic from Monday through Friday. Chronic exposures and recurrent congestion can raise the hazards to one's long-term health. "Incident congestion," which includes traffic jams and disabled vehicles, is often overlooked even though it can have a significant impact on certain acute health consequences, including an exacerbation of asthma (Zhang and Batterman 2013). At the exposure assessment stage, the distribution of population exposure is established. The simplest way to determine population exposure from monitoring data is to compute the arithmetic mean of the concentrations for each time unit using a subset of the city's monitoring stations, if any. The population as a whole is then represented by this average (daily or annual) figure, which serves as a measure of their exposure over a given period. For megacity Delhi, for instance, we used the arithmetic mean of a few chosen concentrations for each year using daily data from seven monitoring stations spread out over the city (Makri and Stilianakis 2008). The data from exposure science and air pollution epidemiology are critical to understanding health impact and burden. Thanks to recent developments in these two fields, health impact assessments may now integrate epidemiological and atmospheric science data, enabling analysts to quantify a higher variety of health consequences in significantly more depth than was previously feasible. In the past few decades, epidemiology of air pollution has identified the risks associated with specific health outcomes for populations exposed to higher levels of pollution compared to those exposed to lower levels of pollution (referred to as "relative risk") (Anenberg et al. 2016). Both indoor and outdoor environments expose people to air pollution, and how much time a person spends indoors and outside, how close they live to sources of ambient air pollution, and how concentrated outdoor pollutants are indoors all affect how exposed they are too ambient urban air pollution. In developing nations, the effects of pollution are found to be substantial when weighed against other leading sources of illness and mortality. This suggests that in order to enhance health and well-being in these areas, appropriate air pollution-related measures are required (Rao et al. 2012).

4.3 Artificial Intelligence

The goal of the diverse focus of artificial intelligence (AI) in computer science is to imitate and recreate human-like cognitive capacities in machines (Holzinger et al. 2019). Artificial intelligence is a field of study in science and technology that develops computer programs and intelligent machines to carry out tasks that call for human intelligence. Artificial intelligence (AI) has many uses that lead to increased productivity and performance because of its capacity to execute tasks that humans cannot (Karthikeyan et al. 2021). AI offers a comparable opportunity: it can make something abundant and inexpensive that has previously been relatively expensive. Prediction, or the capacity to use the knowledge you already have to create new information, is the task that artificial intelligence (AI) performs well and affordably (Agrawal et al. 2017). Artificial intelligence (AI) has become the most extensively utilized technology instrument for controlling and averting the harmful effects of various air pollutants. It has garnered significant interest within the atmospheric and medical sciences fields (Mo et al. 2019). Artificial intelligence has been recognized as a critical instrument that helps regional environmental protection authorities make well-informed judgments about mitigating air pollution to lower the danger of public exposure (Jerrett et al. 2005). AI is able to better capture the pollution episode projections because it can handle the intricate and nonlinear interactions that exist between air quality factors (Fernández and Vico 2013). The field of forecasting ambient air pollution concentrations using AI-based methods has seen a renaissance of attention. The development of sophisticated and precise air pollution forecasting systems has caught the interest of academics due to the quick advancements in big data analytics technology, such as enhanced computing platforms, scalable storage systems, and high-speed parallel processing machines (Bai et al. 2019; Masood and Ahmad 2021).

4.3.1 Use of AI for Monitoring and Forecasting

Artificial intelligence is distinct from computer science and psychology in that it places more of a focus on perception, thinking, and action. It also differs from computer science in that regard. It increases the intelligence and utility of machines. Artificial intelligence has several advantages over natural intelligence. It can carry out some activities far more quickly and effectively than humans and is more permanent, consistent, less expensive, and easy to duplicate and distribute (Pannu 2015).

4.3.1.1 Environmental Monitoring

In order to assess and forecast the effects that various activities will have on the environment, satellite imagery and geographical information systems (GIS) data are essential elements of environmental monitoring. The amount of data collected through various airborne and space-borne sensors for remote sensing (RS) is growing exponentially. To improve the performance of environmental monitoring solutions for the given application, effective AI-based algorithms, big data analytics, data fusion mechanisms, and sufficient computing resources must be employed (Himeur et al. 2022). Amidst the environmental upheaval, artificial intelligence appears to be a ray of hope. Artificial intelligence (AI) has the potential to transform environmental monitoring and conservation through its ability to recognize patterns, process large amounts of data, and do predictive analytics. It alters our understanding of, reactions to, and interpretations of environmental data, offering a flexible and adaptable framework for dealing with the problems of the Anthropocene (Chisom et al. 2024).

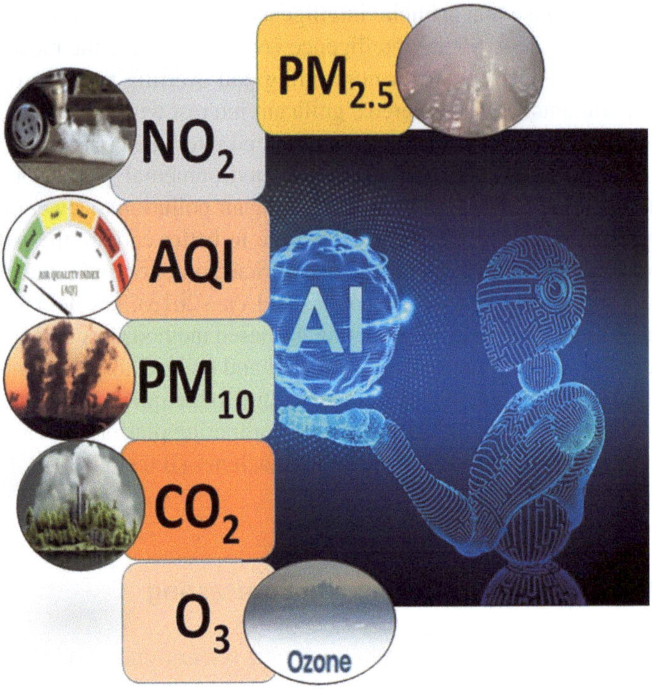

Fig. 4.2 Use of AI in air quality monitoring

4.3.1.2 Air Monitoring

We examine the application of artificial intelligence (AI) in the widely adopted notion of the smart city; few pollutants used in AI modeling have been shown in Fig. 4.2. In actuality, smart cities are frequently commended for being able to modernize rural areas. Air pollution is one of the major issues facing modern cities. The emergence of smart cities has led to an annual upsurge in the application of AI to anticipate air quality and to monitor air conditions. On the other hand, there are stark inconsistencies in the local and global context of environmental justice (Krupnova et al. 2022). Data and technical assistance for environmental public governance can be obtained by integrating big data thinking and artificial intelligence (AI) diagnostic technologies into environmental governance. Furthermore, environmental governance, through data collecting, real-time monitoring, and citizen participation management, can offer precise and scientific suggestions for government decision-making in public environmental monitoring and early warning (Chen et al. 2020; Shneiderman 2020). Artificial intelligence has the capability to collect a wide range of ecological environment monitoring data, including pollution sources, ecological circumstances, and environmental quality. AI is also capable of creating an intelligent ecological environment monitoring brain that supports the whole ecological environment monitoring industry, including all vertical multilevel analytical viewpoints as well as all horizontal monitoring elements (Fu et al. 2023).

4.3.2 Air Quality Forecasting

The chemical composition of the atmosphere at a specific location and time is referred to as air quality. Everyone is impacted by air quality, much as weather (Zhang et al. 2012) . Precise estimates of air quality are necessary for effective planning by the many economic performance-related sectors. Accurate forecasting is also crucial for strategic management that preserves air quality. Since time series analysis is the primary work for researchers utilized in development, the time series modeling and forecasting approach is chosen to be used in air quality management to help make future planning and helps to form the better air quality (Du et al. 2019). Utilizing sensor data to inform decisions is one of smart cities' primary goals. Still, while gathering the data, sensors can occasionally be mistakes and failures. Predictive modeling for air quality forecasting in smart cities appears to be a possible solution to these issues (Harish kumar et al. 2020). When comparing air quality predictions to conventional statistical methods, AI algorithms—more especially, machine learning and deep learning—have shown greater predictive accuracy. Planning for policy and prompt intervention depend on improved prediction accuracy. AI algorithms are able to learn and adapt in response to fresh inputs. AI models, for instance, are able to adjust to shifting patterns and circumstances in air quality because they are able to learn from new data as well as from past data. Advances in IoT and data streaming have made it possible for AI systems to process real-time data and produce instantaneous forecasts and updates on air quality (Kulikova et al. 2023). According to epidemiological research, there may be negative health impacts from being exposed to particles with an aerodynamic diameter of less than 10 μm (PM_{10}) and, more recently, less than 2.5 μm ($PM_{2.5}$). Evidence also points to the possibility that such exposure could raise hospitalization and death rates in older adults with preexisting cardiopulmonary disease (Pope et al. 1995; Chapman et al. 1997; Karaca et al. 2005).

4.4 Role of AI in Environmental Pollution Forecasting

One of the main causes of the worldwide environmental burden of disease is now air pollution. The World Health Organization (WHO) estimates that 92% of people on Earth are exposed to pollution at levels above that at which air quality is considered hazardous to human health. Numerous studies have indicated that long-term exposure to low-quality air increases a person's chance of heart attacks, lung cancer, strokes, and other major, potentially fatal illnesses (Wang et al. 2019). Due to this, there has been a great deal of pressure on the scientific community to produce precise and dependable remedies to counteract the negative health impacts of air pollution.

Furthermore, according to Jerrett et al. (2005), artificial intelligence (AI) is a crucial tool that local environmental protection agencies use to make well-informed judgments on air pollution mitigation methods that lower the risk of public exposure. Artificial intelligence (AI) can better capture pollution episode projections because it can handle the complex and nonlinear interactions that exist between air quality indicators (Fernández and Vico 2013). It seems to be a precise and quick method for locating pollution hotspots and offers great tracking capabilities for the baseline pollution levels. Prior studies have contrasted AI-based methods for air pollution forecasting with alternative methods. McKendry (2002) conducted a comparison between artificial neural networks (ANN) and linear regression (MLR), a statistical technique that uses several explanatory factors to forecast response variable values in particulate matter (PM_{10} and $PM_{2.5}$) simulations. In order to estimate PM_{10} concentrations, Dutta and Jinsart (2021) assessed the effectiveness of decision tree and artificial neural network models. Similarly, for the purpose of forecasting carbon monoxide (CO), suspended particulate matter (SPM), and sulfur dioxide (SO_2) in an industrialized area, Turias et al. (2008) compared the autoregressive integrated moving average (ARIMA), another statistical time series forecasting tool, with the back-propagation neural network (BPNN). In order to anticipate hourly $PM_{2.5}$ concentrations, Shang and He (2018) developed a unique forecasting method based on random forest and ANN approaches. A comparative study of ANN, LASSO, SVR, RF, KNN, and xGBoost modeling techniques for simulating PM_{10} pollution concentrations was given by Bozdağ et al. (2020) and more.

4.5 Models of AI in Environmental Forecasting and Monitoring Air Quality

4.5.1 $PM_{2.5}$

The important and useful task of $PM_{2.5}$ concentration forecasting is to safeguard public health by offering an early warning. An ensemble long short-term memory neural network (E-LSTM) is suggested in this paper for forecasting hourly $PM_{2.5}$ concentrations. Three stages are used to implement the model that is presented: The three steps involved in multimodal feature extraction are as follows: (1) ensemble empirical mode decomposition (EEMD), (2) long short-term memory method (LSTM) for multimodal feature learning, and (3) inverse EEMD computation for multimodal feature estimated integration. The next mode information of the $PM_{2.5}$ concentration is forecasted in each modeling process using the mode information of the $PM_{2.5}$ and the relevant meteorological factors 1 h in advance. Two datasets gathered from two environmental monitoring stations in Beijing, China, are examined in order to assess the effectiveness of the E-LSTM model. One of the main causes of the worldwide environmental burden of disease is now air pollution. The World Health Organization (WHO) estimates that 92% of people on Earth are exposed to pollution at levels above that at which air quality is considered hazardous to human health (WHO 2017). Numerous studies have indicated that long-term exposure to low-quality air increases a person's chance of heart attacks, lung cancer, strokes, and other major, potentially fatal illnesses (Wang et al. 2019). Because of this, there is a great need for the scientific community to offer trustworthy and precise methods to counteract the negative health impacts of air pollution (Bai et al. 2019).

Furthermore, according to Jerrett et al. (2005), artificial intelligence (AI) is a crucial tool that local environmental protection agencies use to make well-informed judgments on air pollution mitigation methods that lower the risk of public exposure. Artificial intelligence (AI) can better capture pollution episode projections because it can handle the complex and nonlinear interactions that exist between air quality indicators (Fernández and Vico 2013). Moreover, Jerrett et al. (2005) state that local environmental protection agencies rely heavily on artificial intelligence (AI) to make well-informed decisions about air pollution mitigation strategies that reduce the danger of public exposure. Because artificial intelligence (AI) can handle the intricate and nonlinear connections that exist between air quality indicators, it can better capture pollution episode projections (Fernández and Vico 2013).

4.5.2 PM_{10}

Over the past 40 years, the source of suspended particulate matter in London has shifted. Particulate emissions have decreased as a result of nearly all London power plants closing and less coal being used for space heating. On the other hand, the transportation sector's particle emissions have increased significantly (Goodwin 1999; Beevers et al. 2009; Brimblecombe 1999). European-wide PM_{10} air quality guidelines have been established. PM_{10} restrictions were established by the first Daughter Directive under the Air Quality Framework Directive 96/62/EC. The objectives of the Air Quality Strategy for England, Scotland, Wales, and Northern Ireland (AQS) (Bell et al. 2000) incorporate them into UK government policy. According to the EU Daughter Directive, by the end of 2004, the annual mean PM_{10} concentration should not exceed 40 μg m^{-3} and the daily mean should not exceed 50 μg m^{-3} on more than 35 days annually. By December 31, 2009, the annual mean PM_{10} should be less than 20 μg m^{-3}, and the daily mean shall not exceed 50 μg m^{-3} on more than 7 days per year. PM_{10} is measured gravimetrically for these reasons. Understanding the sources of PM_{10} now and having the ability to forecast future concentrations of PM_{10} are essential for developing management strategies and policies to meet these limitations. First, measurements of PM_{10} and black smoke were correlated to determine the primary component, and particulate sulfate data were used to determine the secondary component. Future PM_{10} levels were then estimated using estimates of the anticipated changes in these components. Eventually, measured NO_X concentrations were utilized by Stedman et al. (2001) instead of black smoke to determine the main component.

4.5.3 O_3

China's tropospheric ozone (O_3) pollution is getting worse, and lowering the O_3 peak level requires an accurate forecast. China's ambient air quality has significantly improved since the Air Pollution Prevention and Control Action Plan was put into effect in 2013. Nonetheless, despite the implementation of several control measures, the amount of ground-level ozone (O_3) pollution has increased (Li et al. 2020). Using numerical models, deterministic approaches replicate the intricate physical and chemical mechanisms controlling the behaviors of O_3 in the atmosphere. But in order to represent the intricate processes, they are computationally demanding and have problems with uncertainty in the emission inventory and parameterization methods (Abdi-Oskouei et al. 2020; Hu et al. 2021). Because they are excellent at solving nonlinear problems and are computationally efficient, machine learning techniques like support vector machines (SVMs) (Lu and Wang 2008), random forests (RF) (Stafoggia et al. 2020), and artificial neural networks (ANNs) (Hoshyaripour et al. 2016) have become more and more common in O_3 prediction. Recurrent neural networks (RNNs) are able to learn from the long-term dependencies of air pollution concentrations and model the associated sequence of time series patterns (Yafouz et al. 2021). As a result, they are thought to be superior to ANNs when it comes to modeling the variation and evolution of air quality. Research has shown that, in various climatic scenarios, RNNs outperform SVM and ANN in the prediction of hourly O_3 concentrations (Lu et al. 2021). However, the quantity of observation data that is fed into the RNNs determines their high prediction accuracy (Kocijan et al. 2018).

4.5.4 CO_2

The world's cities are experiencing a sharp rise in air pollution, which is seriously endangering not only the environment but also the health of all living beings. Since humans are one of the sources of this pollution, carbon dioxide (CO_2) is the primary cause of indoor pollution. CO_2 testing and monitoring are time-consuming, are expensive, and require sophisticated sensors. Therefore, machine learning (ML) has been applied to forecast the CO_2 concentration inside an office space in order to overcome these restrictions (Wang et al. 2022). The data used as input parameters for this study includes the number of occupants, area per person, exterior temperature, relative humidity, outer wind speed, and air quality index, all of which were measured in real time. Ten algorithms—ensemble learning (EL), artificial neural network (ANN), decision tree (DT), support vector machine (SVM), Gaussian process regression (GPR), linear regression (LR), optimized SVM, optimized GPR, optimized DT, and optimized EL—were used in this study.and enhanced SVMs were employed to forecast the CO_2 concentration. It has been discovered that, in terms of prediction accuracy, the optimized GPR model outperforms the other chosen models (Kapoor et al. 2022). The following is a list of this study's primary contributions in the research associated with CO_2 assessment: (i) determining the crucial elements that are inputted into the CO_2 concentration prediction. (ii) The number of occupants, the area per person, the exterior temperature, the outer wind speed, the relative humidity, and the air quality index are the recognized important characteristics that are used as input for forecasting the CO_2 concentration. (iii) Gathered data on the current CO_2 concentration from the office building. (iv) Six machine learning algorithms were modeled, including ensemble learning

(EL), ANN, SVM, DT, GPR, LR, and GPR. Additionally, four machine learning algorithms—GPR, EL, DT, and SVM—were optimized for this purpose in order to predict the CO_2 concentration inside the office building (Kapoor et al. 2022).

4.5.5 NO_2

To effectively reduce air pollution, precise estimates of urban NO_2 concentrations must be achieved. In order to predict the daily average NO_2 concentration, this paper focused on a DWT-LSTM predictive model based on the discrete wavelet transform and long- and short-term memory network. The concentration of NO_2 in Tianjin was chosen as the research object. As the input indices, five significant air pollutants, important meteorological data, and historical data were chosen, enabling an accurate prediction of the NO_2 concentration for the following day. To enhance the data dimension, the input data were first broken down using the discrete wavelet transform. Moreover, the features of the dissected data were learned using the LSTM network model (Liu et al. 2021). The growth of industrialization and urbanization is causing an annual rise in the per capita energy consumption. Apart from organic sources including dust storms, wildfires, and volcanic eruptions, elevated nitrogen dioxide (NO_2) originating from automobile and boiler exhaust has emerged as a significant environmental issue affecting the majority of global nations. As a matter of fact, nitrogen dioxide reacts with UV light to generate photochemical smog and dissolves in airborne water to form acids, which may cause acid rain. Furthermore, exposure to varying amounts of NO_2 in humans can cause varying degrees of lung function loss, which can have a negative impact on social and industrial activities as well as industrial productivity (Pandey et al. 2005). Thus, early warning of air pollution emergencies and guidance for the release of NO_2 control measures and public health protection activities may be obtained by the effective detection and accurate prediction of NO_2 and the development of a highly accurate and stable prediction model. Moreover, Mirzaei et al. (2019) investigated the spatiotemporal variability between satellite aerosol optical depth (AOD) data and $PM_{2.5}$ concentration at ground monitoring stations using the GTWR model. However, throughout the warm season, there were flaws in the retrieval algorithm when it came to identifying low particulate matter values. As a result, the model's forecast accuracy decreased, which had an impact on the simulation's results. The nonlinear relationship is taken into account and prediction accuracy is increased by the use of machine learning methods, such as random forests, support vector machines (SVMs), and artificial neural networks (ANNs) (Oprea et al. 2016; Gutiérrez et al. 2016).

4.5.6 Air Quality Index

With 342 monitoring stations spread across 240 cities, the National Air Monitoring Program (NAMP) was launched in India by the Central and State Pollution Control Boards. The air quality index (AQI) is divided into various sections. The dataset was gathered and then preprocessed to replace missing values and eliminate redundant data in order to forecast the AQI in Chennai city. The gray-level co-occurrence matrix (GLCM) is used to obtain the mean, mean square error, and standard deviation (Janarthanan et al. 2021). The quality of life and air quality are closely correlated. Forecasting air pollution is essential for issuing air pollution alerts and managing air pollution. However, because the original data are chaotic and nonstationary, it is challenging to produce accurate projections for air pollution indexes. According to Kurt and Oktay (2010), air pollution is a significant environmental issue in many parts of the world. On the one hand, the release of inferior industrial pollutants is a primary cause of severe air pollution in industrialized areas, combined with rapid development. On the other side, there has been a sharp rise in the population and quickening economic growth. China has extremely high levels of air pollution, which is mostly manifested as a high concentration of suspended particles in the atmosphere of Chinese cities. For instance, there is a high concentration of particulate matter (PM) and sulfur dioxide (SO_2), a sharp rise in vehicle exhaust emissions, and a tendency for nitrogen oxides (NO_x) pollution to worsen. Among pollution indices, the air quality index (AQI) and $PM_{2.5}$—particulate matter having a diameter of less than or equal to 2.5 μm—are two significant indicators. The Air Quality Index (AQI) is a metric that assesses and represents the current state of air quality by condensing the amounts of various contaminants into a single numerical value. The new ambient air quality standards (GB3095-2012) are used to calculate the AQI. These standards cover six pollutants: carbon monoxide (CO), sulfur dioxide (SO_2), nitrogen dioxide (NO_2), sulfur dioxide ($PM_{2.5}$), PM_{10} (particulate matter with a diameter less than or equal to 10 μm), ozone (O_3), and $PM_{2.5}$ (Sheng and Tang 2016). Several definitions of the air quality index (AQI) have been proposed. For example, the fuzzy-based air quality index (FAQI) defined the AQI by using a fuzzy criterion to weight the CO, PM_{10}, SO_2, NO_2, and O_3 (Sowlat et al. 2011). The air pollution index only included the CO, PM_{10}, SO_2, NO_2, and O_3. The globe has been aware of air pollution warnings in the last several decades (Chen et al. 2016). In order to prevent the negative impacts of atmospheric pollution on human health, a lot of individuals currently base their decisions on what to do outside on air pollution forecasts. To the best of our knowledge, a few research studies have used hybrid models for AQI forecasting (Wang et al. 2017). Nonetheless, there are no mistake revisions in the EEMD-ANN hybrid models, two-phase decomposition technique, or modified extreme learning machine (Zhu et al. 2017).

4.6 Prediction of Climate Change Using AI and ML Techniques

The damage that climate change and global warming pose to long-term development has made them contentious issues in the international community. The overuse of natural resources, coupled with the combustion of fossil fuels and nonrenewable energy sources, presents significant environmental hazards, such as the scarcity of drinking water, deforestation, and climate change (Maiwada et al. 2014). Because human activity causes the weather to change, it becomes increasingly important to accurately predict weather events that may have an impact on our own surroundings and well-being. As a result of anthropogenic weather change, major oceanic heating events, often known as "marine heatwaves," are typically occurring more frequently and with more ferocity (Subramaniam et al. 2022). As global temperatures rise, heatwaves are becoming more powerful, prolonged, and frequent. Overexposure to heatwaves can harm people's health, the environment, livestock, and agriculture. More attention has been paid to land-based heatwaves than to marine heatwaves (MHWs), which cause significant financial burdens and detrimental impacts on marine life and ecosystems. Acute heat-related exposure of marine life and ecosystems may result in irreversible damage or even the near extinction of species, as seen in the case of widespread bleaching of coral reefs, die-offs in biodiversity, and financial losses in the aquaculture and fisheries industries (Wernberg et al. 2013; Hobday et al. 2016). Deep learning, a type of AI, can also be used to forecast extreme weather events and the environment. Khan et al. (2021) developed a reliable heatwave forecast system for Pakistan based on climate change using machine learning (ML) techniques such artificial neural networks, support vector machines (SVM), and random forests. With an R2 of 0.87 and a normalized root mean square error (NRMSE) of 36%, SVM algorithms performed better in predicting (heatwave days) HWDs than other ML models during the validation period. Benthuysen et al.'s (2021) research suggests that the Australian Community Climate and Earth System Simulator Seasonal version 1 (ACCESS-S1) ocean-atmosphere model should be able to predict the 2020 marine heatwaves in the Great Barrier Reef on a sub-seasonal scale. The model could predict the start of MHWs 1 week ahead of time with accuracy, but it could not predict when the MHW would terminate. They also looked into the use of ACCESS-S2 for monthly MHW forecasts, which had a 40% hit rate when projected 4 months ahead of time.

4.7 Conclusion

In conclusion, the issue of air quality and pollution is a critical global concern that demands immediate attention and innovative solutions. Artificial intelligence (AI) has emerged as a powerful tool in the monitoring and management of air quality, offering significant advancements in accuracy, efficiency, and real-time data analysis. Through AI models, we can achieve more precise predictions of air pollution levels, identify sources of contamination, and strategize effective interventions. Several AI models have been reported for air quality monitoring including pollutants like $PM_{2.5}$, PM_{10}, O_3, NO_2, CO_2, and AQI. These models leverage vast datasets and sophisticated algorithms to continuously improve their predictive capabilities, enabling authorities to make informed decisions swiftly and proactively. By embracing these technologies responsibly, we can pave the way for a cleaner and safer future for generations to come. Collaborative efforts among scientists, policymakers, and technology experts are essential to harnessing AI's full potential in safeguarding public health and the environment.

References

Abdi-Oskouei M, Carmichael G, Christiansen M, Ferrada G, Roozitalab B, Sobhani N, Stanier C (2020) Sensitivity of meteorological skill to selection of WRF-Chem physical parameterizations and impact on ozone prediction during the Lake Michigan Ozone Study (LMOS). J Geophys Res Atmos 125(5):e2019JD031971

Agrawal A, Gans J, Goldfarb A (2017) What to expect from artificial intelligence. In: What the digital future holds. Massachusetts Institute of Technology

Anenberg SC, Belova A, Brandt J, Fann N, Greco S, Guttikunda S, Heroux ME, Hurley F, Krzyzanowski M, Medina S, Miller B (2016) Survey of ambient air pollution health risk assessment tools. Risk Anal 36(9):1718–1736

Ayala A, Brauer M, Mauderly JL, Samet JM (2012) Air pollutants and sources associated with health effects. Air Qual Atmos Health 5:151–167

Bai Y, Zeng B, Li C, Zhang J (2019) An ensemble long short-term memory neural network for hourly PM2.5 concentration forecasting. Chemosphere 222:286–294

Beevers S, Carslaw D, Westmoreland E, Mittal H (2009) Air pollution and emissions trends in London. King's College London report to Defra

Bell W, Davies N, Butterfield D, Blakley K, Lancaster K, Quincey P, Henderson M (2000) Final Report on DETR contract EPG 1/3/155

Benthuysen JA, Smith GA, Spillman CM, Steinberg CR (2021) Subseasonal prediction of the 2020 Great Barrier Reef and Coral Sea marine heatwave. Environ Res Lett 16(12):124050

Bernstein JA, Alexis N, Barnes C, Bernstein IL, Nel A, Peden D, Diaz-Sanchez D, Tarlo SM, Williams PB (2004) Health effects of air pollution. J Allergy Clin Immunol 114(5):1116–1123

Bozdağ A, Dokuz Y, Gökçek ÖB (2020) Spatial prediction of PM10 concentration using machine learning algorithms in Ankara, Turkey. Environ Pollut 263:114635

Briffa J, Sinagra E, Blundell R (2020) Heavy metal pollution in the environment and their toxicological effects on humans. Heliyon 6(9):e04691

Brimblecombe P (1999) Air pollution and health history. In: Air pollution and health. Academic Press, pp 5–18. Elsevier

Chapman RS, Watkinson WP, Dreher KL, Costa DL (1997) Ambient particulate matter and respiratory and cardiovascular illness in adults: particle-borne transition metals and the heart–lung axis. Environ Toxicol Pharmacol 4(3–4):331–338

Chen TM, Kuschner WG, Gokhale J, Shofer S (2007) Outdoor air pollution: nitrogen dioxide, sulfur dioxide, and carbon monoxide health effects. Am J Med Sci 333(4):249–256

Chen W, Tang H, Zhao H (2016) Urban air quality evaluations under two versions of the national ambient air quality standards of China. Atmos Pollut Res 7(1):49–57

Chen J, Wang B, Huang S, Song M (2020) The influence of increased population density in China on air pollution. Sci Total Environ 735:139456

Cheng Y, Li X, Li Z, Jiang S, Li Y, Jia J, Jiang X (2014, November) AirCloud: a cloud-based air-quality monitoring system for everyone. In: Proceedings of the 12th ACM conference on embedded network sensor systems, pp 251–265

Chisom ON, Biu PW, Umoh AA, Obaedo BO, Adegbite AO, Abatan A (2024) Reviewing the role of AI in environmental monitoring and conservation: a data-driven revolution for our planet. World Journal of Advanced Research and Reviews 21(1):161–171

Doherty RM, Wild O, Shindell DT, Zeng G, MacKenzie IA, Collins WJ et al (2013) Impacts of climate change on surface ozone and intercontinental ozone pollution: a multi-model study. J Geophys Res Atmos 118(9):3744–3763

Dolk H, Vrijheid M (2003) The impact of environmental pollution on congenital anomalies. Br Med Bull 68(1):25–45

Dominici F, Sheppard L, Clyde M (2003) Health effects of air pollution: a statistical review. Int Stat Rev 71(2):243–276

Du S, Li T, Yang Y, Horng SJ (2019) Deep air quality forecasting using hybrid deep learning framework. IEEE Trans Knowl Data Eng 33(6):2412–2424

Dutta A, Jinsart W (2021) Air pollution in Indian cities and comparison of MLR, ANN and CART models for predicting PM10 concentrations in Guwahati, India. Asian J Atmos Environ 15(1):2020131

Faustini A, Rapp R, Forastiere F (2014) Nitrogen dioxide and mortality: review and meta-analysis of long-term studies. Eur Respir J 44(3):744–753

Fernández JD, Vico F (2013) AI methods in algorithmic composition: a comprehensive survey. J Artif Intell Res 48:513–582

Fishbein L (1990) Sources, nature and levels of air pollutants. In: Air pollution and human cancer. Springer Berlin Heidelberg, Berlin/Heidelberg, pp 9–34

Fu L, Li J, Chen Y (2023) An innovative decision making method for air quality monitoring based on big data-assisted artificial intelligence technique. J Innov Knowl 8(2):100294

Ghosh S, Mohanty S, Akcil A, Sukla LB, Das AP (2016) A greener approach for resource recycling: manganese bioleaching. Chemosphere 154:628–639

Ghosh S, Gandhi M, van Hullebusch ED, Das AP (2021) Proteomic insights into Lysinibacillus sp.-mediated biosolubilization of manganese. Environ Sci Pollut Res 28:40249–40263

Goodwin P (1999) Transformation of transport policy in Great Britain. Transp Res A Policy Pract 33(7–8):655–669

Gutiérrez L, Mena RH, Ruggiero M (2016) A time dependent Bayesian nonparametric model for air quality analysis. Comput Stat Data Anal 95:161–175

Hamanaka RB, Mutlu GM (2018) Particulate matter air pollution: effects on the cardiovascular system. Front Endocrinol 9:412302

Harish Kumar KS, Yogesh KM, Gad I (2020) Forecasting air pollution particulate matter (PM2. 5) using machine learning regression models. Proc Comput Sci 171:2057–2066

Hester RE, Harrison RM (eds) (1995) Volatile organic compounds in the atmosphere, vol 4. Royal Society of Chemistry

Himeur Y, Rimal B, Tiwary A, Amira A (2022) Using artificial intelligence and data fusion for environmental monitoring: a review and future perspectives. Inf Fusion 86:44–75

Hobday AJ, Alexander LV, Perkins SE, Smale DA, Straub SC, Oliver EC, Benthuysen JA, Burrows MT, Donat MG, Feng M, Holbrook NJ (2016) A hierarchical approach to defining marine heatwaves. Prog Oceanogr 141:227–238

Holman C (1999) Sources of air pollution. In: Air pollution and health. Academic Press, pp 115–148. Elsevier

Holzinger A, Langs G, Denk H, Zatloukal K, Müller H (2019) Causability and explainability of artificial intelligence in medicine. Wiley Interdisciplinary Reviews: Data Mining and Knowledge Discovery 9(4):e1312

Hoshyaripour G, Brasseur G, Andrade MDF, Gavidia-Calderón M, Bouarar I, Ynoue RY (2016) Prediction of ground-level ozone concentration in São Paulo, Brazil: deterministic versus statistic models. Atmos Environ 145:365–375

Hu C, Kang P, Jaffe DA, Li C, Zhang X, Wu K, Zhou M (2021) Understanding the impact of meteorology on ozone in 334 cities of China. Atmos Environ 248:118221

Jaffe LS (1968) Ambient carbon monoxide and its fate in the atmosphere. J Air Pollut Control Assoc 18(8):534–540

Janarthanan R, Partheeban P, Somasundaram K, Elamparithi PN (2021) A deep learning approach for prediction of air quality index in a metropolitan city. Sustain Cities Soc 67:102720

Jerrett M, Arain A, Kanaroglou P, Beckerman B, Potoglou D, Sahsuvaroglu T, Morrison J, Giovis C (2005) A review and evaluation of intraurban air pollution exposure models. J Expo Sci Environ Epidemiol 15(2):185–204

Kampa M, Castanas E (2008) Human health effects of air pollution. Environ Pollut 151(2):362–367

Kapoor NR, Kumar A, Kumar A, Kumar A, Mohammed MA, Kumar K, Kadry S, Lim S (2022) Machine learning-based CO2 prediction for office room: a pilot study. Wirel Commun Mob Comput 2022(1):9404807

Karaca F, Alagha O, Ertürk F (2005) Statistical characterization of atmospheric PM10 and PM2. 5 concentrations at a non-impacted suburban site of Istanbul, Turkey. Chemosphere 59(8):1183–1190

Karthikeyan J, Hie TS, Jin NY (2021) Learning outcomes of classroom research, I ordine nuovo publication

Khan N, Shahid S, Ismail TB, Behlil F (2021) Prediction of heat waves over Pakistan using support vector machine algorithm in the context of climate change. Stoch Env Res Risk A 35:1335–1353

Kocijan J, Gradišar D, Stepančič M, Božnar MZ, Grašič B, Mlakar P (2018) Selection of the data time interval for the prediction of maximum ozone concentrations. Stoch Env Res Risk A 32:1759–1770

Krupnova TG, Rakova OV, Bondarenko KA, Tretyakova VD (2022) Environmental justice and the use of artificial intelligence in urban air pollution monitoring. Big Data Cogn Comput 6(3):75

Kulikova E, Sulimin V, Shvedov V (2023) Artificial intelligence for ambient air quality control. In: E3S web of conferences, vol 419. EDP Sciences, p 03011

Kurt A, Oktay AB (2010) Forecasting air pollutant indicator levels with geographic models 3 days in advance using neural networks. Expert Syst Appl 37(12):7986–7992

Li K, Jacob DJ, Shen L, Lu X, De Smedt I, Liao H (2020) Increases in surface ozone pollution in China from 2013 to 2019: anthropogenic and meteorological influences. Atmos Chem Phys 20(19):11423–11433

Liu X, Ouyang W, Shu Y, Tian Y, Feng Y, Zhang T, Chen W (2019) Incorporating bioaccessibility into health risk assessment of heavy metals in particulate matter originated from different sources of atmospheric pollution. Environ Pollut 254:113113

Liu B, Zhang L, Wang Q, Chen J (2021) A novel method for regional NO2 concentration prediction using discrete wavelet transform and an LSTM network. Comput Intell Neurosci 2021:6631614

Lu WZ, Wang D (2008) Ground-level ozone prediction by support vector machine approach with a cost-sensitive classification scheme. Sci Total Environ 395(2–3):109–116

Lu X, Ye X, Zhou M, Zhao Y, Weng H, Kong H, Li K, Gao M, Zheng B, Lin J, Zhou F (2021) The underappreciated role of agricultural soil nitrogen oxide emissions in ozone pollution regulation in North China. Nat Commun 12(1):5021

Maiwada NA, Abdulkarim HEL, Usman A, Abdullahi S (2014) The role of renewable energy in mitigating deforestation and climate change in Nigeria. J Nat Sci Res 4(24):81–85

Makri A, Stilianakis NI (2008) Vulnerability to air pollution health effects. Int J Hyg Environ Health 211(3–4):326–336

Masood A, Ahmad K (2021) A review on emerging artificial intelligence (AI) techniques for air pollution forecasting: fundamentals, application and performance. J Clean Prod 322:129072

McKendry IG (2002) Evaluation of artificial neural networks for fine particulate pollution (PM10 and PM2. 5) forecasting. J Air Waste Manage Assoc 52(9):1096–1101

Mirzaei M, Amanollahi J, Tzanis CG (2019) Evaluation of linear, nonlinear, and hybrid models for predicting PM2. 5 based on a GTWR model and MODIS AOD data. Air Qual Atmos Health 12(10):1215–1224

Mo X, Zhang L, Li H, Qu Z, (2019) A Novel Air Quality Early-Warning System Based on Artificial Intelligence. Int J Environ Res Public Health 16(19):3505

Mohammed AS, Kapri A, Goel R (2011) Heavy metal pollution: source, impact, and remedies. In: Biomanagement of metal-contaminated soils, pp 1–28. Springer

Montalvo L, Fosca D, Paredes D, Abarca M, Saito C, Villanueva E (2022) An air quality monitoring and forecasting system for Lima city with low-cost sensors and artificial intelligence models. Front Sustain Cities 4:849762

Oprea M, Popescu M, Mihalache SF (2016) A neural network based model for PM 2.5 air pollutant forecasting. In: 2016 20th international conference on system theory, control and computing (ICSTCC). IEEE, pp 776–781

Pandey JS, Kumar R, Devotta S (2005) Health risks of NO2, SPM and SO2 in Delhi (India). Atmos Environ 39(36):6868–6874

Pannu A (2015) Artificial intelligence and its application in different areas. Artif Intell 4(10):79–84

Pénard-Morand C, Annesi-Maesano I (2004) Air pollution: from sources of emissions to health effects. Breathe 1(2):108–119

Pope CA, Thun MJ, Namboodiri MM, Dockery DW, Evans JS, Speizer FE, Heath CW (1995) Particulate air pollution as a predictor of mortality in a prospective study of US adults. Am J Respir Crit Care Med 151(3):669–674

Rao S, Chirkov V, Dentener F, Van Dingenen R, Pachauri S, Purohit P, Schoepp W (2012) Environmental modeling and methods for estimation of the global health impacts of air pollution. Environ Model Assess 17:613–622

Shang Z, He J (2018) Predicting hourly PM2.5 concentrations based on random forest and ensemble neural network. In: Chinese automation congress (CAC). IEEE, pp 2341–2345

Sharma SB, Jain S, Khirwadkar P, Kulkarni S (2013) The effects of air pollution on the environment and human health. Indian J Res Pharm Biotechnol 1(3):391–396

Sheng N, Tang UW (2016) The first official city ranking by air quality in China—a review and analysis. Cities 51:139–149

Shneiderman B (2020) Human-centered artificial intelligence: reliable, safe & trustworthy. Int J Human–Comput Interact 36(6):495–504

Sierra-Vargas MP, Teran LM (2012) Air pollution: impact and prevention. Respirology 17(7):1031–1038

Sowlat MH, Gharibi H, Yunesian M, Mahmoudi MT, Lotfi S (2011) A novel, fuzzy-based air quality index (FAQI) for air quality assessment. Atmos Environ 45(12):2050–2059

Stafoggia M, Johansson C, Glantz P, Renzi M, Shtein A, de Hoogh K, Kloog I, Davoli M, Michelozzi P, Bellander T (2020) A random forest approach to estimate daily particulate matter, nitrogen dioxide, and ozone at fine spatial resolution in Sweden. Atmos 11(3):239

Stedman JR, Goodwin JW, King K, Murrells TP, Bush TJ (2001) An empirical model for predicting urban roadside nitrogen dioxide concentrations in the UK. Atmos Environ 35(8):1451–1463

Subramaniam S, Raju N, Ganesan A, Rajavel N, Chenniappan M, Prakash C, Pramanik A, Basak AK, Dixit S (2022) Artificial intelligence technologies for forecasting air pollution and human health: a narrative review. Sustainability 14(16):9951

Turias IJ, González FJ, Martin ML, Galindo PL (2008) Prediction models of CO, SPM and SO 2 concentrations in the Campo de Gibraltar Region, Spain: a multiple comparison strategy. Environ Monit Assess 143:131–146

Wang D, Wei S, Luo H, Yue C, Grunder O (2017) A novel hybrid model for air quality index forecasting based on two-phase decomposition technique and modified extreme learning machine. Sci Total Environ 580:719–733

Wang M, Aaron CP, Madrigano J, Hoffman EA, Angelini E, Yang J, Laine A, Vetterli TM, Kinney PL, Sampson PD, Sheppard LE (2019) Association between long-term exposure to ambient air pollution and change in quantitatively assessed emphysema and lung function. JAMA 322(6):546–556

Wang P, Zhong Y, Yao Z (2022) Modeling and estimation of CO2 emissions in China based on artificial intelligence. Comput Intell Neurosci 2022(1):6822467

Wei J, Li Z, Li K, Dickerson RR, Pinker RT, Wang J et al (2022) Full-coverage mapping and spatiotemporal variations of ground-level ozone (O3) pollution from 2013 to 2020 across China. Remote Sens Environ 270:112775

Wernberg T, Smale DA, Tuya F, Thomsen MS, Langlois TJ, De Bettignies T, Bennett S, Rousseaux CS (2013) An extreme climatic event alters marine ecosystem structure in a global biodiversity hotspot. Nat Clim Chang 3(1):78–82

Wiedinmyer C, Yokelson RJ, Gullett BK (2014) Global emissions of trace gases, particulate matter, and hazardous air pollutants from open burning of domestic waste. Environ Sci Technol 48(16):9523–9530

Yafouz A, Ahmed AN, Zaini NA, El-Shafie A (2021) Ozone concentration forecasting based on artificial intelligence techniques: a systematic review. Water Air Soil Pollut 232:1–29

Yang Z, Wang J (2017) A new air quality monitoring and early warning system: air quality assessment and air pollutant concentration prediction. Environ Res 158:105–117

Zhang K, Batterman S (2013) Air pollution and health risks due to vehicle traffic. Sci Total Environ 450:307–316

Zhang Y, Bocquet M, Mallet V, Seigneur C, Baklanov A (2012) Real-time air quality forecasting, part I: history, techniques, and current status. Atmos Environ 60:632–655
Zheng K, Zhao S, Yang Z, Xiong X, Xiang W (2016) Design and implementation of LPWA-based air quality monitoring system. IEEE Access 4:3238–3245
Zhou Y, Zhao X, Lin K-P, Wang C-H, Li L (2019) A Gaussian process mixture model-based hard-cut iterative learning algorithm for air quality prediction. Appl Soft Comput 85:105789
Zhu S, Lian X, Liu H, Hu J, Wang Y, Che J (2017) Daily air quality index forecasting with hybrid models: a case in China. Environ Pollut 231:1232–1244

Artificial Intelligence and Internet of Things: Collaborative Steps for Next-Generation Biomedical Waste Management

5

Swati Shrivastava, Nimisha Jadon, Shaily Varshney, Keisham Radhapyari, and Bayartungalag Batsaikhan

Abstract

Biomedical waste refers to pathogenic or infectious waste generated by healthcare facilities, medical centers, and research institutions. In the current COVID-19 era, the use of personal protective equipment, testing kits, and surgical facemasks has become essential. The improper disposal of this new biomedical waste poses significant global concerns for public health and environmental sustainability. It can lead to the transmission of infections and environmental contamination, while handled inappropriately, particularly in developing countries. Effective management of this waste requires careful segregation, collection, and recycling to mitigate risks. While established protocols exist, public understanding of disposal methods is often lacking, highlighting the need for a systematic approach that involves thorough planning and evaluation. The use of artificial intelligence (AI) and the Internet of Things (IoT) for material recognition and robotic automation in solid waste management has been extensively studied. AI offers various machine learning methods, such as artificial neural networks (ANN), adaptive neurofuzzy inference systems (ANFIS), and support vector machines (SVM), to analyze data, make decisions, and perform tasks that typically require human intelligence. While AI continues to advance, the IoT pres-

S. Shrivastava · N. Jadon (✉)
School of Studies in Environmental Chemistry, Jiwaji University, Gwalior, India

S. Varshney
Applied Chemistry and Chemicals Technology Department, SGSITS, Indore, India

K. Radhapyari
Regional Chemical Laboratory, Central Ground Water Board, North Eastern Region, Guwahati, Assam, India

B. Batsaikhan
University of Mongolian Academy of Sciences, Ulaanbaatar, Mongolia

B. Thokchom et al. (eds.), *Artificial Intelligence Enabled Real Time Environmental Monitoring*, https://doi.org/10.1007/978-981-95-3767-9_5

ents significant opportunities in waste management by addressing critical challenges such as proper segregation, tracking, disposal, and compliance with regulatory standards. Despite the potential for a more digitalized future for waste collection, sorting, and recycling, the current level of digitalization among waste management firms remains unclear. Additionally, the application of AI and IoT in biomedical waste management is not as widespread as in other areas of environmental engineering. Therefore, this chapter will focus on the management of biomedical waste and explore the potential use of artificial intelligence, robotic automation, and the Internet of Things to efficiently oversee the entire process, from collection and sorting to recycling.

Keywords

Biomedical waste · Solid waste management · Artificial intelligence · Internet of Things · Recycling of waste

Abbreviations

BMW	biomedical waste
ARGs	antibiotic-resistance genes
ARB	antibiotic-resistant bacteria
WHO	World Health Organization
PPE	personal protective equipment
AI	artificial intelligence
IoT	Internet of things
SVM	support vector machine
GA	genetic algorithm
NN	neural network
DT	decision tree
LSTM	long short-term memory networks
RF	random forest
ANN	artificial neural network
CNN	convolutional neural network
ANFIS	artificial neurofuzzy inference system
AHP	analytical hierarchy process
GIS	geographic information system
RFID	radio-frequency identification
GPS	global positioning system
GPRS	general packet radio system
GLCM	gray-level co-occurrence matrix
KNN	k-nearest neighbor models
GLAM	gray-level aura matrix
MLPs	multilayer perceptions

FFNN	feed forward neural network
DTW	dynamic time warping
GW	Gabor wavelet
AIS	artificial immune system
BP	backpropagation
DOY	day of the year
MOY	month of the year
HHV	higher heating value
LHV	lower heating value

5.1 Introduction

Thriving advancement in the field of medical research and technology has facilitated the galloping growth in human life span, thereby decreasing mortality; recent evidence of COVID-19 vaccine invention has proven this fact. However, there is a growing alarm that such transition is somehow becoming a significant contributor to infectious waste along with the general waste generated from healthcare facilities (Goswami et al. 2021; Ezeudu et al. 2022). Most of the medical waste is generated by hospitals, medical centers, research laboratories, and at-home patient care. Contrary to municipal waste, it includes hazardous waste, such as chemicals, radioactive substances, sharp objects, pharmaceuticals, diagnostic samples, blood, heavy metals, pathogens, and/or genotoxic substances (Chartier et al. 2014; Komilis et al. 2012). The medical waste management market is growing at a compound annual growth rate (CAGR) of 8% due to the increasing production of biomedical waste. It is expected to grow from its current size of USD 16.25 billion to over USD 36 billion by the end of 2036 (Research Nester 2024).

The term "biomedical waste" (BMW) is defined as any solid or liquid waste produced during the diagnosis, treatment, or immunization of human beings or animals (Patil and Bohara 2020). BMW is a great concern as it tends to carry a higher potential for infection and injury than other categories of waste and its poor management causes severe health risks. Open dumping of improperly treated medical waste may also cause severe environmental pollution including ground and surface water contamination, land deterioration, development of antibiotic-resistance genes (ARGs), and antibiotic-resistant bacteria (ARB) (Chi et al. 2020; Iyer et al. 2021). Therefore, huge waste generation calls for proper segregation according to waste type, collection, and recycling to minimize the risk of infection spread through aerosols.

Based on physicochemical properties, waste is normally classified as nonhazardous or hazardous. It's been stated by the guidelines of the US Environmental Protection Agency and World Health Organization that 15% of BMW is hazardous, whereas the remaining 85% is nonhazardous (WHO 2018). Hazardous waste could pose physical, chemical, and/or microbiological risks to the healthcare staff who is associated with handling, treatment, and disposal of waste. Currently, this hazardous waste has increased about five times since the COVID-19 pandemic hit globally. In Wuhan there was an increase from 40 tons/day to 240 tons/day and a drastic increase from 5 million tons/year to 2.5 million tons/month was estimated in the USA (Ilyas et al. 2020). This volume increase can be attributed to the increased use of personal protective equipment (PPE), COVID-19 testing kits, etc., but their disposal remains a big question mark regarding the environmental impact. To preserve the environment and public health, several systematic national and international guidelines were established and successfully revised frequently to regulate the disposal of hazardous waste (WHO 2020; CDC 2003; OSHA 2022; SWANA 2024; EU 2020).

However, for better management of such hazardous waste, there is a tremendous potential for utilizing the benefits of artificial intelligence, robotic automation with computational advancements to avoid human touch points, and increased time efficiency in the collection, segregation, and disposal of waste (Fang et al. 2023). Along with AI, the Internet of Things can significantly improve biomedical waste management as recently few studies utilized the IoT for estimating real-time data on bin levels, waste quantities, types, and locations (Ishaq et al. 2023; Uppala et al. 2023). Together, they could offer a powerful solution for managing biomedical waste efficiently, safely, and in compliance with regulations. These technologies are within the realm of possibility for BMW management, as they are already widespread in other fields of environmental engineering. To date, there is limited evidence about the prediction of biomedical waste generation. As adequate management of BMW is a need of the hour for keeping the current and future generations of human beings healthy, these technological advancements promise an increasingly digitalized future for BMW management. However, little is known about the current extent of digitalization by waste management firms. Hence, this chapter focuses on the (i) categories of biomedical waste and (ii) scope of artificial intelligence in the effective management of BMW, which covers the implications of innovative segregation, collection, and disposal methods equipped with technology such as artificial intelligence (AI) and Internet of Things (IoT).

5.2 Sources and Classification of Biomedical Waste

Biomedical waste growing through the years in its kinds and volume creates a menace to the environment and public health particularly, to the hazardous for patients and workers who deal with it. The sources of biomedical waste are classified into two categories: prime and subsidiary sources.

Prime Sources: It included the sources from which biomedical waste is generated in large quantities, e.g., government and private hospitals, nursing homes, healthcare centers, medical and veterinary colleges, biotechnology institutes and research centers, blood banks, mortuaries, manufacturing units, and so on (Franchini et al. 2004).

Minor Sources: These include those sources that generate small portions of biomedical waste, e.g., physicians, dental clinics, animal and slaughterhouses, blood donation camps, vaccination centers, acupuncturists, cosmetic piercing, and institutes for disabled persons and many others (Kumari et al. 2013; Zimmermann 2017).

Based on the sources and nature of waste, biomedical waste management is divided into four steps: Step 1, segregation; Step 2, collection and storage; Step 3, transportation; and Step 4, treatment and disposal. These four steps of waste management are the key elements of biomedical waste classification. In this chapter it has been classified based on the following:

5.2.1 Types of Waste

The World Health Organization (WHO 2017) categorizes biomedical waste (BMW) into eight categories that are described briefly in the following subsections:

Infectious Waste. These types of waste include body fluids, blood, and pathogens, for example, laboratory waste, discarded diagnostic samples, blood, tissues, cotton swabs, bandages, excreta, waste from isolation wards, and infected equipment.

Pathology Waste. Waste that contains human tissues, body parts, blood samples, contaminated animal carcasses, urine samples, and other body fluids is known as pathology waste.

Genotoxic Waste. Genotoxic waste contains cytotoxic drugs used in cancer therapy and genotoxic chemicals.

Pharmaceutical Waste. Waste containing expired medicine or no longer needed products, drugs, and contaminated pharmaceutical chemicals is referred to as pharmaceutical waste.

Heavy Metals Waste. Unused batteries, broken thermometers, gauges of blood pressure, gas cylinders, and pressurized containers have high content of heavy metals.

Radioactive Waste. Waste that contains infected glassware, urine, and excreta of patients treated with radionuclides is referred to as radioactive waste.

Chemical Waste. Chemical waste includes laboratory chemicals, solvents, and disinfectants that are expired.

Sharps. Sharps include broken glasses, blades, syringes, scalpels, needles, and other objects that can cause a puncture.

5.2.2 Category of Waste

The biomedical waste (management and handling) rules (1998) categorized BMW into ten categories that are summarized in Table 5.1.

5.2.3 Based on Color Coding

The biomedical waste (management and handling) rules were amended in 2016 in which the biomedical waste was divided into four categories instead of ten to improve the segregation of waste at source based on color coding. It has been summarized in Table 5.2:

5.3 Scope of Artificial Intelligence in Effective Management of Biomedical Waste

Conceptually, BMW management is segmented into three major steps: (a) segregation of BMW, (b) collection and transportation of BMW, and (c) disposal of BMW. A brief overview of some essential AI techniques to boost segregation, transportation, and disposal for reducing adverse environmental impact is presented below. To serve this purpose, several models are used such as adaptive neurofuzzy inference system (ANFIS), support vector machine (SVM), genetic algorithm (GA), neural network (NN), decision tree (DT), random forest (RF), and long short-term memory (LSTM) networks that mimic human traits such as learning, problem-solving, understanding, perception, reasoning, and awareness of surroundings. In the last two decades, AI has been extensively used to forecast waste generation rates, optimize waste collection, and truck routes, locate waste management facilities, and simulate waste conversion processes.

5.3.1 AI in Segregation

To effectively manage biomedical waste, segregation of the waste at the point of generation is always encouraged no matter what final strategy for treatment and disposal of waste is selected. Despite available guidelines, healthcare waste segregation practice among healthcare workers is overlooked in developing countries. A study has been conducted to assess the awareness of waste segregation among various categories of healthcare workers; the goal is to determine the rate of mismanagement of waste. Over three years, there has been an increased number of accidental incidents involving needles and sharp objects reported by doctors, paramedics, and housekeepers (A Glance at the World 2010). Currently, as the increased generation of BMW is inevitable during the COVID-19 outbreak, it is not only limited to healthcare facilities but also comes from municipal solid waste and hence it requires safe handling, treatment, and disposal to minimize the contamination of land, water,

Table 5.1 Categories of biomedical waste (Biomedical Waste (Management and Handling) Rules 1998)

Category	Waste content	Components	Method of treatment and disposal
Category No. 1	Human anatomical waste	Human tissues, body parts	Incineration/deep burial
Category No. 2	Animal waste	Animal tissues, organs, body parts carcasses, bleeding parts, fluid, blood, and experimental animals used in research, waste generated by veterinary hospitals colleges, discharge from hospitals, animal, houses	Incineration/deep burial
Category No. 3	Microbiology and biotechnology waste	Wastes from laboratory cultures, stocks or specimens of microorganisms live or attenuated vaccines, human and animal cell cultures used in research and infectious agents and industrial laboratories, wastes from production of biological from research toxins, dishes and devices used for transfer of cultures	Local autoclaving/ microwaving/ incineration
Category No. 4	Waste sharps	Needles, syringes, scalpels, blades, glass, etc. that may cause punctures and cuts. This includes both used and unused sharps	Disinfection chemical treatment/ autoclaving/ microwaving and mutilation shredding
Category No. 5	Discarded medicines and cytotoxic drugs	Wastes comprising outdated, contaminated, and discarded medicines	Incineration/ destruction and drug disposal in secured landfills
Category No. 6	Solid waste	Items contaminated with blood, and body fluids including cotton, dressings, soiled plaster casts, lines, beddings, and other materials contaminated with blood	Incineration, autoclaving/ microwaving
Category No. 7	Solid waste	Waste generated from disposable items other than the waste sharps such as tubing, catheters, and intravenous sets	Disinfection chemical treatment/ autoclaving/ microwaving and mutilation shredding
Category No. 8	Liquid waste	Waste generated from laboratory and washing, cleaning, housekeeping, and disinfecting activities	Disinfection by chemical treatment and discharge into drains
Category No. 9	Incineration ash	Ash from incineration of any biomedical waste	Disposal in municipal landfill
Category No. 10	Chemical waste	Chemicals used in the production of biological, chemicals used in disinfections such as insecticides	Chemical treatment and discharges into drains

Table 5.2 Color coding for common biomedical waste (Guidelines for Management of Healthcare Waste as per Biomedical Waste Management Rules 2016)

Color Coding	Type of Container	Waste Category	Treatment Options (Schedule I)
Yellow	Plastic bag	Cat.1, Cat.2, Cat.3, and Cat.6	Incineration/Deep burial
Red	Disinfected container /Plastic bag	Cat.3, Cat.6, and Cat.7	Autoclaving/ microwaving/ chemical treatment
Blue	Plastic bag/ puncture-proof container	Cat.4 And Cat.7	Auto calving / microwaving/ Chemical treatment and destruction / Shredding
Black	Plastic bag	Cat.5, Cat.9, and Cat.10 (solid)	Disposal in a secured landfill

and air. Waste generated from COVID-19 patients is like any other infectious waste, so its segregation from solid waste is essential. Therefore, creating public awareness about the COVID-19 waste hazards and segregation at source is highlighted in all guidelines as a recommendation. To boost automated segregation, AI helps in reducing human contact with potentially infectious waste and could imply in the following ways. This segment highlights the role of AI in waste classification and the generation rate prediction.

5.3.1.1 Waste Classification

Waste classification based on an automated sorting system could eliminate manual waste segregation and protect sanitary workers from unnecessary exposure to infection. Deep learning-based algorithms are a kind of artificial intelligence that imitates the way humans gain certain types of knowledge and have promoted significant advances in sorting waste material. Waste materials entering the sorting systems are usually dirty, containing the dust and such intelligent system needs to be able to recognize the same material under various conditions in the presence of dirt and other tainting agents. Two deep learning methods named artificial neural network (ANN) and convolutional neural network (CNN) are widely studied by researchers in waste sorting. Numerous studies have focused on classifying solid waste and have utilized CNNs to identify various waste categories. One such study used CNN for the identification and classification of eight different kinds of medical waste based on a dataset of 3480 images; 97.2% accuracy was achieved using the proposed methodology (Zhou et al. 2022). Another CNN-based; multitask learning framework is used for sorting four categories of waste with a huge dataset of 56,964 waste images. The proposed framework showed promising performances for multi-label waste classification tasks and can help in improving future automation sorting goals (Liang and Gu 2021). Similarly, deep CNNs were also used to segregate different types of paper and cardboard using various models, and their accuracy ranged between 61.9 and 77.5%; these low values were attributed to the limited size of the training dataset of only 24 images (Vrancken et al. 2019).

Another study combined the cloud computing technique with seven different deep learning CNN models to classify the nine categories of waste at source using a mobile phone application. Among seven CNN models, MobileNetV3 architecture showed the best performance, with 94.26% accuracy, a comparatively small storage size of 49.5 MB with a fast computing speed of 261.7 ms (Wang et al. 2021). A study comparing CNN-based automated sorting with manual sorting revealed that the automated process significantly improved waste classification while reducing computation time (Sudha et al. 2016). Similarly, several other studies utilized the different CNN models for automated sorting of various other waste products such as glass and metals with a classification accuracy of 98% (Funch et al. 2021) and another model used transfer learning with CNN for waste classification and demonstrated 82.80% accuracy with their proposed methodology (Zhang et al. 2021). Few studies also utilized the multilayer hybrid approach to enhance the accuracy of their developed methodology; one such study reported a multilayer hybrid method with CNN to perform waste classification in public areas (Chu et al. 2018). Another study utilized hyperspectral imaging and multilayer ANNs to recognize three different types of plastics in e-waste (Tehrani and Karbasi 2017). For both studies, the multilayer hybrid approach demonstrated higher accuracy more than 90% with the CNN model and 99% with the ANN model in identifying their respective materials.

5.3.1.2 Waste Generation Rate

Forecasting waste generation is crucial for effective waste management. Biomedical waste is typically heterogeneous due to its complexity, and accurately estimating waste generation rates—whether in the short or long term—is essential for optimizing the design and operation of the biomedical waste management system. The waste generation rate can be influenced by many variables. For a general healthcare facility, these variables can vary from the number of available wards, active wards, total hospital beds, active beds, number of hospital staff, type of wards (such as maternity, emergency, etc.), type of hospital ownership (government or private), the total number of patients who come for diagnosis and treatment, number, and type of hospital's activity in a year, etc. The pandemic of COVID-19 has also increased the number of variables as the infected wastes are generated from isolation wards, quarantine camps, homes of suspected patients, testing centers, laboratories, common treatment, and disposal facilities during the testing of people. Therefore, predicting the waste generation rate for such a complex system requires advanced machine learning methods. Several studies have been reported that implied the artificial intelligence techniques for BMW management. Among them, Golbaz et al. conducted a study on hospital solid waste generation rates, categorizing the waste into three groups: infectious, general, and total solid waste. They employed two machine learning methods, specifically neural networks and kernel-based methods. Their findings indicated that ANN-based models delivered high predictability with minimal errors when compared to the observed data. Additionally, they discovered that kernel-based methods outperformed conventional regression models (Golbaz et al. 2019). Another study compared the ANNs and multiple linear regression models for predicting the rate of hospital waste generation, where ANN-based methods demonstrated superiority in assessing the quality in terms of accuracy (Jahandideh et al. 2009). Recently, Altin et al. compared kernel-based SVM and deep learning methods to predict the amount of medical waste. They used inputs such as the number of surgeries, outpatients, inpatients, intensive care patients, and the number of days in intensive care. To evaluate the performance of these methods, they utilized metrics such as root mean square error (RMSE), mean absolute error (MAE), and R-squared (R^2). The results indicated that the deep learning method outperformed the kernel-based SVM, with better results: (RMSE = 0.094, MAE = 0.079, R^2 = 0.466) compared to SVM (RMSE = 0.264, MAE = 0.202, R^2 = 0.221) (Altin et al. 2023). A variety of machine learning methods have been implemented for waste generation forecasting and predicting solid waste characteristics using different input and output variables. Among these methods, ANN was widely implemented followed by support vector regression (SVMs), adaptive neurofuzzy inference system (ANFIS), decision tree (DT), random forest (RF), long short-term memory networks (LSTM) (Kumar et al. 2018; Younes et al. 2015; Adeleke et al. 2020; Kannangara et al. 2018; Cubillos 2020). A pictorial presentation of the uses of AI in the BMW segregation process is depicted in Fig. 5.1.

5.3.2 AI in Collection

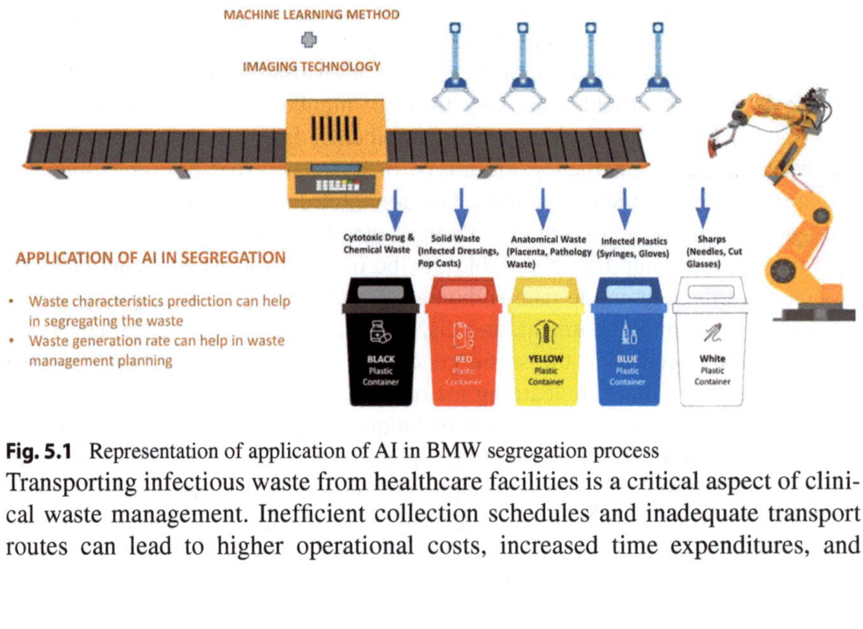

Fig. 5.1 Representation of application of AI in BMW segregation process

Transporting infectious waste from healthcare facilities is a critical aspect of clinical waste management. Inefficient collection schedules and inadequate transport routes can lead to higher operational costs, increased time expenditures, and unnecessary vehicle emissions. To implement an effective BMW management strategy, two key steps can enhance the overall collection process: vehicle routing and bin-level detection. Proper waste collection routing is essential to ensure that infectious waste is contained and processed correctly according to its category and level of hazardousness. Additionally, bin-level detection can assess the status of collection bins and determine whether a site is ready for waste collection. While both of these steps have been previously explored in the context of solid waste management using various AI-based algorithms, this discussion will focus on the different types of AI techniques that can be applied to improve the collection of infectious waste.

5.3.2.1 Vehicle Routing

Smart waste collection techniques have been developed to effectively manage the transportation of waste and to replace the conventional practices in terms of time, pollution, and cost. The ever-increasing BMW generation, especially widely impacted due to COVID-19, has presented an even more challenging situation for the collection system. Recent studies suggested new approaches to improve the waste collection process by optimizing the waste collection frequency and route planning (Fig. 5.2). These methods are based on mostly genetic algorithm (GA) and its hybrid versions; Krol et al. developed a GA-based procedure for optimizing routes used during the collection of electrical and electronic household waste. This methodology reduced collection costs by minimizing route distances, the number of vehicles needed, and the required collection staff (Krol et al. 2016). The results indicated that while the AI-optimized route length was shorter, the average service time was longer compared to the non-optimized solution.

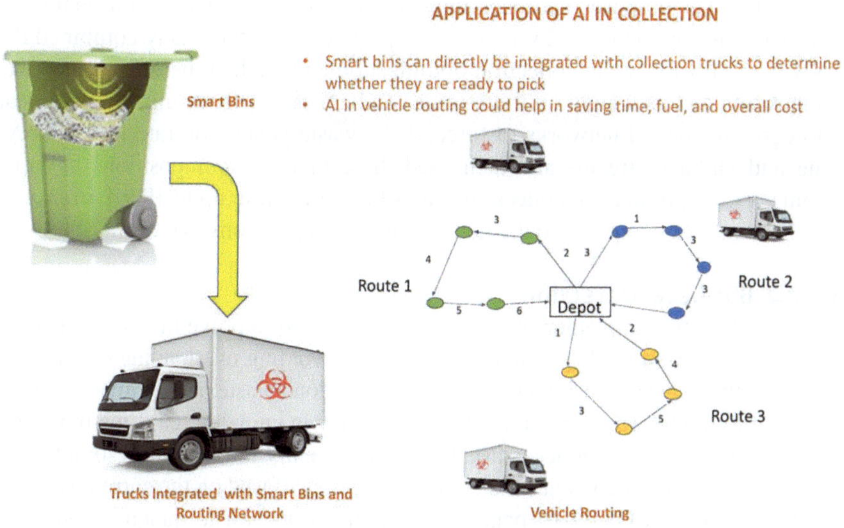

Fig. 5.2 Application of AI in BMW collection

Identification of dumping location is another challenge for waste management firms; to solve this issue, Ramaswami and Velumani (2016) proposed an optimized GA-based chromosome-modeled method for finding all possible location information (Ramaswami and Velumani 2016). Similarly, a study has been conducted in a metropolitan area of Barcelona to locate the collection areas using GA-based methodology (Bautista and Pereira 2006). Test analysis proved the better quality of the proposed method for real instances as compared to the literature and GRASP heuristics with a shorter running time, though their method was compromised on instance randomness. There are a few studies where the genetic algorithm and fuzzy logic together have also been used to solve vehicle routing problems. A study compared a hybrid fuzzy analytical hierarchy process (AHP) and goal programming with hybrid GA for disposing of infectious waste (Wichapa and Khokhajaikiat 2018). This comparison concluded that hybrid GA was efficient at optimizing the minimum transportation cost and a limited number of vehicles for the systematic collection and disposal of waste.

Several studies integrated geographic information system (GIS) tools into the collection route optimization for better management of geospatial data. One such study developed software that integrated GIS with hybrid GA to optimize vehicle routing by addressing path constraints such as road inclinations, traffic directions, and U-turns during collection points (Duzgun et al. 2016). However, optimization resulted in a longer route to avoid higher elevations and served flat regions for the loaded trucks while the loading percentage remained relatively low; model favored the higher elevations and steep roads to reduce fuel consumption. Validation results indicated that the routes proposed by GA were closely correlated to the optimal values, with low error values up to 3.15%. Another study integrated GA with spatial GIS and utilized a modified Dijkstra algorithm in GIS for optimal route options in terms of traveling time and distances of vehicles during waste collection (Amal et al. 2018). The proposed approach improved traveling distance, operational time, and fuel consumption by 8, 28, and 3%, respectively. Another study combined the ANN with GIS for waste collection optimization and studied the impact of waste composition and its weight on truck routes and emissions. They utilized nonlinear autoregressive neural networks to forecast the waste generation rate for the recycling and garbage streams and then used these modified compositions as input parameters to optimize the collection routes that resulted in up to 19.9% improved travel distance as compared to the non-modified compositions (Vu et al. 2019).

5.3.2.2 Bin-Level Detection

Bin-level detection implementation has the potential to improve the overall performance of collection models in multiple ways: (i) avoiding overloading of bins that will limit the exposure of infection through hazardous waste, (ii) estimated arrival and bin pick-up information for a collection agency in knowing real-time bin status helping cut down on the unnecessary drive and fuel emission, and (iii) estimation of weight and density of content for a collection agency. Based on the current requirements due to the COVID-19 pandemic, the infectious waste quantity generated globally will continue to rise at par and thus requires a systematic approach to

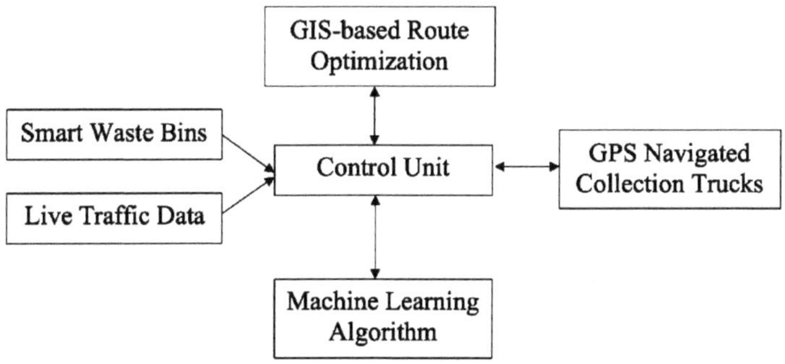

Fig. 5.3 Architecture of the proposed smart waste collection system (Abdallah et al. 2019)

provide actual quantitative data to enhance the quality of current collection practices. This may be achieved by using smart bins installed with level or image sensors attached via AI and other communication technologies such as GIS, radio-frequency identification (RFID), global positioning system (GPS), and general packet radio system (GPRS) (Arebey et al. 2010; Vu et al. 2019).

Few studies have improvised the collection models by incorporating AI to monitor the real-time bin status. One such study implemented a smart waste collection system (Fig. 5.3) to determine the filling rate of bins for different types of households; obtained data were then used to simulate the smart route optimization, resulting in a 19% cost reduction as compared to conventional practices (Abdallah et al. 2019). Rovetta et al. (2009) developed a set of sensors and mounted them in two-wheel 240 L bins to estimate the waste content. The performance accuracy and consistency were evaluated by calculating waste weight and volume and then they used those parameters for routing optimization of collection trucks. Arebey et al. (2010) mounted a tag on bins and trucks that were attached with an RFID reader, a camera associated with an on-board GPS, and a GSM module to make the complete process fully automated for a solid waste-bin monitoring system. Additionally, in another study, they assessed the effectiveness of an image extraction technique known as the gray-level co-occurrence matrix (GLCM) for detecting solid bin levels using ANN and k-nearest neighbor (KNN) models. Their findings indicated that the KNN classifier outperformed the ANN when applied to the same dataset (Arebey et al. 2012).

Hannan et al. (2012) have implemented an advanced image processing technique to assess waste bin levels using the gray-level aura matrix (GLAM). They evaluated the performance of this approach using multilayer perceptron (MLP) and k-nearest neighbor (KNN) classifiers. The results showed that the MLP classifier achieved classification accuracies of 98.98% for class and 90.19% for grade, while the KNN classifier attained 96.91% for class and 89.14% for grade. Further, they employed a feedforward neural network (FFNN) in conjunction with the Hough transform image extraction technique to categorize the amount of solid waste within the bin. The output was classified into five categories: empty, medium, full, flow, and overflow (Hannan et al. 2014). However, the positioning of the camera remains a challenge in these studies, as it is crucial to capture images of the bin that provide a properly centered view of the bin area. To overcome this issue, Islam et al. have introduced dynamic time warping (DTW) for detecting and cropping the bin area and combined it with Gabor wavelet (GW) filters. The performance accuracy evaluated with the MLP classifier for the waste level estimation was 98.50% with the proposed methodology (Islam et al. 2014). Further, a study proposed the use of an artificial immune system (AIS) for the optimization of route planning and bin location to reduce the overall cost and vehicle emission. Various constraints were assessed during optimization, including parking facilities, busy streets, and narrow pavements (Nowakowski and Mrowczynska 2018). The implication of these constraints could also be valuable in designing the BMW collection models as generally healthcare facilities are busy in terms of crowds and vehicles. Recent studies indicate that the Internet of Things (IoT) has the potential to revolutionize the waste collection industry by integrating cloud data systems for more reliable and efficient operations (Singh and Misra 2022). IoT technology enables real-time monitoring of waste levels through smart bins equipped with sensors, which alert employees when bins are full, thereby reducing manual labor and operational costs (Mohamed et al. 2023). Qureshi et al. demonstrated the use of IoT for real-time biomedical waste monitoring and collection. This study explains the use of strategically placed bins with sensors in hospital departments and wards to collect biomedical waste, and the data from these sensors is stored, processed, and analyzed in the cloud. Data from these sensors are stored in the cloud, processed, and analyzed. When a bin reaches capacity, it sends a notification to a central waste management system, which alerts waste collectors to collect, segregate, and transport the waste to a centralized facility (Qureshi et al. 2023). Similarly, Ishaq et al. described the use of IoT sensors in hospital bins to facilitate timely waste collection and analyze data to improve waste management practices (Ishaq et al. 2023).

5.3.3 AI in Disposal

Biomedical waste disposal is the last segment of the waste management process where AI techniques may contribute to process output prediction and parameter forecasting. Once the biomedical waste is collected and segregated, it is then treated using various methods including incineration, autoclaving, irradiation, chemical disinfection, and landfilling as per their category assigned by regulatory boards. Landfilling, also known as sanitary landfilling, may be used for the disposal of previously treated BMW (autoclaved/microwaved/chemically treated) or incineration ash. This segment will focus on the role of AI in quantifying the by-products such as biogas generation, leachate formation, heavy metal removal from leachate, energy recovery, and monitoring the process parameters like equipment parameters, heating value, co-melting temperature of waste (Fig. 5.4).

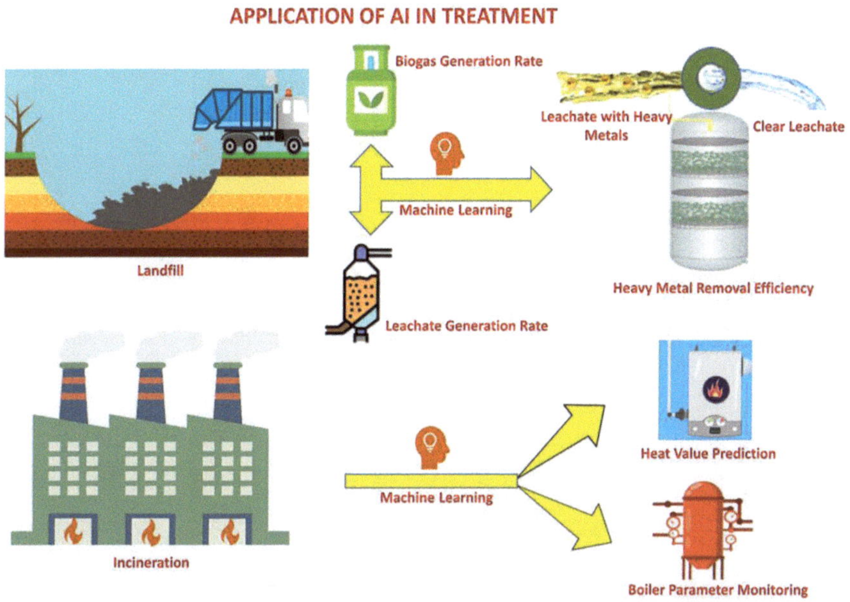

Fig. 5.4 Representation of application of AI in BMW treatment

5.3.3.1 Process Output Prediction

Since landfill gas and leachate formation is one of the major concerns of permanent land disposal, it is necessary to quantify these by-products for optimal waste management. Several models like ANN, ANFIS, and SVM are used for predicting composition and quantifying different by-products generated from waste management methods, as shown in Table 5.3. Among these, ANN is one of the most popular networks developed by many researchers for forecasting the outputs; backpropagation (BP) is the extensively used algorithm for neural networks, consisting of processing elements called neurons and connections arranged in three or more layers: an input layer, a single or more hidden layer(s), and an output layer. Several studies also examined the various BP algorithms to select the most appropriate fitting algorithm, and mostly Levenberge-Marquardt algorithm was proven best compared to the results of different algorithms in predicting the process outputs. For developing these models, various meteorological, climatic, and periodic parameters have been studied as input variables. However, optimization of the number of neurons for a hidden layer is crucial in the design and operation. A smaller number of neurons in the hidden layer could lead to lower predictive accuracy, while too many also result in problems. A highly complex model can suffer from overfitting the training dataset and takes much more computational time to train large networks. Hence, a substitution needs to be found to determine the number of hidden layers, the number of neurons in each layer, and the transfer function used in the hidden layer. Table 5.3 consists of the features of optimized models for predicting methane and leachate generation rates.

Table 5.3 Summary of AI models used for process output predictions

Application	ML Model	Optimized model specifications	Data size	Input	Output	Performance evaluation	References
Landfill gas prediction	ANN	Two bioreactors with (C1) or without (C2) leachate circulation Two-layer network: hidden layer; output layer 12 and 15 neurons in a hidden layer for each bioreactor	–	pH, alkalinity, COD, sulfate, conductivity, chloride, waste temperature, refuse age (day)	Methane percentage	*R*: 0.951 (C1); 0.957 (C2) MSE: 0.00263 (C1); 0.00250 (C2)	Ozkaya et al. (2007)
	ANFIS	Five-layer network	50, 100, 150, 200, 250, 300	Time, sludge addition rate, leachate recirculation rate	Biogas generation rate per week	R^2: 0.97	Abdallah (2011)
	ANN	Two hidden layers with 25 neurons in each layer, one output layer	177	Temperature, total solids, total volatile solids, and pH	Methane percentage	R^2: 0.87 MSE: 6×10^{-5}	Qdias et al. (2010)
	ANN	Two-stage time series model Missing data imputation stage Auto-regressive neural network model with exogenous inputs (NARX) Includes 3 layers, 9 neurons in the hidden layer with sigmoidal function and an output layer with linear function	2344	Temperature, dew point, humidity, wind speed, air pressure	Daily methane generation rate	*R*: 0.85; AIA: 0.92; MAPE: 3.08% (testing stage)	Fallah et al. (2020)

	ANN	9 ANN (NARX) models The optimized model includes 4 inputs temperature, air pressure, humidity, month of the year	2344	Climatic parameters: Temperature, dew point, humidity, wind speed, air pressure Periodic parameters: Month of the year (MOY) and day of the year (DOY)	Daily methane generation rate	R^2: 0.83; MAPE: 2.05% (testing stage)	Fallah et al. (2020)
	ANN	Input layer, log-sigmoid function at 2 hidden layers with 12 and 9 neurons, 1 output layer	121	Soil and air temperature, soil moisture content, methane and oxygen concentration	Methane oxidation	R^2: 0.937 MSE: 0.0082	Abushammala et al. (2013)
Landfill leachate prediction	ANN, SVM	ANN-MLP1 M2 with single hidden layer, 3 inputs ANN-MLP2 M1 with double hidden layer, 9 inputs SVM-M2	120	Dumped waste quantity, rainfall level, emanated gases	Monthly leachate generation rate	R^2: 0.956; RMSE: 398.5 (MLP1-M2) R^2: 0.964; RMSE: 354.7 (MLP2-M1) R^2: 0.889; RSME: 692.1 (SVM-M2)	Abunama et al. (2018b)
	ANFIS	3 inputs, 1 output, 2 triangular membership functions (MFs), and 8 rules	120	Dumped waste quantity, rainfall level, emanated gases	Monthly leachate generation rate	R: 0.952	Abunama et al. (2018a)
	ANN	Two-layer neural network, tan-sigmoidal function at hidden layer with 22 neurons and linear function at output layer	–	pH, temperature, conductivity, months, air temperature, pressure, cloudiness, humidity, precipitation, maximum and minimum temperature	Leachate daily discharge	R^2: 0.847	Karaca and Ozkaya (2006)

(continued)

Table 5.3 (continued)

Application	ML Model	Optimized model specifications	Data size	Input	Output	Performance evaluation	References
Heavy-metal removal prediction from leachate	ANN	Two neurons in each layer i.e., input layer, hidden layer, and output layer	–	Liner material such as bentonite (B), natural zeolite (NZ), expanded vermiculite (EV), and pumice (P)	Cu(II) and Zn(II) removal efficiency	Removal (%): 95% Cu(II) Removal (%):89.19% Zn(II): with (60% NZ + 40% B)	Turan et al. [2013]
	ANFIS,2^3	3 inputs, each has 2 input MF with 8 rules and 8 output MF, one output layer	–	pH, adsorbent dosage, and contact time	Cu(II) removal efficiency with Clinoptilolite	Removal (%): 90.43 (ANFIS) Removal (%): 93.65 (2^3)	Turan and Ozgonenel (2013)
	ANN	Five inputs, 2 hidden layers with 7 neurons in each layer, one output layer	50	pH, temperature, feed flux, substrate flow, retention time	Removal % for CO(II), Ni(II), Zn(II), and turbidity	R^2: 96.56 CO(II)% R^2: 97.63 Ni(II)% R^2: 98.38 Zn(II)% R^2: 95.28 turbidity% (testing stage data)	Maurya et al. (2021)

2^3: Full factorial experimental design, *ANN* artificial neural network, *ANFIS* adaptive-network fuzzy inference system, *MAPE* mean absolute percentage error, *MLP* multilinear perceptron, *RMSE* root mean square error, *MSE* mean squared error, R^2 R squared, *SVM* support vector machine

Ozkaya et al. (2007) have developed a neural network model to predict methane generation in landfill gas from two bioreactors where they used previously recorded leachate components as input parameters. Abdallah et al. (2011) developed a hybrid artificial neural networks-fuzzy logic (ANN-FL) model that utilizes knowledge-based FL to describe and predict the process of landfill gas production at a given time based on operational parameters qualitatively and implements the learning algorithm of ANN. Another study developed the ANN and ANN integrated GA models, where the sigmoid function was observed as the most appropriate BP algorithm method for the prediction of methane yield. Their integrated model showed a 6.9% increase in methane yield by optimizing the operational variables as compared to the ANN model alone (Abu Qdais et al. 2010). Recently, Fallah et al. (2020) developed three landfill gas prediction ANN models, namely, MLP 1, MLP 2 (multilayer perceptrons), and a two-stage neural network autoregressive model with exogenous inputs (NARX) to predict the methane generation rate, and found NARX as an optimum model. They also performed the data preprocessing by removing collinearity and filtering the outliers and observed improved ANN model performance during preprocess steps. Similarly, the work was continued using the NARX model with preprocessing steps as previously used to predict the daily collection rate of methane using 3-year training data. In addition to climatic parameters, periodic parameters were also studied as input variables, and sensitivity analysis was performed using periodic parameters, i.e., day of the year (DOY) and month of the year (MOY); favorable precision was achieved by DOY-based model with MAPE of 2.12% (Fallah and Torabi 2021). As leachate is another by-product coming from landfills, it is a prerequisite to predict the leachate generation rate too. Various AI models have been implemented to predict and analyze the leachate formation, included in Table 5.3. Similarly, the application of AI models in heavy metal removal from landfill leachate and predicting the removal efficiency of liner material has also been explored, presented in Table 5.3. As heavy metal is a major concern of biomedical waste, these applications could also be useful in BMW management.

5.3.3.2 Process Parameter Predictions

Heating value predictions are crucial for determining the amount of energy that can be extracted from waste during conversion processes such as gasification, pyrolysis, and combustion. The operational stability of waste-to-energy plants is affected by variations in heating value. Therefore, effective prediction models have the potential to improve plant efficiency, reduce maintenance costs, and lower emissions. Traditionally, the heating value has been experimentally determined using a bomb calorimeter. However, with advancements in information technology, several studies have focused on developing predictive models for the heating value of solid waste (Table 5.4); among them, ANN is the most commonly used. For such models, input variables are often based on ultimate analysis, physical composition, or proximate analysis, while higher heating value (HHV) or lower heating value (LHV) is recorded as an output; however, a few studies focused on predicting multiple outputs. One such study predicted single/multiple outputs, namely, lower heating values of gas, gasification products (tars, entrained char), and synthetic gas yielding a fluidized bed reactor gasification process using single/double hidden layer neural network models (Pandey et al. 2016). The models were trained using the Levenberg-Marquardt BP algorithm. The predictive performance of multiple output models was higher compared to single output models with mean squared error (MSE) as low as 0.00074. The study reported longer computational time for ANNs with double layers (8 days), which was more than ten times higher than that of single-layer ANN models. As discussed in the previous section about the complexity of optimization of ANN models, You et al. compared the different AI methods, namely, ANN, SVM, ANFIS, and RF, to analyze the variation trend in heating value, where ANFIS showed the best performance although it took longer computation time (You et al. 2017). Another study developed the SVM model to predict the HHV; their optimized model achieved a correlation coefficient of 1.00 with 100 HHV data points (Rostami and Baghban 2018). Additionally, AI methods can also be utilized in monitoring equipment parameters and providing engineering assistance in real-time conditions, for such application few studies developed the ANN model to evaluate the operational performance of the boiler in terms of temperature, pressure, and flow rate of the steam. Among these, Behera et al. (2013) observed higher predictive capabilities than the normal fitted plant data and Shaha et al. (2020) compared their ANN model with random forest (RF) and long short-term memory (LSTM) models; results are presented in Table 5.4. Intelligent control and real-time prediction can maximize energy production and minimize the adverse environmental impact.

5.4 Conclusion

The implication of AI models in BMW management has a wide scope as these algorithms have been applied in all segments for handling waste, from predicting the waste generation rate, waste sorting, vehicle routing, bin-level detection, forecasting of process output, and process parameters involved in disposal steps. In the previous 3 to 4 years, there has sudden increase in the application of AI models in every segment of waste handling from collection, segregation, treatment, and disposal. These predictive models can reduce the cost, workload, and most importantly exposure to hazardous waste. Despite the widespread use of intelligent applications in the current scenario, the adoption of AI in waste management remained challenging due to some commonly observed challenges using AI models, which include overfitting, normalization, validation against classical AI models, and lack of learning/technology transfer, a requirement of high-quality datasets, the rapid evolution of numerous AI models, difficulty in replication of most AI model, etc. However, advanced AI prediction systems have shown advantages over traditional models—such as regression analysis, material flow analysis, and time series analysis—in waste management research, especially for modeling complex nonlinear behavior. This chapter deals with the inclusion of AI models in each segment of waste handling and finds out that neural network is the most widely studied AI model, among which CNN models are widely implemented in the classification of waste, whereas other ANN models with different logarithms such as backpropagation, feedforward, radical basis function

Table 5.4 Summary of AI models used for process parameter predictions

Application	ML model	Optimized model specifications	Data size	Input	Output	Performance evaluation	References
HV prediction	ANN-MLP, SVM, ANFIS, and RF	–	2200	Feeding rate of MSW and coal, bed temperature, furnace outlet gas temperature, steam- (flow, temperature, pressure), primary and secondary airflow.	LHV	Prediction precision: ANFIS (94%) > RF (90%) > SVM (87.5%) > ANN-MLP (73.5%)	You et al. (2017)
	ANN	Five inputs, one hidden layer, and one output layer	60	Plastic, paper, glass, textile, food	LHV	R^2: 0.992	Ogwueleka and Ogwueleka (2010)
	ANN	Seven inputs, one hidden layer with five neurons, and one output layer with three neurons	–	Wood, paper, kitchen garbage, plastic, textile, equivalence ratio, temperature	LHV of gas, gasification products (GP), and gas yields	Relative error (%)-LHV of gas: 20.5 LHV of GP: 8.7–22.3 Gas yield: 12.9	Xiao et al. (2009)
	ANN	Six inputs, three-layer network: two hidden layers with 16 and 4 neurons and output layer ((6-16-4-1)	89	Food, paper, plastic, textile, wood, moisture	LHV	R: 0.9500	Ozveren (2016)

(continued)

Table 5.4 (continued)

Application	ML model	Optimized model specifications	Data size	Input	Output	Performance evaluation	References
	SVM, ANN	ANN: five inputs, two-layer network: one hidden layer with five neurons and one output layer (5-5-1)	252	C, H, O, S, H	HHV	SVM- R^2: 0.978; RMSE: 1.25 ANN- R^2: 0.972 RMSE: 1.44	Bagheri et al. (2019)
	ANN-MLP	Seven inputs, three-layer network: two hidden layers with two and three neurons and output layer ((7-2-3-1)	123	C, O, H, N, S, ash and moisture content	HHV	*R*: 0.986 RMSE: 3.095	Olatunji et al. (2019)
RPF-fired boiler monitoring	RF, LSTM, ANN	–	215	Conveyor speed, feed water temperature and pressure, and incinerator exit temperature	Steam temperature, pressure, and flow rate	MAPE: RF- 1.12, 5.54, 15.6LSTM- 1.14, 4.21, 9.63 ANN- 6.93, 1.49, 18.77	Shaha et al. (2020)

ANFIS adaptive-network fuzzy inference system, *ANN* artificial neural network, *HHV* higher heating value, *LHV* lower heating value, *LSTM* long short-term memory networks, *MAPE* mean absolute percentage error, *MSW* municipal solid waste, R^2 R squared, *RF* random forest, *RMSE* root mean square error, *SVM* support vector machine

(RBF), autoregressive, etc. showed outstanding performances in almost every segment of waste handling from prediction to simulation and optimization of waste management. The development of hybrid models was also noticed in various studies by combining either two AI models or an AI model with other smart applications such as the Internet of Things (IoT), to utilize the benefits of involved tools mutually and to eliminate the shortcomings of each model, when used separately. Adaption of these hybrid models showed improvement in performance in the reported studies, but there is a drawback that there are only a few studies on hybrid models and almost every study uses a different hybridized combination, which leaves no room to compare any two models for waste management applications.

5.5 Future Prospects

While comparing the developed AI models with conventional waste management methods, an improved performance was noted in each study demonstrating the potential of such advanced tools to make an impactful change in handling hazardous waste safely. Additionally, since various factors influence the waste generation rates in healthcare facilities, there is a clear need for robust AI-powered tools that can more accurately predict both waste generation rates and composition. Such tools would facilitate more efficient management of biomedical waste. Internet of Things (IoT) systems can further improve waste management by enabling healthcare facilities to monitor and analyze waste generation patterns in real time. This capability allows for the identification of opportunities to reduce waste at the source and promotes sustainability. When AI and IoT are integrated, they complement each other: IoT provides the necessary infrastructure for real-time monitoring and data collection, while AI analyzes this data to optimize processes, make informed decisions, and enhance overall system performance. To effectively implement AI models in biomedical waste management, waste management associations should prioritize raising awareness and educating organizations about the benefits, challenges, and best practices for utilizing digital technologies. Investing in education and quality data generation will lead to more efficient and sustainable waste management practices.

References

A Glance at the World (2010) Waste Manag 30(1):162–164. https://doi.org/10.1016/j.wasman.2009.10.007

Abdallah M, Warith M, Narbaitz R, Petriu E, Kennedy K (2011) Combining fuzzy logic and neural networks in modeling landfill gas production. World Acad Sci Eng Technol 78:559–565

Abdallah M, Adghim M, Maraqa M, Aldahab E (2019) Simulation and optimization of dynamic waste collection routes. Waste Manag Res 37(8):793–802. https://doi.org/10.1177/0734242x19833152

Abu Qdais H, Bani Hani K, Shatnawi N (2010) Modeling and optimization of biogas production from a waste digester using artificial neural network and genetic algorithm. Resour Conserv Recycl 54(6):359–363. https://doi.org/10.1016/j.resconrec.2009.08.012

Abunama T, Othman F, Younes MK (2018a) Predicting sanitary landfill leachate generation in humid regions using ANFIS modeling. Environ Monit Assess 190(10). https://doi.org/10.1007/s10661-018-6966-y

Abunama T, Othman F, Ansari M, El-Shafie A (2018b) Leachate generation rate modeling using artificial intelligence algorithms aided by input optimization method for an MSW landfill. Environ Sci Pollut Res 26(4):3368–3381. https://doi.org/10.1007/s11356-018-3749-5

Abushammala MFM, Basri NEA, Elfithri R, Younes MK, Irwan D (2013) Modeling of methane oxidation in landfill cover soil using an artificial neural network. J Air Waste Manage Assoc 64(2):150–159. https://doi.org/10.1080/10962247.2013.842510

Adeleke O, Akinlabi SA, Jen T-C, Dunmade I (2020) Prediction of municipal solid waste generation: an investigation of the effect of clustering techniques and parameters on ANFIS model performance. Environ Technol 43(11):1634–1647. https://doi.org/10.1080/09593330.2020.1845819

Altin FG, Budak I, Özcan F (2023) Predicting the amount of medical waste using kernel-based SVM and deep learning methods for a private hospital in Turkey. Sustain Chem Pharm 33:101060. https://doi.org/10.1016/j.scp.2023.101060

Amal L, Son LH, Chabchoub H (2018) SGA: spatial GIS-based genetic algorithm for route optimization of municipal solid waste collection. Environ Sci Pollut Res 25(27):27569–27582. https://doi.org/10.1007/s11356-018-2826-0

Arebey M, Hannan MA, Basri H, Begum RA, Abdullah H (2010) Integrated technologies for solid waste bin monitoring system. Environ Monit Assess 177(1–4):399–408. https://doi.org/10.1007/s10661-010-1642-x

Arebey M, Hannan MA, Begum RA, Basri H (2012) Solid waste bin level detection using gray level co-occurrence matrix feature extraction approach. J Environ Manag 104:9–18. https://doi.org/10.1016/j.jenvman.2012.03.035

Bagheri M, Esfilar R, Golchi MS, Kennedy CA (2019) A comparative data mining approach for the prediction of energy recovery potential from various municipal solid waste. Renew Sust Energ Rev 116:109423. https://doi.org/10.1016/j.rser.2019.109423

Bautista J, Pereira J (2006) Modeling the problem of locating collection areas for urban waste management. An application to the metropolitan area of Barcelona. Omega 34(6):617–629. https://doi.org/10.1016/j.omega.2005.01.013

Behera SK, Rene ER, Kim MC, Park HS (2013) Performance prediction of an RPF-fired boiler using artificial neural networks. Int J Energy Res 38(8):995–1007. https://doi.org/10.1002/er.3108

Bio-medical waste (management and handling) rules 1998. [online] Available at: https://hspcb.gov.in/BMW%20Rules.pdf

Chartier Y, Emmanuel J, Pieper U, Prüss A, Rushbrook P, Stringer R, Townend W, Wilburn S, Zghondi R (eds) (2014) World Health Organization. Safe management of wastes from health care activities. World Health Organization Blue Book

Chi T, Zhang A, Zhang X, Li A, Zhang H, Zhao Z (2020) Characteristics of the antibiotic resistance genes in the soil of medical waste disposal sites. Sci Total Environ 730:139042. https://doi.org/10.1016/j.scitotenv.2020.139042

Chu Y, Huang C, Xie X, Tan B, Kamal S, Xiong X (2018) Multilayer hybrid deep-learning method for waste classification and recycling. Comput Intell Neurosci 2018:1–9. https://doi.org/10.1155/2018/5060857

Cubillos M (2020) Multi-site household waste generation forecasting using a deep learning approach. Waste Manag 115:8–14. https://doi.org/10.1016/j.wasman.2020.06.046

Düzgün HŞ, Uşkay SO, Aksoy A (2016) Parallel hybrid genetic algorithm and GIS-based optimization for municipal solid waste collection routing. J Comput Civ Eng 30(3). https://doi.org/10.1061/(asce)cp.1943-5487.0000502

DIRECTIVE (EU) 2024/3019 OF THE EUROPEAN PARLIAMENT AND OF THE COUNCIL of 27 November 2024 concerning urban wastewater treatment. https://eur-lex.europa.eu/legal-content/EN/TXT/PDF/?uri=CELEX:32024L3019

Ezeudu OB, Ezeudu TS, Ugochukwu UC, Tenebe IT, Ajogu AP, Nwadi UV, Ajaero CC (2022) Healthcare waste management in Nigeria: a review. Recycling 7(6):87–102. https://doi.org/10.3390/recycling7060087

Fallah B, Torabi F (2021) Application of periodic parameters and their effects on the ANN landfill gas modeling. Environ Sci Pollut Res 28(22):28490–28506. https://doi.org/10.1007/s11356-021-12498-5

Fallah B, Ng KTW, Vu HL, Torabi F (2020) Application of a multi-stage neural network approach for time-series landfill gas modeling with missing data imputation. Waste Manag 116:66–78. https://doi.org/10.1016/j.wasman.2020.07.034

Fang B, Yu J, Chen Z, Osman AI, Farghali M, Ihara I, Hamza EH, Rooney DW, Yap P (2023) Artificial intelligence for waste management in smart cities: a review. Environ Chem Lett 21:1959–1989. https://doi.org/10.1007/s10311-023-01604-3

Franchini M, Rial M, Buiatti E, Bianchi F (2004) Health effects of exposure to waste incinerator emissions: a review of epidemiological studies. Annali dell'Istituto superiore di sanita 40(1):101–115

Funch OI, Marhaug R, Kohtala S, Steinert M (2021) Detecting glass and metal in consumer trash bags during waste collection using convolutional neural networks. Waste Manag 119:30–38. https://doi.org/10.1016/j.wasman.2020.09.032

Golbaz S, Nabizadeh R, Sajadi HS (2019) Comparative study of predicting hospital solid waste generation using multiple linear regression and artificial intelligence. J Environ Health Sci Eng 17(1):41–51. https://doi.org/10.1007/s40201-018-00324-z

Goswami M, Goswami PJ, Nautiyal S, Prakash S (2021) Challenges and actions to the environmental management of bio-medical waste during the COVID-19 pandemic in India. Heliyon 7(3):e06313. https://doi.org/10.1016/j.heliyon.2021.e06313

Guidelines for Management of Healthcare Waste as per Biomedical Waste Management Rules (2016) [online] Available at: https://cpcb.nic.in/uploads/Projects/Bio-Medical-Waste/Guidelines_healthcare_June_2018.pdf. Accessed 26 Aug 2024

Hannan MA, Arebey M, Begum RA, Basri H (2012) An automated solid waste bin level detection system using a gray-level aura matrix. Waste Manag 32(12):2229–2238. https://doi.org/10.1016/j.wasman.2012.06.002

Hannan MA, Zaila WA, Arebey M, Begum RA, Basri H (2014) Feature extraction using Hough transform for solid waste bin level detection and classification. Environ Monit Assess 186(9):5381–5391. https://doi.org/10.1007/s10661-014-3786-6. https://apps.who.int/iris/handle/10665/259491

Ilyas S, Srivastava R, Kim H (2020) Disinfection technology and strategies for COVID-19 hospital and bio-medical waste management. Sci Total Environ 749:141652. https://doi.org/10.1016/j.scitotenv.2020.141652

Ishaq A, Mohammad SJ, Bello AAD, Wada SA, Adebayo A, Jagun ZT (2023) Smart waste bin monitoring using IoT for sustainable biomedical waste management. Environ Sci Pollut Res. https://doi.org/10.1007/s11356-023-30240-1

Islam MS, Hannan MA, Basri H, Hussain A, Arebey M (2014) Solid waste bin detection and classification using dynamic time warping and MLP classifier. Waste Manag 34(2):281–290. https://doi.org/10.1016/j.wasman.2013.10.030

Iyer M, Tiwari S, Renu K, Pasha M, Pandit S, Singh B, Raj N, Krothapalli S, Kwak H, Balasubramanian V, Jang S, G. D, Uttpal A, Narayanasamy A, Kinoshita M, Subramaniam M, Nachimuthu S, Roy A, Valsala Gopalakrishnan A, Ramakrishnan P, Cho S, Vellingiri B (2021) Environmental survival of SARS-CoV-2 – A solid waste perspective. Environ Res 197:111015. https://doi.org/10.1016/j.envres.2021.111015

Jahandideh S, Jahandideh S, Asadabadi EB, Askarian M, Movahedi MM, Hosseini S, Jahandideh M (2009) The use of artificial neural networks and multiple linear regression to predict the

rate of medical waste generation. Waste Manag 29(11):2874–2879. https://doi.org/10.1016/j.wasman.2009.06.027
Kannangara M, Dua R, Ahmadi L, Bensebaa F (2018) Modeling and prediction of regional municipal solid waste generation and diversion in Canada using machine learning approaches. Waste Manag 74:3–15. https://doi.org/10.1016/j.wasman.2017.11.057
Karaca F, Özkaya B (2006) NN-LEAP: a neural network-based model for controlling leachate flow-rate in a municipal solid waste landfill site. Environ Model Softw 21(8):1190–1197. https://doi.org/10.1016/j.envsoft.2005.06.006
Komilis D, Fouki A, Papadopoulos D (2012) Hazardous medical waste generation rates of different categories of health-care facilities. Waste Manag 32(7):1434–1441. https://doi.org/10.1016/j.wasman.2012.02.015
Król A, Nowakowski P, Mrówczyńska B (2016) How to improve WEEE management? Novel approach in mobile collection with the application of artificial intelligence. Waste Manag 50:222–233. https://doi.org/10.1016/j.wasman.2016.02.033
Kumar A, Samadder SR, Kumar N, Singh C (2018) Estimation of the generation rate of different types of plastic wastes and possible revenue recovery from informal recycling. Waste Manag 79:781–790. https://doi.org/10.1016/j.wasman.2018.08.045
Kumari R, Srivastava K, Wakhlu A, Singh A (2013) Establishing biomedical waste management system in the Medical University of India – A successful practical approach. Clin Epidemiol Global Health 1(3):131–136. https://doi.org/10.1016/j.cegh.2012.11.004
Liang S, Gu Y (2021) A deep convolutional neural network to simultaneously localize and recognize waste types in images. Waste Manag 126:247–257. https://doi.org/10.1016/j.wasman.2021.03.017
Maurya AK, Reddy BS, Theerthagiri J, Narayana PL, Park CH, Hong JK, Yeom J-T, Cho KK, Reddy NS (2021) Modeling and optimization of process parameters of biofilm reactor for wastewater treatment. Sci Total Environ 787:147624. https://doi.org/10.1016/j.scitotenv.2021.147624
Mohamed NH, Khan S, Jagtap S (2023) Modernizing medical waste management: unleashing the power of the Internet of Things (IoT). Sustainability 15(13):9909. https://doi.org/10.3390/su15139909
Nowakowski P, Mrówczyńska B (2018) Towards sustainable WEEE collection and transportation methods in the circular economy – comparative study for rural and urban settlements. Resour Conserv Recycl 135:93–107. https://doi.org/10.1016/j.resconrec.2017.12.016
Ogwueleka TC, Ogwueleka FN (2010) Modeling energy content of municipal solid waste using artificial neural network. Iranian J Environ Health Sci Eng 7(3):259–266
Olatunji O, Akinlabi S, Madushele N, Adedeji P, Felix I (2019) Multilayer perceptron artificial neural network for the prediction of the heating value of municipal solid waste. AIMS Energy 7(6):944–956. https://doi.org/10.3934/energy.2019.6.944
Ozkaya B, Demir A, Bilgili M (2007) Neural network prediction model for the methane fraction in biogas from field-scale landfill bioreactors. Environ Model Softw 22(6):815–822. https://doi.org/10.1016/j.envsoft.2006.03.004
Ozveren U (2016) An artificial intelligence approach to predict a lower heating value of municipal solid waste. Energy Sour Part A Recov Util Environ Effects 38(19):2906–2913. https://doi.org/10.1080/15567036.2015.1107864
Pandey DS, Das S, Pan I, Leahy JJ, Kwapinski W (2016) Artificial neural network-based modeling approach for municipal solid waste gasification in a fluidized bed reactor. Waste Manag 58:202–213. https://doi.org/10.1016/j.wasman.2016.08.023
Patil P, Bohara R (2020) Nanoparticles impact in biomedical waste management. Waste Manag Res J Sustain Circ Econ 38(11):1189–1203. https://doi.org/10.1177/0734242X20936761
Qureshi KN, Khan A, Jamil SUU, Sharma B, Jeon G (2023) Internet of Things enables smart solid waste bin management systems for a sustainable environment. Environ Sci Pollut Res 30:125188–125196. https://doi.org/10.1007/s11356-023-28611-9

Ramasami K, Velumani B (2016) Location prediction for solid waste management – a genetic algorithmic approach. In: IEEE international conference on Computational Intelligence and Computing Research (ICCIC)

Rostami A, Baghban A (2018) Application of a supervised learning machine for accurate prognostication of higher heating values of solid wastes. Energy Sour Part A Recov Util Environ Effects 40(5):558–564. https://doi.org/10.1080/15567036.2017.1360967

Rovetta A, Xiumin F, Vicentini F, Minghua Z, Giusti A, Qichang H (2009) Early detection and evaluation of waste through sensorized containers for a collection monitoring application. Waste Manag 29(12):2939–2949. https://doi.org/10.1016/j.wasman.2009.08.016

Shaha AP, Singamsetti MS, Tripathy BK, Srivastava G, Bilal M, Nkenyereye L (2020) Performance prediction and interpretation of a refuse plastic fuel-fired boiler. IEEE Access 8:117467–117482. https://doi.org/10.1109/access.2020.3004156

Singh AP, Misra DC (2022) Waste management issues and solutions using IoT. Int J Sci Res Arch 7(1):260–282. https://doi.org/10.30574/ijsra.2022.7.1.0209

Sudha S, Vidhyalakshmi M, Pavithra K, Sangeetha K, Swaathi V (2016) An automatic classification method for the environment: friendly waste segregation using deep learning. In: 2016 IEEE technological innovations in ICT for Agriculture and Rural Development (TIAR). [online]. https://doi.org/10.1109/tiar.2016.7801215

Tehrani A, Karbasi H (2017) A novel integration of hyper-spectral imaging and neural networks to process waste electrical and electronic plastics. In: 2017 IEEE conference on Technologies for Sustainability (SusTech). https://doi.org/10.1109/sustech.2017.8333533

Turan NG, Ozgonenel O (2013) The design and implementation of adsorptive removal of Cu(II) from leachate using ANFIS. Sci World J 2013:1–9. https://doi.org/10.1155/2013/590267

Turan NG, Gümüşel EB, Ozgonenel O (2013) Prediction of heavy metal removal by different liner materials from landfill leachate: modeling of experimental results using artificial intelligence technique. Sci World J 2013:1–5. https://doi.org/10.1155/2013/240158

Uppala R, Ramya V, Murugan VS (2023) Classification, collection, and notification of medical waste using IoT-based smart dust bins. Ingenierie Des Systemes d'Inform 28(1):149–154. https://doi.org/10.18280/isi.280115

Vrancken C, Longhurst P, Wagland S (2019) Deep learning in material recovery: development of a method to create training database. Expert Syst Appl 125:268–280. https://doi.org/10.1016/j.eswa.2019.01.077

Vu HL, Bolingbroke D, Ng KTW, Fallah B (2019) Assessment of waste characteristics and their impact on GIS vehicle collection route optimization using ANN waste forecasts. Waste Manag 88:118–130. https://doi.org/10.1016/j.wasman.2019.03.037

Wang C, Qin J, Qu C, Ran X, Liu C, Chen B (2021) A smart municipal waste management system based on deep-learning and the Internet of Things. Waste Manag 135:20–29. https://doi.org/10.1016/j.wasman.2021.08.028

WHO.int (2018) Health-care waste. [online] Available at: https://www.who.int/news-room/fact-sheets/detail/health-care-waste. Accessed 25 Aug 2024

WHO.int (2020) Water, sanitation, hygiene, and waste management for SARS-CoV-2, the virus that causes COVID-19. [online] Available at: https://www.who.int/publications/i/item/WHO-2019-nCoV-IPC-WASH-2020.4. Accessed 25 Aug 2024

Wichapa N, Khokhajaikiat P (2018) Solving a multi-objective location routing problem for infectious waste disposal using hybrid goal programming and hybrid genetic algorithm. Int J Ind Eng Comput:75–98. https://doi.org/10.5267/j.ijiec.2017.4.003

World Health Organization (2017) Safe management of wastes from health-care activities: a summary. World Health Organization [online]. Available at: https://www.who.int/publications/i/item/WHO-FWC-WSH-17.05. Accessed 25 Aug 2024

www.cdc.gov (2003) Medical waste | Background | Environmental guidelines | Guidelines library | Infection control | CDC. [online] Available at: https://www.cdc.gov/infection-control/hcp/environmental-control/regulated-medical-waste.html?CDC_AAref_Val=https://www.cdc.gov/infectioncontrol/guidelines/environmental/background/medical-waste.html. Accessed 25 Aug 2024

www.osha.gov (2022) Hazardous waste – control and prevention | Occupational safety and health administration. [online] Available at: https://www.osha.gov/hazardous-waste/control-prevention. Accessed 25 Aug 2024

www.researchnester.com (2024) Healthcare & pharmaceuticals | Medical waste management market | Global market size, forecast, and trending highlights over 2024–2036. [online] Available at: https://www.researchnester.com/reports/medical-waste-management-market/6054?gad_source=1&gbraid=0AAAAACxz_SOdyOj6Isldq3bBwk7PUYCv4&gclid=Cj0KCQjwz7C2BhDkARIsAA_SZKb1ocO9hb4feN573ahbCwl9OtQ82PgRbXjJhZsPUyQa2a3aPIDZzq4aArY7EALw_wcB#. Accessed 26 Aug 2024

www.swana.org (2024) Safety first: handling biohazardous waste. [online] Available at: https://swana.org/news/blog/swana-post/swana-blog/2024/03/05/safety-first-handling-biohazardous-waste. Accessed 25 Aug 2024

Xiao G, Ni M, Chi Y, Jin B, Xiao R, Zhong Z, Huang Y (2009) Gasification characteristics of MSW and an ANN prediction model. Waste Manag 29(1):240–244. https://doi.org/10.1016/j.wasman.2008.02.022

You H, Ma Z, Tang Y, Wang Y, Yan J, Ni M, Cen K, Huang Q (2017) Comparison of ANN (MLP), ANFIS, SVM, and RF models for the online classification of the heating value of burning municipal solid waste in circulating fluidized bed incinerators. Waste Manag 68:186–197. https://doi.org/10.1016/j.wasman.2017.03.044

Younes MK, Nopiah ZM, Basri NEA, Basri H, Abushammala MFM, KNA M (2015) Solid waste forecasting using modified ANFIS modeling. J Air Waste Manage Assoc 65(10):1229–1238. https://doi.org/10.1080/10962247.2015.1075919

Zhang Q, Yang Q, Zhang X, Bao Q, Su J, Liu X (2021) Waste image classification based on transfer learning and convolutional neural network. Waste Manag 135:150–157. https://doi.org/10.1016/j.wasman.2021.08.038

Zhou H, Yu X, Alhaskawi A, Dong Y, Wang Z, Jin Q, Hu X, Liu Z, Kota VG, Abdulla MHAH, Ezzi SHA, Qi B, Li J, Wang B, Fang J, Lu H (2022) A deep learning approach for medical waste classification. Sci Rep 12(1). https://doi.org/10.1038/s41598-022-06146-2

Zimmermann K (2017) Microwave as an emerging technology for the treatment of biohazardous waste: a mini-review. Waste Manag Res 35(5):471–479. https://doi.org/10.1177/0734242x16684385

Artificial Intelligence for Environmental Decision Support Systems

6

Geeta Aggarwal and Seema Aggarwal

Abstract

Artificial intelligence (AI) is a technology that empowers computer systems and machines to act, think, and behave like humans for problem-solving. It can perform tasks that would otherwise require human intelligence. AI along with machine learning and deep learning is a field of computer science, where algorithms are used to model the decision-making processes of the human brain. These models can "learn" from available data and make accurate predictions over time. One of the main advantages of utilizing AI tools is their ability to analyze and process large amounts of data efficiently.

Human life is most vulnerable to climatic disasters. The natural environment is deteriorating at an unexpected pace. The need of the hour is to take urgent collective action for environmental protection. We want to develop an intelligent and smart system to monitor, analyze, and guide environmental protection. Environmental decision support systems (EDSS) will help in analyzing environmental issues. It will help to find the interrelated aspects of problems related to the environment and find optimum solutions for existing alternatives. It is a rapidly progressing field. We talk about various algorithms and methods to develop such systems. We look at the use of AI for handling the issues. It is a challenging task due to the dynamic nature and variation in the environment. To address the challenges, AI and environmental professionals are collaborating for effective design of solutions.

G. Aggarwal
PGDAV College, University of Delhi, New Delhi, India
e-mail: geeta.gupta@pgdav.du.ac.in

S. Aggarwal (✉)
Miranda House, University of Delhi, New Delhi, India
e-mail: seema.aggarwal@mirandahouse.ac.in

B. Thokchom et al. (eds.), *Artificial Intelligence Enabled Real Time Environmental Monitoring*, https://doi.org/10.1007/978-981-95-3767-9_6

Keywords

Artificial intelligence · Environmental decision support systems · Environment · Protection

6.1 Introduction

The development of mankind is leading to the deterioration of the environment, which in turn will limit the growth of human beings. Heat waves, floods, earthquakes, global warming, and many other environmental issues are occurring more and more due to increasing population, urbanization, and other human activity. Weather patterns over the years are getting disrupted and worsening. Extreme weather conditions like very hot summers and cold winters are becoming prevalent, which results in pressure on energy supplies to fulfil requirements for heating and air conditioning. Additionally, as a result of the melting of glaciers and polar ice caps brought on by rising temperatures, coastal populations are at risk of flooding. In addition to upsetting ecosystems, this causes species to struggle to adapt, which results in a loss of biodiversity.

There has been much advancement in information technology and the quick creation of new, faster hardware. The immediate concern of environmental degradation can be solved by the right use of the advanced technology for betterment of both mankind and the environment. The connections between computer scientists and environmental scientists could be very beneficial.

AI is a technology that empowers computer systems and machines to act, think, learn, and behave like humans for problem-solving. AI allows the machine to perform tasks that would otherwise require human intelligence. One of the main advantages of utilizing AI tools is their ability to analyze and process large amounts of data efficiently. Since computers are essential to modern environmental protection for jobs like data processing, communication, information storing and retrieval, and monitoring, it makes sense to consider integrating artificial intelligence into these functions to improve and integrate them all. AI-based decision support system (DSS) is an approach to human reasoning in contrast to binary Boolean logic, fuzzy logic (FL), etc. Interactive computer programs known as DSS (Rahamathunnisa and Chellappa 2018) assist decision-makers in using models and data to address unstructured problems, hence enhancing the decision-making process.

AI-based DSS are being used in all walks of life these days. Using such DSS, one can assess options to investigate potential outcomes, weigh costs and benefits, and suggest suitable courses of action. The purpose of DSS is to assist decision-makers in using models and data to address unstructured problems, hence enhancing the quality of decisions. DSSs are interactive computer-based systems. The DSS Quality Assessment Tool is intended to evaluate AI-based decision support systems comprehensively, with an emphasis on factors like human engagement, data similarity, calibration, and robustness.

Medical professionals are using AI systems more frequently to help with illness diagnosis and patient safety. The efficacy and accuracy of clinical decision support systems depend on the underlying AI techniques used, aiming to provide the best outcomes in medical decision-making. AI methodologies are utilized to implement intelligent decision support systems in serious games, addressing complex decision-making environments. AI decision-making systems are incorporated into management, encompassing diverse applications like optimization, predictive modeling, and decision support systems.

Information technologies in general and artificial intelligence, in particular, have been successful in creating tools that are suitable for planning, modeling, simulation, prediction, and design of systems that aid in decision-making for environmental preservation and management. The intersection of AI and environmental study represents one of the most promising yet challenging frontiers in modern science and technology. AI's potential and ability to significantly impact environmental change is being realized through innovative applications in multiple sectors, from renewable energy optimization to enhancing climate models.

Concepts of AI along with those of machine learning and deep learning are used to model the decision-making processes in the environmental domain leading to the development of EDSS. EDSS assists decision-makers in addressing complex environmental issues. The capability of EDSS has greatly improved because of AI (Cortés et al. 2000). It can process data in real time, make predictions that are more accurate, and manage enormous volumes of data. EDSS is different from regular DSS in terms of the size and complexity of the datasets and problems they handle.

However, this journey and this collaboration do not and will not come without its own set of challenges and obstacles. Through this chapter, we aim to dive deep into the progress made so far, explore the benefits, and discuss the challenges being faced as far as leveraging AI to tackle the problem of environmental change is concerned. The chapter also explores the intersection of AI and EDSS, highlighting key technologies, applications, and future directions. We discuss more about the integration of AI and EDSS with special emphasis on important technologies, uses, and potential future developments. An overview of how AI can be incorporated into environmental decision support systems to address a range of environmental concerns is discussed in detail. This is intended for researchers, practitioners, and policymakers who are interested in utilizing AI for environmental management.

6.2 Impact of Mankind on the Environment

Environmental concerns have become increasingly prominent due to the impact of human activities on the planet. Rapid urbanization leads to the loss of natural habitats, increased pollution, and higher demands on resources. Sustainable urban planning and green infrastructure are necessary to mitigate these impacts. Transitioning to renewable energy sources like solar, wind, and hydropower is essential to reduce environmental impacts. The increasing amount of waste, particularly nonbiodegradable waste like plastics, poses significant challenges for waste management systems. Landfills, incineration, and littering are causing environmental and health issues.

The most pressing environmental issue is climate change, driven by greenhouse gas emissions from activities like burning fossil fuels, deforestation, and industrial processes. The effects include global warming, rising sea levels, and more extreme weather events. Human activities such as deforestation, habitat destruction, pollution, and climate change are causing a rapid decline in biodiversity. Species are becoming extinct at an alarming rate, which disrupts ecosystems and reduces genetic diversity.

Pollution of air, water, and soil is a major concern. Industrial processes, vehicle emissions, and improper waste disposal contribute to pollution, affecting human health and the environment. Plastic pollution has become a significant problem in oceans and waterways. Increased CO_2 levels are causing the oceans to become more acidic, which affects marine life, particularly organisms with calcium carbonate shells or skeletons, such as corals and shellfish. Over-extraction of groundwater and pollution are contributing to water scarcity in many regions. This affects agriculture, drinking water supplies, and ecosystems. Contaminated water, air, and soil can lead to serious health conditions, highlighting the need for environmental protection to safeguard public health.

The clearing of forests for agriculture, logging, and urban development is leading to habitat loss and disruption of water cycles. Forests are vital for maintaining biodiversity and mitigating climate change. Intensive agriculture, deforestation, and industrial activities are degrading soil quality. This leads to reduced agricultural productivity, loss of arable land, and increased desertification. Unsustainable fishing practices are depleting fish stocks and harming marine ecosystems. This affects food security and the livelihoods of communities that depend on fishing.

Addressing all these concerns requires global cooperation, innovative technologies, sustainable practices, and strong policies to protect the environment and ensure a healthy planet for future generations.

6.3 Protecting the Environment

Ensuring the health and welfare of all life on Earth, including people, depends on environmental protection. By lowering exposure to contaminants that can lead to respiratory illnesses, cardiovascular problems, and other major health concerns, a clean and stable environment promotes human health. Moreover, pollination, water purification, climate management, and soil fertility are just a few of the critical services that healthy ecosystems offer and are necessary for both food production and stable economies. By lowering greenhouse gas emissions and protecting natural carbon sinks like forests and oceans, environmental preservation also plays a crucial part in lessening the effects of climate change.

Protecting the environment also keeps biodiversity intact, clean water and air available, and our planet's natural beauty and recreational value intact. It is essential for the well-being and progress of people. It is a moral requirement. Protecting the environment is crucial for several interconnected reasons that impact the well-being of all life forms on Earth, including humans. A clean environment reduces exposure to pollutants that can cause respiratory diseases, cardiovascular problems, and other health issues. In summary, protecting the environment is essential for maintaining the health and well-being of humans, preserving biodiversity, ensuring economic stability, and fulfilling our ethical responsibilities to future generations and the natural world. It is the premise for social development and the precondition for further economic development.

The problems are extremely complicated and require all our technological know-how to create both short-term and long-term answers (Duan et al. 2019). Technological innovations and international cooperation play crucial roles in addressing environmental protection, emphasizing the need for immediate international legal regulation and cooperation between countries to combat global environmental security and sustainable development.

6.4 Artificial Intelligence and Its Role

AI comprises various technologies, such as computer vision, natural language processing (NLP), deep learning (DL), and machine learning (ML). Besides these, robotics, self-learning, and decision support take care of the thinking capability of AI. ML algorithms of AI are algorithms that learn from data to forecast or do trend analysis. A few examples of the common techniques used are regression, classification, clustering, and reinforcement learning (García et al. 2019). DL is a branch of machine learning that uses multilayered neural networks to extract intricate patterns and representations from massive datasets (Shrestha and Mahmood 2019).

NLP refers to methods for producing and comprehending human language. These methods help handle textual material and promote communication between humans and computers (Kurdi 2017). Natural language processing, or ML, has been used in extensive biodiversity research. It has modeled social media data to forecast public opinion, employed NLP techniques to anticipate ecosystem services, and modeled habitat suitability and availability for biodiversity conservation (Bagstad et al. 2015; Tuuli et al. 2019). The primary areas of concentration in energy research are fuzzy logic models, expert systems, ANN, and pattern recognition.

Computer vision (CV) refers to methods for evaluating and deciphering visual information from the outside world, such as pictures and satellite photos. Major CV applications include object detection and classification, including face recognition (Kumar et al. 2019), feature extraction in photos and videos (Buch et al. 2011), and how autonomous cars employ the perception of their surroundings (Elliott et al. 2019).

Thus we see AI includes a wide variety of technologies each having special qualities that can be used to help environmental decision-making. This facilitates real-time processing and interpretation of environmental data, aiding in data-driven decision-making and strategic planning for environmental management to address various environmental concerns and their protection. Also, technology has been widely recognized as a feasible solution for environmental protection. Technologies such as Internet of Things (IoT), drones, and Lidar (light detection and ranging) allow for intelligent monitoring of the environment, measuring parameters, modularization, real-time data transmission, etc.

Thus, it is important to use AI capabilities to discover solutions to mitigate the environmental catastrophe. To do this, thorough research is necessary to determine how AI solutions can be integrated with human feelings, ideas, social conventions, and actions. AI has the potential to overcome information asymmetries and the bias of human emotions, two factors that create barriers to developing solutions for environmental sustainability (Cullen-Knox et al. 2017).

6.5 Environment Decision Support Systems

Environmental decision support systems (EDSS) are developed by modeling the decision-making processes in the environment domain using concepts from artificial intelligence, machine learning, and deep learning. It can handle massive volumes of data, process data in real time, and produce more accurate forecasts. EDSS facilitates the resolution of intricate environmental challenges by decision-makers. Artificial intelligence, machine learning, and deep learning concepts are utilized to mimic the decision-making processes that lead to the creation of EDSS. Most EDSS models have been constructed and tested utilizing methods from artificial neural network (ANN), support vector machine (SVM), genetic algorithm (GA), linear regression, and autoregressive moving average model (ARMA).

Development of an EDSS for any application will require building of its various components shown in Fig. 6.1. First and foremost we have to decide about the data which is required. A lot of high-quality data is required. To guarantee accuracy and relevance, data from sensors, satellite photography, and historical records are gathered and preprocessed. Data management component handles all the tasks related to it. Model management component creates environment/predictive models using AI

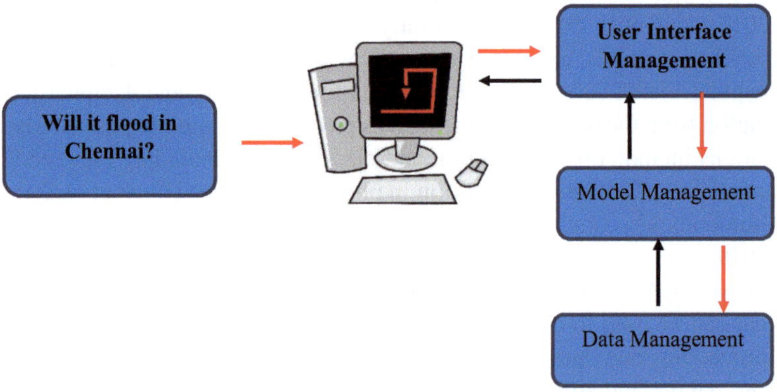

Fig. 6.1 Components of EDSS

methods. In order to do this, suitable methods are chosen. Models are trained using historical data, and their performance is verified. Further, developed models are integrated into the EDSS architecture to guarantee smooth communication with other system elements, which is known as system integration. Finally, there is a need to develop user-friendly interfaces that allow users communicate with the AI-powered EDSS, analyze data and come to wise decisions. Thus, environment decision support systems make considerable use of AI approaches in a variety of components, including the data, knowledge base, model base, and user interface. Essentially, a geographical information systems (GIS) spatial data management capabilities are frequently used to realize the various coupled environmental models, databases, and assessment tools that make up an environmental decision support system (EDSS). All of these components are integrated under a graphical user interface (GUI) (Matthies et al. 2007).

Using EDSS, a vast amount of the environment can be analyzed at a speed unattainable by human capabilities. This rapid processing of vast datasets helps in obtaining real-time insights and thus makes us capable to take timely action or at least respond better to environmental threats. EDSS can predict the risks and problems that might occur. They allow the decision-makers, the scientists, governments, and policymakers to act proactively and nip the threat in the bud to as much extent as possible.

6.6 Applications of EDSS

Waste management, energy efficiency, water pollution control, climate modeling, and biodiversity preservation are just a few of the environmental sectors that employ EDSS. They are used to enhance natural resource management decision-making processes, predict ecosystem services, conserve and optimize water resources, and provide an understanding of ecosystems. Their use has had a significant impact on environmental sustainability and resource management. The use of EDSS is instrumental in optimizing energy systems to reduce carbon footprints. They help in optimizing the distribution and consumption of electricity, integrating renewable energy sources like wind and solar more effectively. The systems manage demand and supply fluctuations, enhancing energy efficiency and reducing dependency on fossil fuels. Satellite imagery combined with the identification of emission hotspots, monitoring deforestation, and tracking urban expansion can better assess the impact on the environment of these factors. More so, once identified, AI also assists in optimizing and bettering the processes to be able to reduce these emissions through predictive maintenance of machinery and by improving resource efficiency. EDSS has been used in many areas as shown in Fig. 6.2. The subsequent section discusses these applications one by one.

Fig. 6.2 Application of EDSS

6.6.1 Modeling of Climate Change

Climatic systems are modeled and future climatic scenarios are predicted using EDSS. Large volumes of data from many sources may be processed by these models, which increases the precision of climate projections and aids in the creation of practical mitigation plans by policymakers.

AI is seen as a game-changing technology (Dwivedi et al. 2024; Khaleel et al. 2024) that, by using AI-based EDSS, can quicken humanity's fight against climate change. The use of AI-based EDSS in climate change research signifies a significant leap toward proactive adaptation and sustainable decision-making, providing invaluable insights for policymakers and researchers striving to address the global climate crisis. It has significant potential in tackling climate change and supporting efforts to adapt to it. It presents new opportunities for understanding and mitigating climate risks, especially in urban areas.

EDSS using ML has revolutionized climate modeling and prediction. Traditional climate models, although informative, are often complex and computationally demanding. Using AI, EDSS has not only been able to perform faster but also is observed to be giving precise predictions through its ability to process huge amounts of datasets, thus making it more efficient, accurate, and reliable. Neural networks, for instance, have been utilized to improve the accuracy of weather forecasting, predict extreme weather events, and model complex climate dynamics. AI's usage in the prediction and mitigation of climate change can undoubtedly prove to be useful.

6.6.2 Agricultural Management

EDSS that incorporates AI is revolutionizing the agriculture sector by increasing sustainability, production, and efficiency. EDSS uses data from multiple sources, including market trends, soil sensors, weather forecasts, and satellite imagery, to give farmers useful information (Nain and Singh 2016; Rinaldi and He 2014).

Systems are able to evaluate soil properties and weather trends to suggest the best times to sow and harvest, increasing productivity while consuming fewer resources. Doing trend analysis of climate variables also helps the farmers to decide about the choice and time of crop to grow. EDSS also can detect disease and pest outbreaks early on, facilitating prompt and focused responses that lower crop losses and reduce the need for chemical treatments. In order to solve the issues of water scarcity and diminishing water quality, EDSS is also essential for field and district water management.

By evaluating data on crop health, weather trends, and soil conditions, AI-driven EDSS enhances precision agriculture. By optimizing irrigation, fertilization, and pest management, these systems assist farmers in increasing sustainability and productivity. By lowering food waste and strengthening the circular economy, artificial intelligence (AI) supports sustainable resource management in the agri-food industry. In order to assist farmers in selecting crops and developing market strategies that would maximize revenue, machine learning models can also evaluate historical data and market patterns.

Using AI-based EDSS for agricultural management, farmers are encouraged to adopt sustainable practices by lowering carbon emissions, increasing land efficiency, and preserving water. In reaction to climate change, it helps farmers to use more robust and adaptable farming practices.

EDSS has been used to help farmers identify and monitor herbivorous pests. With the use of EDSS, farmers in India have been able to increase groundnut yields per hectare by 30% by receiving guidance on when to plant, apply fertilizer, and prepare the field.

A novel decision support system for tomatoes in high tunnels uses fuzzy logic and an adaptive neuro-fuzzy inference system to control climate variables, such as temperature, humidity, and solar radiation (Habib et al. 2017). Another environmental decision support system for the greenhouse culture plan uses a temperature integration model and market data to predict the marketing period and price and control the greenhouse climate in real time (Wang et al. 2013).

6.6.3 Management of Natural Disasters

EDSS has the potential to significantly improve disaster management at every stage, including preparedness, response, recovery, and prediction. Using EDSS in disaster management, decision-makers can allocate resources under different crisis scenarios, such as allocating temporary disaster response facilities for the delivery of relief supplies (Çavdur et al. 2021).

Authorities can increase their capacity to safeguard people, assets, and property by using EDSS, strengthening society's resistance to natural calamities. Even though, crucial challenges exists with data quality, system integration, ethical considerations, and striking a balance between reliance on technology and human judgment. In order to identify weak points in disaster preparation, ascertain the most successful disaster response tactics, and facilitate real-time disaster response coordination, EDSS can also simulate meteorological events and natural disasters.

Natural calamities including floods, hurricanes, and wildfires can be predicted by an AI-powered EDSS. These systems assist in evacuation and disaster response planning by evaluating real-time data from sensors and satellite pictures. They also provide early warnings. Artificial intelligence has proven to be useful in a number of environmental management applications, including wastewater treatment, ecosystem monitoring, and water conservation. Prediction, optimization, and environmental sustainability have all benefited from it. The information system that the Philippines is building for disaster risk reduction and management includes a decision support system with real-time data console data, historical information visualization, and the compilation of anticipated evacuation lists for potential evacuees (Goh et al. 2019).

"Meeting the resource and service needs of current and future generations without compromising the health of the ecosystems" is the definition of environmental sustainability (Morelli 2011). EDSS is actively used in over 80 areas of environmental protection initiatives, including waste reduction, biodiversity conservation, and clean energy. It has proven effective in executing routine tasks with higher accuracy and faster completion rates and is progressing in unsupervised decision-making.

AI technologies present significant opportunities for investigating renewable energy, water, and the environment, empowering sustainable human development, and promoting sustainable development. Techniques have been developed for green security, including wildlife protection, and aim to optimally allocate scarce resources under uncertainty for environmental conservation.

6.6.4 Conservation of Biodiversity

EDSS analyzes information from satellite photos, audio recordings, and camera traps to assist in the monitoring of biodiversity. Conservation efforts can be aided by machine learning algorithms that can identify species, follow their movements, and evaluate changes to their habitat. EDSS has proven instrumental in monitoring biodiversity. The use of AI in the forestry and wildlife sectors can aid in the effective monitoring, management, and conservation of biodiversity and forest resources. AI has been applied to predict ecosystem services and optimize water resource conservation, demonstrating its relevance in biodiversity research and environmental sustainability. AI facilitates real-time processing and interpretation of environmental data, aiding in informed decision-making and strategic planning in conservation initiatives.

6.6.5 Monitoring of Water and Air Quality

EDSS uses data from sensors measuring air and water quality to identify pollution levels and forecast patterns in the future. By using this information, companies and regulatory bodies can minimize environmental pollution more quickly. Furthermore, to improve data and knowledge interoperability and enhance urban water resource management within a river basin, AI approaches are applied to environmental decision support systems.

Artificial intelligence can track the quality of drinking water, control household water consumption, find subterranean leaks in water supply lines, and forecast when water plants require maintenance. In the Tigris Basin water contamination case, DSS addressed various sources of water pollution and provided support to various decision-makers (Salah et al. 2014).

6.6.6 Weather Forecasting

EDSS is used for weather forecasting and the anticipation of extreme events. Predictive models have the potential to improve decisions and future results. Enhancing forecasting models is essential to the supply chain operation as a whole. Developing an EDSS for forecasting is a challenging endeavor since numerous interrelated aspects affect demand, adding to its complexity. Farmers would be able to make more educated judgments about their farming operations with the use of the data's insights. Tsunami and cyclones are intricate natural phenomena driven by forces beyond human comprehension. EDSS is used to assist in predicting their pattern, enabling one to plan.

6.6.7 Energy Efficiency and Renewable Energy Management

AI presents new avenues for sustainable urban development through its application in energy optimization that helps in improving energy efficiency. For example, the system could determine when users were most likely to view data-sucking YouTube videos and adjust the cooling accordingly. Google cut its energy consumption by 40% as a result. In the energy sector, DSS is used to determine the optimal energy pricing and the quantity of excess capacity required to satisfy periods of peak energy demand. DSS is used not just to determine the best locations for energy-generating infrastructure but also to plan energy production, determine whether or not to build capacity above local demand, estimate the environmental cost of installing infrastructure at certain locations, and more.

6.6.8 Smart Cities

These days, every city experiences congestion, air pollution, traffic delays, and the inefficient utilization of already-existing services. The goal of smart cities, a future city plan, is to improve the quality of life for city inhabitants by providing solutions to various problems related to urban administration. Papastamatiou et al. (2017) unveiled a decision support system in 2017 to optimize Smart City energy use. Park et al. (2015) have developed a social network-based DSS for smart UParking planning that addresses several urban problems associated with high-density development and population increase.

DSS is an essential part of smart cities, supporting decision-making across a range of industries, including finance, healthcare, the environment, and security. Referred to as artificial intelligence DSS (AIDSS) or IDSS, the systems possess full control over semi-structured and unstructured situations and are very adaptive and versatile. AI can make cities more livable and sustainable. With the development of vehicle-to-vehicle and infrastructure communication, artificial intelligence will assist drivers in avoiding collisions and gridlock.

6.7 Challenges and Opportunities

Despite all its advantages, unparalleled progress, and numerous benefits, there are various challenges that this technology is facing that are restricting its full integration and limiting its potential. Artificial intelligence (AI) applications aim to tackle environmental sustainability; however, they face several obstacles such as a reliance on historical data in machine learning (ML), ambiguous human behavioral reactions to AI-based interventions, heightened cybersecurity risks, unfavorable effects of AI applications, and insufficient performance measurements or intervention tactics.

Most environmental data is not created by humans, unlike data related to a financial transaction or a property boundary, Rather, it is data that has been taken out of a difficult physical process that we can never fully comprehend. Mental representations of physical processes provide the basis of our comprehension. When several sources are used, several datasets are frequently dealt with, each of which is connected to a physical model that could not be readily put in challenges incorporated into a shared structure. One dataset, for instance, might identify "contaminating" compounds in rivers that are linked to higher bacterial growth. Another dataset may identify a whole different collection of compounds and link "contamination" to heavy metals.

Real-time, high-quality data is necessary for effective AI models, yet it's not always readily available. Sometimes, it can be difficult to integrate AI solutions with the current infrastructure for disaster management. Ethical Concerns: It is imperative to guarantee the impartiality and transparency of AI decision-making processes. Relying too much on artificial intelligence (AI) technologies may result in a decline in human knowledge and judgment.

6.7.1 Availability of Large Quantity of Good Quality Data

A prerequisite for EDSS and its applications to perform effectively, and efficiently and to give the useful, desired outcome is a large quantity of high-quality data. Data on climate change however is either limited or biased or of poor quality that hinders the right training of models and thus might lead to faulty or undesirable outcomes.

The success of an environmental decision support system would be greatly affected by the quality of the information that is provided as input. The climate data has to be collected from a wide variety of sources. Weather monitoring equipment is placed at geographically distant locations to monitor and collect climate data. Also in order to make some decisions the data has to be collected and analyzed for a long period of time, maybe hundreds of years. Since the data is vast and heterogeneous (i.e., derived from a variety of fields or sources) and its quality cannot be predicted in advance, issues can occur.

The quality and availability of data determine how well AI works with EDSS. Among the difficulties are:

- Availability of data: The efficacy of AI models may be restricted in certain places due to a lack of available or sparse data.
- Data quality: Poor model performance and untrustworthy forecasts might result from missing or inaccurate data.
- Data integration: Merging information from many sources and formats necessitates sophisticated preparation methods.

6.7.2 Energy Consumption of AI

EDSS using AI technologies, especially deep learning, are resource-intensive and consume significant amounts of energy. Given that the objective in this case is to make use of AI and its technologies in mitigating climate change, this consumption poses a problem because instead of reducing the emissions, these technologies themselves end up being a contributors to these.

6.7.3 Ethical and Social Implications

Like a two-edged sword, AI too comes with its own set of perils. The first and the concern of paramount importance is AI displacing jobs, especially in traditional industries. Also, climate change is a global problem that is being addressed. However certain not so technologically developed and sound nations might not be able to put AI to proper use. This again is a concern that needs to be addressed and figured out if climate change is to be completely tackled.

6.7.4 Implementation Barriers

Lastly, implementing EDSS can be complex and costly, requiring substantial investments in technology and expertise. Resistance to change within industries and governance structures can also impede the adoption of EDSS. Because of their complexity, EDSS is frequently referred to as "black boxes." Gaining the trust of decision-makers requires ensuring that these systems are transparent and easy to understand. EDSS demands large computational resources. One of the main obstacles to their widespread adoption is making sure these materials are inexpensive and readily available models used in EDSS may carry over biases from training sets, producing results that are inequitable or discriminatory. Gathering and examining environmental data might give rise to concerns about privacy, especially when it comes to human activity.

6.7.5 Disadvantages of AI

It might be hard to assess if the conclusions drawn by EDSS are desirable or correct because they are "black box" results that are incomprehensible to humans. For applications where greater assurance is required, such as early warning systems for natural disasters, deep learning may be dangerous. Security of the systems is very important as they can be hacked, allowing malicious actors to disrupt critical systems including early warning, transportation, energy, and others. EDSS may result in more bias since it makes decisions based on assumed assumptions about communities and organizations. Privacy concerns are also brought up by data collection.

6.8 Road Ahead

Artificial intelligence holds great promise to transform environmental decision support systems by offering insights that are more precise, timely, and comprehensive. Even if there are still many obstacles to overcome, continuous developments in AI techniques and technologies hold the potential to improve EDSS's capabilities and produce better environmental results and more knowledgeable decision-making.

The success of an EDSS depends on the quality of the input data. Future EDSS and AI research and development ought to concentrate on enhancing data-gathering means and methods. To get high-quality environmental data, new sensors and data-gathering techniques should be developed. Focus should be made on developing techniques to improve the interpretability and transparency of AI models' judgments. Efforts should be made to lower the price and complexity of AI technology so that a wider spectrum of consumers can utilize it. Policies and procedures should be created to guarantee the moral use of AI in environmental decision-making. Better utilization of interdisciplinary data, integration, and display of temporal and spatial outcomes should be the main focus in the future. Efforts should be also made to enhance user-friendliness and better representation in models of EDSS.

EDSS holds transformative potential for combating environmental concerns, with notable progress in modeling, energy optimization, and emissions control. The benefits of AI, including its scalability and predictive capabilities, are significant and can be roped in EDSS to improve them. However, the road ahead involves addressing challenges ranging from data quality and energy consumption to ethical and socioeconomic considerations. These challenges need to be jointly addressed and tackled with a united, global front committed to sustainable and inclusive technological advancement.

References

Bagstad KJ, Reed JM, Semmens DJ, Sherrouse BC, Troy A (2015) Linking biophysical models and public preferences for ecosystem service assessments: a case study for the Southern Rocky Mountains. Reg Environ Change 16(7):2005–2018

Buch N, Velastin SA, Orwell J (2011) A review of computer vision techniques for the analysis of urban traffic. IEEE Trans Intell Transp Syst 12(3):920–939

Çavdur F, Sebatlı-Sağlam A, Köse-Küçük M (2021) A scenario-based decision support system for allocating temporary-disaster-response facilities. J Fac Eng Archit Gazi Univ 36:1499–1514

Cortés U, Sànchez-Marrè M, Ceccaroni L, Rodríguez-Roda I, Poch M (2000) Artificial intelligence and environmental decision support systems. Appl Intell 13:77–91

Cullen-Knox C, Eccleston R, Haward M, Lester E, Vince J (2017) Contemporary challenges in environmental governance: technology, governance, and the social license. Environ Policy Gov 27(1):3–13

Duan Y, Edwards JS, Dwivedi YK (2019) Artificial intelligence for decision making in the era of Big Data–evolution, challenges, and research agenda. Int J Inf Manag 48:63–71

Dwivedi DN, Mahanty G, Dwivedi VN (2024) Intelligent conservation: a comprehensive study on AI-enhanced environmental monitoring and preservation. In: The convergence of self-sustaining systems with AI and IoT. IGI Global Scientific Publishing

Elliott D, Keen W, Miao L (2019) Recent advances in connected and automated vehicles. J Traffic Transp Eng (English Edition) 6(2):109–131

García CG, Valdez ERN, Díaz VG, Bustelo BCPG, Lovelle JMC (2019) A review of artificial intelligence in the Internet of Things. Int J Interact Multimedia Artif Intell 5(4):9–20

Goh JEE, Goh MLI, Baccay MA (2019) Community-based disaster risk reduction and management information system in the Philippines. In: Proceedings of 2019 international conference on computational intelligence and knowledge economy, ICCIKE 2019

Habib S, Akram M, Ashraf A (2017) Fuzzy climate decision support systems for tomatoes in high tunnels. Int J Fuzzy Syst 19(3):751–775

Khaleel M, Murtaza N, Mueen QH, Qadri SF (2024) Use of AI in conservation and for understanding climate change. In: A biologist's guide to artificial intelligence. Academic Press, pp 201–240

Kumar A, Kaur A, Kumar M (2019) Face detection techniques: a review. Artif Intell Rev 52(2):927–948

Kurdi MZ (2017) Natural language processing and computational linguistics, semantics, discourse and applications, vol 2. Wiley

Matthies M, Giupponi C, Ostendorf B (2007) Environmental decision support systems: Current issues, methods and tools. Environ Model Softw 22(2):123–127

Morelli J (2011) Environmental sustainability: a definition for environmental professionals. J Environ Sustain 1(1):2

Nain AS, Singh KK (2016) Conceptualization of a framework of decision support system for agriculture in a hilly region. Mausam 67:195–204

Papastamatiou I, Marinakis V, Doukas H, Psarras J (2017) A decision support framework for smart cities energy assessment and optimization. Energy Procedia 111:800–809

Park J, Nam K, Hong S, Moon T, Lee S, Leem Y (2015) Social network based decision support system for smart U-parking planning. Int J Archit Environ Eng 9(3):288–292

Rahamathunnisa U, Chellappa B (2018) Decision support systems – an overview. Int J Mech Eng Technol. 9(7):252–255

Rinaldi M, He Z (2014) Decision support systems to manage irrigation in agriculture. Adv Agron 123:229–279

Salah HA, Mocanu M, Florea A (2014) Towards an integrated decision support system for the evaluation of water pollution in Tigris Basin (DSSWAPIT). In: Proceedings IEEE 10th international conference on intelligent computer communication and processing, ICCP 2014

Shrestha A, Mahmood A (2019) Review of deep learning algorithms and architectures. IEEE Access Pract Innov Open Solut 7:53040–53065

Tuuli T, Vuokko H, Christoph F, Anna H, Tuomo H, Olle J, Enrico DM (2019) Social media data for conservation science: a methodological overview. Biol Conserv 233:298–315

Wang J, Li P, Zhao Q, Ban JDXK (2013) Decision support system of greenhouse production planning based on temperature integration model. J Jiangsu Univ (Natural Science Edition) 34(5):543–547

Exploring the Intersection of Artificial Intelligence and Climate Change Action

7

Milu Maria Jose, Bashabi Gupta, and Seema Aggarwal

Abstract

Climate change poses an unprecedented threat to the planet, requiring urgent and coordinated action across all sections of society (Intergovernmental Panel on Climate Change (IPCC). Climate change 2022—impacts, adaptation and vulnerability. In Climate change 2022—impacts, adaptation and vulnerability. Cambridge University Press, 2023. https://doi.org/10.1017/9781009325844). In recent years, artificial intelligence (AI) has emerged as a potential game-changer in the fight against climate change, offering innovative solutions and insights across multiple domains (Rolnick et al. ACM Comput Surv, 55(2), 2019. http://arxiv.org/abs/1906.05433). However, integrating AI into climate change mitigation and adaptation efforts is not without its challenges and complexities (Clutton-Brock et al. Climate change & AI: recommendations for government, 2021; Kaack et al. Aligning artificial intelligence with climate change mitigation, 2021. https://hal.science/hal-03368037). This chapter looks into the intersection of AI and climate change, examining both the promising opportunities and potential pitfalls associated with AI applications in climate analysis and action. Drawing on a comprehensive review of existing literature and case studies, we explore the diverse ways in which AI is being utilized to address climate-related issues, from enhancing renewable energy efficiency, optimizing resource management, and monitoring environmental changes to monitoring deforestation patterns. The chapter also critically assesses the ethical, social, and ecological implications of AI deployment in the context of climate change, highlighting the instances where AI technologies may inadvertently exacerbate environmental degradation or perpetuate existing inequalities.

M. M. Jose · B. Gupta · S. Aggarwal (✉)
Miranda House, University of Delhi, New Delhi, India
e-mail: milum@mirandahouse.ac.in; Bashabi.Gupta@mirandahouse.ac.in; seema.aggarwal@mirandahouse.ac.in

B. Thokchom et al. (eds.), *Artificial Intelligence Enabled Real Time Environmental Monitoring*, https://doi.org/10.1007/978-981-95-3767-9_7

Keywords

Artificial intelligence · Climate change · Sustainability · Renewable energy · Environmental justice · Ethical considerations

7.1 Introduction

Climate change has been widely accepted as the biggest threat to humankind in the next centuries, mainly caused by anthropogenic causes (IPCC 2023). According to IPCC (2007), climate change is the alteration in climate patterns that persist for an extended period, either due to natural or anthropogenic causes. With the rapidly changing climate, the world has witnessed an intensification of the hydrological cycle (Bekele et al. 2016; Doulabian et al. 2021). It has significantly threatened the quality and quantity of freshwater supply, especially across Asia and Africa. The issue has been compounded by the increased demand due to the agricultural, urban, industrial, and environmental sectors being on the rise globally due to the increasing population and poor management practices, posing a major challenge given the large population (Dahal et al. 2016).

Climate change has contributed to worsening the sustainability of global food systems (V. Owino et al. 2022; Owino and Mouratidou 2019). It has directly impacted food production with variations in soil fertility by altering temperature and rainfall patterns, reducing growing periods, decreasing crop yields and food production, and food-nutrient and anti-nutrient composition (Owino et al. 2022). Climate change has also indirectly impacted the food systems with increasing pest occurrences that increase the spoilage and food safety hazards at all stages of the food chain from production to post-harvest and consumption (Hodges et al. 2011; Owino et al. 2022; Parfitt et al. 2010).

The rising sea level is another major effect of global warming that has exacerbated coastal erosion, damaging the mangroves, salt marshes, and coral reefs. Even a global temperature rise to 1.5 °C will cause a sea-level increase of 1.7–3.2 feet by 2100. A 2 °C rise could be catastrophic, which submerges multiple islands and vast coastal areas, causing saltwater intrusion and increasing soil erosion. Islands like the Maldives, where 80% of the land is less than 1 m above sea level, would be severely affected, and countries like China, Vietnam, Fiji, Japan, Indonesia, India, and Bangladesh are at high risk. Sundarbans National Park, a UNESCO World Heritage Site and the world's largest mangrove forest, has already lost 12% of its shoreline due to rising sea levels and may be entirely submerged (Climate Change and Its Impact—PMC n.d.; Shivanna 2020, 2022).

Ultimately, climate change will and has compounded the impacts of urbanization and economic development, adversely affecting the sustainable development of most nations by increasing pressure on natural resources and the environment. It has impacted human health, security, and livelihoods, particularly for the vulnerable groups. It has affected almost all aspects of human life, as well as that of ecosystems. Climate change has induced habitat and phenological shifts, which have led to decreased biodiversity (Shivanna 2020, 2022).

Considering the extensive and interlinked impacts of anthropogenic climate change, coordinated global efforts are necessary to tackle the challenge. International agreements like the Kyoto Protocol and the Paris Agreement aim to limit global temperature rise to well below 2 degrees Celsius above pre-industrial levels, with efforts to limit the increase to 1.5 degrees Celsius (Government of India 2022; Mohan 2017). Achieving the climate goals is a necessity, but despite global efforts and intensified diplomacy, the desired goals are yet to be achieved (Abbass et al. 2022; Maizland 2023; Murshed et al. 2021).

It is in this background that artificial intelligence (AI) has emerged as a powerful tool that has the potential to significantly contribute to both climate change mitigation and adaptation efforts. AI, in this chapter, refers to a wide range of technologies, including machine learning, neural networks, and advanced data analytics, which enable machines to learn from data, identify patterns, make decisions, and perform tasks that typically require human intelligence (Xu et al. 2021).

Artificial intelligence is a key player that can aid in combating climate change through its rapid, accurate data processing capabilities. It is capable of improving the existing climate modeling and prediction systems, enhancing disaster preparedness, and optimizing energy usage to reduce emissions with a click of a finger (Artificial Intelligence—PMC n.d.; Kaack et al. 2021; Xu et al. 2021). AI applications in smart grids integrate renewable sources efficiently, while in agriculture, it maximizes sustainability and yield by monitoring crop health and optimising resources. AI also helps in managing natural resources, monitoring deforestation via satellite imagery, and supports sustainable practices. Additionally, AI drives behavioral change and raises climate awareness by offering personalized carbon footprint reduction strategies and making scientific information more accessible.

AI is useful in the efficient management of all levels of systems, from small computers and televisions to houses and larger cities with enough development. With AI, a large-scale collection of data from people, devices, and buildings is possible. Then, analyses are performed on the data to optimise control over infrastructure, traffic, and energy. Smart energy infrastructure, along with AI, can monitor energy utilization in the city and reduce costs and carbon emissions (Kumar et al. 2024; Serban and Lytras 2020; Shabandri et al. 2020; Smart Cities Artificial Intelligence Techniques in Smart Grid: A Survey n.d.). Ultimately, AI can, if utilized properly, democratize and revolutionize the data. AI-based analytics can find hidden, useful information within large datasets and transform it into actionable outcomes and knowledge to support improved decision-making (Bibri et al. 2024a; Bibri et al. 2024b).

This chapter explores the role of AI in addressing climate change, outlining its opportunities and challenges in climate action. It covers AI's application in renewable energy, resource optimization, and emissions monitoring for mitigation, as well as environmental monitoring, disaster prediction, and biodiversity management for adaptation. The chapter critically examines the technical, ethical, and social hurdles in leveraging AI for climate efforts, including data issues, computational challenges, AI's carbon footprint, algorithmic bias, and privacy concerns. It advocates for ethical AI use, emphasizing transparency, accountability, and inclusiveness.

7.2 The Role of AI in Climate Change Mitigation

According to UNFCCC (2020), climate change mitigation is an umbrella term for the efforts to reduce emissions and enhance sinks. Artificial intelligence (AI) is a transformative technology with the capability to process vast amounts of data and identify patterns by optimizing systems and processes; it can lead to substantial reductions in greenhouse gas emissions and more efficient use of resources, ultimately assisting in climate change mitigation. AI plays different roles across many critical environmental domains. In the area of climate modeling, AI systems allow us to perform accurate and precise simulations that support better-informed policy decisions. An example of this is neural GCM, developed by Google DeepMind in collaboration with ECMWF. This hybrid model combines ML and AI methods to deliver faster and more accurate long-term climate predictions (Kochkov et al. 2024). Management of energy sytems through smart grid management and predictive maintenance enhances the efficiency of resource management, optimizes resource utilization, and ensures a smoother transition to renewable energy systems. Google DeepMind Energy AI models predict renewable energy generation and enhance energy grid efficiency (Google DeepMind n.d.).

Through smart agriculture, AI enables farmers to make better-informed decisions and assists them in an easier transition to sustainable farming and reducing resource waste by optimization. The AI-driven platform "Microsoft Azure Farm Beats" enables sustainable farming by integrating sensors, drones, and predictive analysis (Farm Beats n.d.). By optimizing the operation of carbon capture and storage, AI models reduce carbon emissions. One such AI-driven platform is "Carbon Re" (Carbon n.d.). Using AI tools such as Wildbook (AI-based wildlife conservation tool) (Wildbook—Wild Me n.d.) for constant monitoring and predictive analysis helps in the area of biodiversity protection.

7.2.1 AI and Climate Change Modeling

The key application of artificial intelligence in climate change mitigation is in the arena of climate modeling (Feng et al. 2014; Khwairakpam et al. 2019; Mujumdar and Kumar n.d.). The advanced machine learning algorithms are able to analyze large datasets, which include historical climate data and real-time satellite observations, to create simulations and improve the accuracy and precision of existing climate models (HESS n.d.; Okkan et al. 2021; Ouyang et al. 2021; Scher et al. 2021; Smith et al. 2016). Incorporating AI-driven understandings enables scientists to better understand complex climate interactions, predict changes and refine them, and aid in developing informed policies and mitigation strategies. These systems can increase the accuracy of the existing models with the capability to handle a wider range of data (Gent et al. 2011; Miller et al. 2014; Racah 2017).

Artificial intelligence systems are also crucial in restoring missing values in historical temperature records, which is crucial for global climate datasets like HadCRUT4. Image inpainting techniques combined with data from numerical climate models have effectively filled the data gaps in these datasets with high accuracy (Pearson correlation coefficients ≥0.9941 and root mean squared errors ≤0.0547 °C) compared to the original records. The resulting datasets are better than the ones produced by traditional techniques such as kriging interpolation and principal component analysis-based infilling (Miller et al. 2014; Illing and Cubasch 2014).

7.2.2 AI and Energy Efficiency

Energy efficiency is a key component in reducing emissions. It reduces GHG emissions, direct emissions from fossil fuel combustion, and indirect emissions from electricity generation. AI shows promising capabilities in this with the development and management of smart grids. Compared to the traditional power grids that are designed with the centralized energy production model as the core, in which power is generated in large plants and distributed to consumers, the modern systems, with the increasing integration of renewable energy sources, like wind and solar, which are highly localized and are done on varying scales, higher flexibility and intelligence are required from the grid systems.

The smart grid systems function and coordinate multiple systems consisting of information management, control technologies, digitally based sensing, communication technologies, and field devices (U.S. Department of Energy 2018). These smart grids monitor the processes in real life, communicate data back to operation centres, respond automatically to adjust a process, and process analysis, and help operators access and apply the data from the grid. In smart grid applications, AI mimics the role of a grid operator. The AI systems are often more precise, reliable, and comprehensive in action.

AI significantly enhances smart grids as they enable real-time monitoring and control of the grid, enhancing the efficiency of the grid operation. The machine learning algorithms predict energy demand and supply by analyzing weather forecasts, consumption patterns, and grid performance metrics, thus reducing wastage. This enables the grid to have better short-term, medium-term, and long-term load balancing and seamless integration of renewable energy, ultimately reducing the reliance on fossil fuels. Energy storage plays a crucial role in smart grids, compensating for the intermittent nature of renewable energy sources. AI optimizes energy storage system operations, determining the best times to store and release energy, thus maximizing renewable use and minimizing waste (Ali and Choi 2020; Chung et al. 2021; Yip et al. 2017; Zhang et al. 2018). Thus, the AI-driven systems assist in better integration of the renewable energy systems and better forecasting of renewable energy production, and assist the grid operators in balancing supply and demand more efficiently.

The AI also assists the smart grid system with system stability assessment, fault detection, and ensuring grid security. Predictive maintenance, powered by AI, improves renewable energy system efficiency by analyzing sensor data from equipment like wind turbines and solar panels to anticipate failures (Chung et al. 2021; Sankarananth et al. 2023; Zhang et al. 2018). This proactive approach ensures timely repairs, reduces system downtime, sustains energy production, and cuts the operational costs and energy production losses linked with unexpected equipment failures. AI supports the design and management of smart grids (Ali and Choi 2020; Yip et al. 2017).

Agriculture is another sector where AI and machine learning techniques are used to reduce wastage and contribute to climate change mitigation by optimizing resource usage and improving crop yield predictions. Agriculture is a significant contributor to greenhouse gas emissions, primarily through methane emissions from livestock, nitrous oxide emissions from fertilized soils, and carbon dioxide emissions from land-use changes. However, AI can play a crucial role in optimizing agricultural practices to reduce these emissions and enhance sustainability.

7.2.3 AI in Agriculture

AI can assist farmers with precision agriculture, where efficacy is achieved through optimizing resource allocation, reducing input costs, and maximizing crop yields. With the help of machine learning algorithms, farmers can collect real-time data on crop conditions, soil moisture, and nutrient concentrations, and pest infestations, enabling them to make targeted decisions for optimal results (Nahvi et al. 2016). AI can also be used to optimize irrigation systems, optimize water use, and reduce wastage. It can reduce the wastage of fertilizers and nutrients and avoid leaching of the soil by adjusting the amount as well as the time of application. These decisions make farmers adapt better to the changing climate and shift more smoothly to sustainable practices (Peters et al. 2020; Raza et al. 2023).

Precision agriculture not only improves the yield but also the soil quality by avoiding the salinization of soil due to the overuse of fertilizers and pesticides. They also assist farmers in early crop disease detection through digital images (Liakos et al. 2018). By effectively controlling the level of water, fertilizer, and other resources, it minimizes the number of greenhouse gases emitted from the process. The AI can assist in identifying the optimal time of harvest, and if within greenhouses, it can control the amount of moisture and temperature using IoT and produce optimal results, reducing environmental impact. AI can plan the process of crop rotations, intercropping, and cover cropping more effectively. AI algorithms can analyze historical crop performance and environmental data to recommend optimal crop rotation sequences that improve soil health and reduce emissions (Peters et al. 2020; Sarker et al. 2023).

Methane emissions are one of the important aspects of greenhouse gas emissions, and livestock is a major contributor to methane emissions globally (Altshuler et al. 2023; Ghassemi Nejad et al. 2024). Thus, effective livestock management is another area where AI can contribute to sustainability. AI can be used to advance on-farm emission monitoring and real-time emission monitoring (Ghassemi Nejad et al. 2024). AI-driven tools can monitor the health and behavior of animals, optimize feeding practices, and improve animal welfare.

7.2.4 AI in Wildlife Conservation

One key application of AI is in wildlife protection, which uses machine learning algorithms to analyze data from camera traps, which capture images and videos of animals in their natural habitats. These traps produce vast amounts of data that are difficult to process. AI automates the process with image recognition algorithms that identify species, count individuals, and track animal movements, providing crucial information for assessing population sizes, studying animal behavior, and identifying critical habitats. AI assists in combating poaching and illegal wildlife trade, major threats to endangered species, and identifying poaching activities in real-time. AI can also track and predict illegal wildlife trade patterns online, helping authorities intercept illegal shipments and dismantle trafficking networks.

In marine conservation, AI can analyze the data to monitor marine animal movements and health and map critical marine habitats such as coral reefs and seagrass beds. AI-driven restoration projects benefit from analyzing satellite imagery to identify areas for reforestation and monitoring the health of newly planted forests. AI can model deep-sea resources, monitor marine biodiversity, and predict ocean dynamics on any given day (Ditria et al. 2022; Jiang and Zhu 2022; Marrable et al. 2022). AI-enabled Mobile Marine Protected Areas (MMPAs) are the most recent and significant development in the field, in addition to the usual monitoring and assessment. MMPAs are dynamic geographical regions with mobile boundaries that, rather than just protecting one specific area like the traditional Marine Protected Areas, protect entire tracts of a species, including migration routes (Bakker 2022). AI supports community-based conservation by providing tools for citizen scientists to contribute to biodiversity monitoring, making it easier for nonexperts to participate in scientific research and conservation initiatives.

7.3 AI in Climate Change Adaptation

According to UNDP (2024), climate change adaptation refers to actions that can help reduce vulnerability to climate change's current or expected impacts. Even with the smallest increase in global warming, the impacts of climate change become more frequent and intense, and adaptation becomes more difficult and expensive for people and ecosystems (UNDP 2024).

There is an urgency, especially for developing countries, which are feeling the impacts of climate change at a greater level and are particularly vulnerable to them due to their geographical and climatic conditions, high dependence on natural resources, and limited capacity to adapt to a changing climate because of limited resource availability (Hecht 2014; Zurek et al. 2022). Even in the most positive scenarios of climate change predictions, where greenhouse gas emissions are significantly and swiftly cut, climate change will continue to impact the world for decades due to the energy already trapped in the system. So, adaptation forms the second part of the response to the climate crisis, along with mitigation (UNDP 2024).

7.3.1 AI and Climate Modeling

Climate modeling and weather forecasting are critical components of climate change adaptation. It enables scientists and policymakers to understand future climate scenarios and prepare for potential impacts. It enables communities and stakeholders to effectively plan for the short-term and long-term impact of climate change and effectively adapt to the change (HESS n.d.; Leal Filho et al. 2022). The first step to any adaptive measure is knowing the threat. Traditional climate models relied on complex mathematical equations to simulate the interactions between the atmosphere, oceans, land surface, and ice. These models are essential for predicting long-term climate patterns but often require significant computational resources and time. Machine learning algorithms can analyze vast amounts of climate data and identify patterns and relationships that may not be evident through conventional modeling techniques (Al-Jarrah et al. 2015; Barnes et al. 2019; Chattopadhyay et al. 2020). AI can refine the parameters used in climate models, making them more accurate and reliable. By learning from historical climate data, AI models can also improve the representation of physical processes within the climate system. In weather forecasting, AI can enhance the precision and timeliness of predictions. While the traditional models provide forecasts with varying degrees of accuracy, depending on the atmosphere's conditions and the data's availability, AI can complement these models by analyzing real-time data from satellites, weather stations, and other sensors to provide more accurate and localized forecasts.

AI in weather forecasting can use deep learning techniques to predict extreme weather events. Convolutional neural networks (CNNs) can analyze satellite images to identify patterns associated with hurricanes, tornadoes, and other extreme weather events, as well as locate remote sites where extreme weather events like glacial cloud bursts or landslides at a much faster rate, reducing the time to get help to the region (Flora et al. 2021). By recognizing these patterns early, advanced warnings to communities at risk can be provided, allowing them to take preventive measures and reduce potential damage (Gronquist et al. 2021; McGovern et al. 2017; Slater et al. 2021).

Moreover, AI-driven weather forecasting can assist in managing agricultural practices and water resources. Accurate weather predictions enable farmers to plan their planting and harvesting schedules, optimize irrigation, and protect crops from adverse weather conditions. AI can predict rainfall patterns and water availability, helping ensure sustainable water use and adapt the communities to the impacts of droughts and floods (Chattopadhyay et al. 2020).

7.3.2 AI and Disaster Risk Reduction

Natural disasters, such as hurricanes, floods, earthquakes, and wildfires, are becoming more frequent and severe due to climate change. Early warning systems are crucial for mitigating the impacts of these disasters, providing timely information to communities and authorities to prepare and respond effectively. AI can enhance the capabilities of early warning systems by improving the accuracy of predictions and the speed of information dissemination (Ali et al. 2018; Park et al. n.d.; Larbi et al. 2021; Li and Babovic 2019).

One of the key advantages of AI in early warning systems is its ability to process and analyze large datasets in real time. For instance, AI algorithms can analyze seismic data to detect early signs of earthquakes and provide rapid alerts to at-risk areas (Adeli and Panakkat 2009; Aden-Antoniów et al. 2022; Zhang et al. 2024). Machine learning models can also predict the likelihood of aftershocks, helping authorities plan for ongoing seismic activity (Adeli and Panakkat 2009). In the case of hurricanes and typhoons, AI can improve the prediction of storm paths and intensities (Wei 2021). By analyzing data from satellites, aircraft, and weather stations, AI models can provide more accurate forecasts of where and when a storm will make landfall, its potential impact, and the necessary evacuation measures. This information is vital for emergency responders and communities to prepare and minimize casualties and property damage (Hashemi et al. 2016).

AI technology holds the potential to revolutionize flood prediction and management. By leveraging AI-powered models that parse through extensive datasets—like rainfall patterns, river flow metrics, and soil moisture content—it becomes possible to accurately forecast the probability and intensity of flooding events (Frame et al. 2022; Kumar et al. 2021). Moreover, these advanced models can run simulations across various flood scenarios, enabling planners and authorities to craft strategic flood response plans and more effectively distribute resources. In urban contexts, the integration of AI with sensor data and social media analytics can offer immediate flood warnings and facilitate efficient evacuation strategies, markedly improving emergency responses. AI can help in utilizing new geographical information, enhancing the computing efficiency of models, and enabling the measurement of uncertainty. Impact modeling procedures can be developed and implemented in a modular, flexible manner thanks to the modeling platform. These methods allow for the quick and simple development and implementation of various weather/climate application workflows. The possibility for advanced scientific knowledge in this multidisciplinary subject to be used for a variety of purposes exists with emerging AI techniques. It can facilitate the data-driven identification of dynamical correlations. We still need to investigate AI's full potential in this situation (Haer et al. 2017; Piadeh et al. 2022; Sajjad et al. 2023).

With the growing temperatures across the globe, instances of wildfires have increasingly become common occurrences, particularly in regions experiencing prolonged droughts and heat waves, with more and more regions added to the list. AI, by processing and constantly monitoring satellites, drones, and ground-based sensor data, can combat wildfires more effectively (Gholamnia et al. 2020; Rodrigues and De la Riva 2014; Zhao et al. 2021). For example, AI algorithms that detect heat signatures and smoke patterns associated with wildfires provide early warnings to fire management agencies, reducing the risk and increasing the chances of evacuation. The spread of wildfires can be predicted using AI by analyzing weather conditions, vegetation types, and topography, helping firefighters strategize their response and contain the fires more effectively.

Utilizing machine learning (ML) and AI offers a dependable method for managing and forecasting wildfires. With the data at its disposal, AI can learn and forecast with accuracy. To predict the occurrence and pattern of wildfires, AI connected to imaging satellites can analyze variables like the appearance of smoke, the incidence of fire, and the disturbance of vegetation. These variables can then be correlated with different physical parameters of the forest, such as vegetation type, climate, landscape, fire susceptibility mapping, and soil deposits. Prediction accuracy rises as a result of AI systems learning during the prediction process. AI systems have produced surprisingly accurate experimental wildfire predictions with shorter detection times. By detecting smoke, AI-assisted software-defined cameras (SDCs) can forecast forest fires on a local scale (Sayad et al. 2019; Zhao et al. 2021).

AI-driven early warning systems are not limited to natural disasters but can also be applied to public health emergencies, such as disease outbreaks. By analyzing epidemiological data and monitoring social media for early signs of disease spread, AI can provide timely alerts and help public health authorities implement measures to control outbreaks and protect public health. Spatial modeling, risk prediction, disinformation control, public health surveillance, illness forecasting, pandemic/epidemic modeling, and health diagnosis are a few ways AI has helped with public health delivery. Artificial intelligence (AI) has been used to discover new drug targets, predict the emergence of infectious diseases, and diagnose illnesses. AI has also been applied to drug administration and discovery, as well as the interpretation of medical imaging. AI was essential to the COVID-19 pandemic's forecasting of the virus's propagation, contact tracing, and quick testing and diagnosis (Guerrero n.d.; Kaufhold et al. 2020; Lu et al. 2022).

7.3.3 AI and Conservation and Management of Wildlife

Deforestation and habitat loss significantly impact biodiversity and climate change by releasing stored carbon (Christin et al. 2019). Utilizing AI for monitoring and management is vital for conservation and climate mitigation. AI enhances the analysis of satellite imagery and remote sensing data, detecting deforestation and habitat changes efficiently. CNNs are particularly effective in identifying specific deforestation activities through high-resolution images, offering real-time alerts to prevent further damage. The industry's efforts to achieve sustainable development goals are being made easier by the application of increasingly sophisticated technologies to monitor, manage, and conserve the forest and its resources. Globally, an increasing number of countries are adopting precision forestry and smart forest techniques, which use cutting-edge technology, including sensors, artificial intelligence, satellite imaging, and digital systems, to sustainably manage, protect, and use forest resources. The use of cutting-edge technologies like digital recorders, drones, wireless sensor networks, camera technology, and crowdsourcing data methods like citizen science is part of the continuous expansion of data collection capabilities. Furthermore, data extraction techniques from social media and other Internet sources are being created (Liu et al. 2019; Raihan 2023; Shivaprakash et al. 2022).

AI plays a pivotal role in addressing two major threats to biodiversity: habitat loss and fragmentation. By analyzing satellite and drone imagery, AI identifies and maps endangered habitats, highlighting areas at risk of fragmentation (Zhao et al. 2020). This information is crucial for developing strategies to conserve or restore connectivity between isolated habitat patches, ensuring wildlife can move freely and thrive. AI extends its capabilities to biodiversity monitoring through acoustic sensors, which analyze audio recordings from natural habitats to identify various bird and animal species by their sounds. This noninvasive monitoring provides a clear picture of ecosystem health and the impacts of environmental changes on different species. Beyond these applications, AI significantly contributes to conservation efforts by tracking animal migrations, detecting illegal poaching, and enabling conservationists to make informed, data-driven decisions. Through innovative tools and techniques, AI is at the forefront of global efforts to protect endangered species and preserve biodiversity (Kim and Park 2009; Shivaprakash et al. 2022).

7.4 Challenges and Complexities in Integrating AI

Artificial intelligence, while playing a promising role in assisting climate change mitigation and adaptation strategies, integrating and effectively utilizing AI for these efforts is not without many challenges and complexities. There are technical, ethical, and social challenges, as well as ecological impacts, that need to be considered while rolling out artificial intelligence for wider use.

Artificial intelligence (AI) refers to the ability of machines to learn like humans from experience and adjust, adapt, and perform human-like tasks (Duan et al. 2020; Gibson et al. 2021; Kaplan et al. 1997). AI has become an integral part of day-to-day life. However, AI applications that promote environmental sustainability face challenges when they open themselves to the rebound effects of high-energy-intensive structures. These undermine the efforts to achieve carbon neutrality and often overexploit the primary resources (Leal Filho et al. 2022). Servers that operate and execute essential AI computational tasks and store and process big data require considerable energy for cooling and uninterrupted operation. The production of these servers also requires huge amounts of nonrenewable resources like lithium, nickel, or cobalt, which are often sourced unethically and violate human rights. The systems also require efficient e-waste management; however, oftentimes, the waste is dumped into landfills (Belkhir and Elmeligi 2018).

The carbon footprint of AI is a growing concern as the demand for AI applications increases every day. For example, training a single large neural network can emit as much carbon as five cars over their lifetimes. So, there is a need for more energy-efficient AI technologies and practices that can reduce the footprint, like using renewable energy sources for data centres and developing hardware specifically designed to reduce energy consumption (Anthony et al. 2020; Patterson et al. 2021; Tamburrini 2022).

Beyond its carbon footprint, the deployment of AI at a widespread level in fields of agriculture and resource management could lead to over-reliance on technology and the potential to disrupt traditional practices and local knowledge systems. The constant use of AI-driven machinery and equipment for real-time monitoring and global systems, while it might prove useful in some instances, can contribute to electronic and energy waste if not managed properly (Benda et al. 2021; Fjeld et al. 2020; LaGrandeur 2021).

The use of AI applications could prioritize short-term efficiency gains over long-term sustainability. As the AI systems focus on the optimization of results in fields like agriculture and forestry, it might lead to increased monoculture or other practices that can reduce biodiversity and soil health over time. Similarly, AI applications in industries might focus on productivity maximization without considering the broader environmental impacts. To mitigate risks, adopting a holistic approach that considers long-term sustainability and ecological balance is essential. This includes integrating environmental impact assessments into AI project planning, promoting sustainable technology use, and encouraging practices that support biodiversity and ecosystem health (Benda et al. 2021; Fjeld et al. 2020).

The technical challenge to integrating AI for climate change starts from the availability and quality availability of data. As AI systems heavily rely on large datasets to train models, make predictions, and generate insights, acquiring high-quality, relevant data is often difficult, especially in regions with limited monitoring infrastructure (Cowls et al. 2023). Climate and environmental data are often sparse, incomplete, or inconsistent. This lack of comprehensive data leads to inaccurate models and unreliable predictions. Using historical climate data to predict change in today's environment may not accurately reflect current situations attributed to rapid environmental changes, further complicating the training of AI models. Inaccurate, outdated, or biased data can significantly impact the performance of AI algorithms, leading to faulty insights and flawed recommendations (Ballatore et al. 2017; Nost and Colven 2022; Andrea et al. 2022).

Improving data access and quality for AI in climate analysis involves expanding networks, standardizing collection methods, and encouraging open data. Collaboration across sectors is crucial. However, AI algorithms, especially deep learning, demand high computational resources and face scalability challenges, limiting their efficiency in large-scale or real-time climate applications (Cowls et al. 2023). Advancements in cloud computing and AI-specific hardware are addressing these limitations, but more efficient algorithms are needed (Newton et al. 2021). Bias in AI, arising from training data or algorithm design, presents ethical concerns and risks unfair outcomes, especially in underrepresented regions (Baeza-Yates 2018; Ntoutsi et al. 2020; Raghavan et al. 2020). Addressing bias demands diverse datasets, fairness-aware algorithms, and ongoing bias evaluation. However, the lack of proper evaluation mechanisms for AI is a challenge that needs consideration (Mohseni et al. 2021).

The rolling out of AI requires citizens to be aware of the risks associated with it as a self-progressive technology. There are many privacy concerns with AI due to the use of potentially sensitive data (Benda et al. 2021; Pickering 2021; Weitz et al. 2019). Strong data protection measures and regulations are essential for maintaining trust and ethical standards. There needs to be a clear understanding and full transparency regarding the management of data, especially the datasets used to train the algorithms.

7.5 Ethical Considerations and Responsible AI Use

As AI becomes increasingly integrated into climate change mitigation and adaptation efforts, ensuring its deployment is ethical and responsible is essential. Ethical frameworks are essential to provide guidelines to ensure AI is used responsibly. The ethical guidelines should address fairness, transparency, accountability, and privacy of AI services (Floridi 2018; Yu et al. 2018). One widely recognized framework is the AI Ethics Guidelines by the European Commission's High-Level Expert Group on AI, which emphasizes the principles of lawful, ethical, and robust AI. The guidelines should ensure that AI respects fundamental rights, is ethically sound, and is technically reliable (Carbonell 2016; Lin et al. 2017).

Transparency and accountability are critical to maintaining public trust in AI systems. Transparency involves making AI processes and decision-making understandable and accessible to stakeholders (Cowgill et al. 2020). This can be achieved through explainable AI (XAI), which provides insights into how AI models arrive at their conclusions (Adadi and Berrada 2018; Mohseni et al. 2021; Samek et al. 2017). Accountability requires a clear delineation of responsibility for AI actions and outcomes. AI deployment must be inclusive and address existing inequalities to avoid perpetuating or exacerbating disparities. This involves ensuring that AI systems are designed and implemented in ways that consider the needs and contexts of diverse populations, particularly marginalized and vulnerable communities (Baeza-Yates 2018; Ntoutsi et al. 2020). It also requires using diverse and representative datasets to train AI models, preventing biases that could lead to unfair outcomes. Addressing inequality also involves ensuring that the benefits of AI are equitably distributed.

7.6 Conclusion

Climate change presents an urgent and unprecedented threat to our planet and necessitates coordinated action across all sectors of society. As AI emerges as a powerful tool in the fight against climate change, providing innovative solutions across multiple domains, there are multiple opportunities and challenges that one needs to be conscious of while rolling out AI. The AI systems can help in climate change mitigation by enhancing renewable energy efficiency through smart grids and predictive maintenance of renewable energy sources, optimizing resource management in agriculture and industries, and aiding in monitoring and reducing greenhouse gas emissions. AI can assist in improving environmental monitoring systems, weather forecasting mechanisms, and early warning systems for natural disasters. It supports biodiversity and ecosystem management by monitoring deforestation and habitat loss and bolstering conservation efforts.

However, integrating AI into climate action is filled with technical challenges, such as data availability, quality, and computational limitations. Ethical and social implications, including bias in AI algorithms and privacy concerns, have the potential to exacerbate existing inequalities. The ecological impacts of AI, a system used to tackle carbon footprint, are ironical, but the carbon footprint and potential negative environmental effects cannot be overlooked. To address these challenges, it is crucial to adopt ethical frameworks for AI deployment, ensure transparency and accountability, and promote inclusivity in the management of systems as well as in data usage. Emerging trends and innovations in AI for climate action offer promising avenues for further exploration, but constant checks are necessary. Effective policy recommendations and multistakeholder collaboration are essential for the responsible integration of AI into climate strategies.

Just the use of AI can democratize climate action by making advanced tools and insights accessible to a broader audience and more nations, but it is also a double-edged sword. AI technologies need to be augmented by community-based monitoring and citizen science initiatives that can engage the public in climate action, fostering a sense of ownership and responsibility. This democratization is vital for ensuring that climate solutions are inclusive and equitable. Without this, it is a piece of technology detached from reality. So, coordinated efforts and responsible deployment are imperative to fully realize the potential of AI in addressing climate change. Governments, industry, academia, and civil society must collaborate and cooperate to develop and implement AI solutions that are ethical, transparent, and inclusive.

AI holds immense potential to transform climate change mitigation and adaptation. It is also important to realize that this potential requires coordinated efforts and a commitment to responsible AI deployment. Only by adopting ethical frameworks, enhancing data infrastructure, promoting inclusivity, supporting innovation, implementing robust policy frameworks, and fostering collaboration can we harness the power of AI to build a sustainable and resilient future. The time for action is now; together, we can leverage AI to combat climate change and protect our planet for future generations.

References

Abbass K, Qasim MZ, Song H, Murshed M, Mahmood H, Younis I (2022) A review of the global climate change impacts, adaptation, and sustainable mitigation measures. Environ Sci Pollut Res Int 29(28):42539–42559. https://doi.org/10.1007/S11356-022-19718-6

Adadi A, Berrada M (2018) Peeking inside the black-box: a survey on explainable artificial intelligence (XAI). IEEE Access 6:52138–52160. https://doi.org/10.1109/ACCESS.2018.2870052

Adeli H, Panakkat A (2009) A probabilistic neural network for earthquake magnitude prediction. Neural Netw 22(7):1018–1024. https://doi.org/10.1016/J.NEUNET.2009.05.003

Aden-Antoniów F, Frank WB, Seydoux L (2022) An adaptable random Forest model for the Declustering of earthquake catalogs. J Geophys Res Solid Earth 127(2). https://doi.org/10.1029/2021JB023254

Ali SS, Choi BJ (2020) State-of-the-art artificial intelligence techniques for distributed smart grids: a review. Electronics (Switzerland) 9(6):1–28. https://doi.org/10.3390/ELECTRONICS9061030

Ali D, Yohanna M, Ijasini PM, Garkida MB (2018) Application of fuzzy—neuro to model weather parameter variability impacts on electrical load based on long-term forecasting. Alex Eng J 57(1):223–233. https://doi.org/10.1016/J.AEJ.2016.12.008

Al-Jarrah OY, Yoo PD, Muhaidat S, Karagiannidis GK, Taha K (2015) Efficient machine learning for big data: a review. Big Data Res 2(3):87–93. https://doi.org/10.1016/j.bdr.2015.04.001

Altshuler Y, Chebach TC, Cohen S (2023) From microbes to methane: AI-based predictive modeling of feed additive efficacy in dairy cows. http://arxiv.org/abs/2311.12901

Andrea T, Lorenzo C, Agathe B, Mireia Y, Philip L, Marco B, Jie Y (2022) A.I. Robustness: a human-centered perspective on technological challenges and opportunities. ACM Comput Surv. https://doi.org/10.1145/3665926

Anthony LFW, Kanding B, Selvan R (2020) Carbontracker: tracking and predicting the carbon footprint of training deep learning models. http://arxiv.org/abs/2007.03051

Artificial intelligence: A powerful paradigm for scientific research – PMC. (n.d.). Retrieved June 7, 2024, from https://www.ncbi.nlm.nih.gov/pmc/articles/PMC8633405/

Baeza-Yates R (2018) Bias on the web. Commun ACM 61(6):54–61. https://doi.org/10.1145/3209581

Bakker K (2022) Smart oceans: artificial intelligence and marine protected area governance. Earth Syst Gov 13:100141. https://doi.org/10.1016/J.ESG.2022.100141

Ballatore A, Graham M, Sen S (2017) Digital hegemonies: the localness of search engine results. Ann Am Assoc Geogr 107(5):1194–1215. https://doi.org/10.1080/24694452.2017.1308240

Barnes EA, Hurrell JW, Ebert-Uphoff I, Anderson C, Anderson D (2019) Viewing forced climate patterns through an AI lens. Geophys Res Lett 46(22):13389–13398. https://doi.org/10.1029/2019gl084944

Bekele D, Alamirew T, Kebede A, Zeleke G, Melese AM (2016) Analysis of rainfall trend and variability for agricultural water management in Awash River Basin, Ethiopia. J Water Clim Change 8(1):127–141. https://doi.org/10.2166/wcc.2016.044

Belkhir L, Elmeligi A (2018) Assessing ICT global emissions footprint: trends to 2040 & recommendations. J Clean Prod 177(March):448–463. https://doi.org/10.1016/j.jclepro.2017.12.239

Benda NC, Novak LL, Reale C, Ancker JS (2021) Trust in AI: why we should be designing for APPROPRIATE reliance. J Am Med Inform Assoc 29(1):207–212. https://doi.org/10.1093/JAMIA/OCAB238

Bernstein L, Bosch P, Canziani O, Chen Z, Christ R, & Riahi K (2008) IPCC, 2007: Climate Change 2007: Synthesis Report. Geneva: IPCC. ISBN 2-9169-122-4

Bibri SE, Huang J, Jagatheesaperumal SK, Krogstie J (2024a) The synergistic interplay of artificial intelligence and digital twin in environmentally planning sustainable smart cities: a comprehensive systematic review. Environ Sci Ecotechnol:100433. https://doi.org/10.1016/J.ESE.2024.100433

Bibri SE, Huang J, Krogstie J (2024b) Artificial intelligence of things for synergizing smarter eco-city brain, metabolism, and platform: pioneering data-driven environmental governance. Sustain Cities Soc 108. https://doi.org/10.1016/J.SCS.2024.105516

Carbon Re – Industrial Decarbonisation for Zero-Carbon Materials. (n.d.). Retrieved April 28, 2025, from https://carbonre.com/

Carbonell I (2016) The ethics of big data in big agriculture. Internet Policy Rev 5(1):1–13

Chattopadhyay A, Hassanzadeh P, Pasha S (2020) Predicting clustered weather patterns: a test case for applications of convolutional neural networks to spatio-temporal climate data. Sci Rep 10(1):1317. https://doi.org/10.1038/s41598-020-57897-9

Christin S, Hervet É, Lecomte N (2019) Applications for deep learning in ecology. Methods Ecol Evol 10(10):1632–1644. https://doi.org/10.1111/2041-210X.13256

Chung HM, Maharjan S, Zhang Y, Eliassen F (2021) Distributed deep reinforcement learning for intelligent load scheduling in residential smart grids. IEEE Trans Industr Inform 17(4):2752–2763. https://doi.org/10.1109/TII.2020.3007167

Climate change and its impact on biodiversity and human welfare – PMC. (n.d.). Retrieved June 7, 2024, from https://www.ncbi.nlm.nih.gov/pmc/articles/PMC9058818/

Clutton-Brock P, Rolnick D, Donti PL, Kaack LH, Maharaj T, Luccioni A, Das HP (2021) Climate change & AI: recommendations for government

Cowgill B, Dell'acqua F, Deng S, Hsu D, Verma N, Chaintreau A (2020) Biased programmers? or biased data? A field experiment in operationalizing AI ethics. In: EC 2020 – proceedings of the 21st ACM conference on economics and computation, pp 679–681. https://doi.org/10.1145/3391403.3399545

Cowls J, Tsamados A, Taddeo M, Floridi L (2023) The AI gambit: leveraging artificial intelligence to combat climate change—opportunities, challenges, and recommendations. AI Soc 38(1):283–307. https://doi.org/10.1007/S00146-021-01294-X/FIGURES/11

Dahal V, Shakya NM, Bhattarai R (2016) Estimating the impact of climate change on water availability in the Bagmati Basin, Nepal. Environ Process 3(1):1–17. https://doi.org/10.1007/s40710-016-0127-5

Ditria EM, Buelow CA, Gonzalez-Rivero M, Connolly RM (2022) Artificial intelligence and automated monitoring for assisting conservation of marine ecosystems: a perspective. Front Mar Sci 9. https://doi.org/10.3389/FMARS.2022.918104/FULL

Doulabian S, Golian S, Toosi AS, Murphy C (2021) Evaluating the effects of climate change on precipitation and temperature for Iran using RCP scenarios. J Water Clim Change 12(1):166–184. https://doi.org/10.2166/WCC.2020.114/851726/JWC2020114.PDF

Duan S, Ullrich P, Shu L (2020) Using convolutional neural networks for streamflow projection in California. Front Water 2. https://doi.org/10.3389/FRWA.2020.00028

FarmBeats: AI, Edge & IoT for Agriculture – Microsoft Research (n.d.) Retrieved April 28, 2025, from https://www.microsoft.com/en-us/research/project/farmbeats-iot-agriculture/

Feng S, Hu Q, Huang W, Ho CH, Li R, Tang Z (2014) Projected climate regime shift under future global warming from multi-model, multi-scenario CMIP5 simulations. Glob Planet Chang 112:41–52. https://doi.org/10.1016/j.gloplacha.2013.11.002

Fjeld J, Achten N, Hilligoss H, Nagy A, Srikumar M (2020) Principled artificial intelligence: mapping consensus in ethical and rights-based approaches to principles for AI. SSRN Electron J. https://doi.org/10.2139/SSRN.3518482

Flora ML, Potvin CK, Skinner PS, Handler S, Mcgovern A (2021) Using machine learning to generate storm-scale probabilistic guidance of severe weather hazards in the warn-on-forecast system. Mon Weather Rev 149(5):1535–1557. https://doi.org/10.1175/MWR-D-20-0194.1

Floridi L (2018) Soft ethics, the governance of the digital and the general data protection regulation. Philos Trans R Soc A Math Phys Eng Sci 376(2133). https://doi.org/10.1098/RSTA.2018.0081

Frame JM, Kratzert F, Klotz D, Gauch M, Shelev G, Gilon O, Qualls LM, Gupta HV, Nearing GS (2022) Deep learning rainfall-runoff predictions of extreme events. Hydrol Earth Syst Sci 26(13):3377–3392. https://doi.org/10.5194/HESS-26-3377-2022

Gent PR, Danabasoglu G, Donner LJ, Holland MM, Hunke EC, Jayne SR, Lawrence DM, Neale RB, Rasch PJ, Vertenstein M, Worley PH, Yang ZL, Zhang M (2011) The community climate system model version 4. J Clim 24(19):4973–4991. https://doi.org/10.1175/2011jcli4083.1

Ghassemi Nejad J, Ju MS, Jo JH, Oh KH, Lee YS, Lee SD, Kim EJ, Roh S, Lee HG (2024) Advances in methane emission estimation in livestock: a review of data collection methods, model development and the role of AI technologies. *Animals 14*(3):435. https://doi.org/10.3390/ANI14030435

Gholamnia K, Nachappa TG, Ghorbanzadeh O, Blaschke T (2020) Comparisons of diverse machine learning approaches for wildfire susceptibility mapping. Symmetry 12(4):1–20. https://doi.org/10.3390/SYM12040604

Gibson PB, Chapman WE, Altinok A, Delle Monache L, DeFlorio MJ, Waliser DE (2021) Training machine learning models on climate model output yields skillful interpretable seasonal precipitation forecasts. Commun Earth Environ 2(1). https://doi.org/10.1038/S43247-021-00225-4

Google DeepMind (n.d.) Retrieved April 28, 2025, from https://deepmind.google/

Government of India (2022) India's updated first nationally determined contribution under the Paris agreement. https://unfccc.int/sites/default/files/NDC/2022-08/India%20Updated%20First%20Nationally%20Determined%20Contrib.pdf

Gronquist P, Yao C, Ben-Nun T, Dryden N, Dueben P, Li S, Hoefler T (2021) Deep learning for post-processing ensemble weather forecasts. Philos Trans R Soc A Math Phys Eng Sci 379(2194). https://doi.org/10.1098/RSTA.2020.0092

Guerrero E (n.d.) Science-based approaches to respond to COVID and other public health threats

Haer T, Botzen WJW, de Moel H, Aerts JCJH (2017) Integrating household risk mitigation behavior in flood risk analysis: an agent-based model approach. Risk Anal 37(10):1977–1992. https://doi.org/10.1111/RISA.12740

Hashemi MR, Spaulding ML, Shaw A, Farhadi H, Lewis M (2016) An efficient artificial intelligence model for prediction of tropical storm surge. Nat Hazards 82(1):471–491. https://doi.org/10.1007/S11069-016-2193-4/FIGURES/14

Hecht AD (2014) Past, present and future: urgency of dealing with climate change. Atmos Clim Sci 04(05):779–795. https://doi.org/10.4236/ACS.2014.45069

HESS – Hybrid forecasting: blending climate predictions with AI models (n.d.) Retrieved June 7, 2024, from https://hess.copernicus.org/articles/27/1865/2023/hess-27-1865-2023.html

Hodges RJ, Buzby JC, Bennett B (2011) Postharvest losses and waste in developed and less developed countries: opportunities to improve resource use. J Agric Sci 149(S1):37–45. https://doi.org/10.1017/S0021859610000936

Illing S, Kadow C, Kunst O, Cubasch U (2014) MurCSS: a tool for standardized evaluation of decadal hindcast systems. J Open Res Softw 2:e24

Intergovernmental Panel on Climate Change (IPCC) (2023) Climate Change 2022—impacts, adaptation and vulnerability. In: Climate Change 2022—impacts, adaptation and vulnerability. Cambridge University Press. https://doi.org/10.1017/9781009325844

Jiang M, Zhu Z (2022) The role of artificial intelligence algorithms in marine scientific research. Front Mar Sci 9:920994. https://doi.org/10.3389/FMARS.2022.920994/BIBTEX

Kaack LH, Donti PL, Strubell E, Kamiya G, Creutzig F, Rolnick D (2021) Aligning artificial intelligence with climate change mitigation. https://hal.science/hal-03368037

Kaplan A, Kushnir Y, Cane MA, Blumenthal MB (1997) Reduced space optimal analysis for historical data sets: 136 years of Atlantic Sea surface temperatures. J Geophys Res Oceans 102(C13):27835–27860. https://doi.org/10.1029/97jc01734

Kaufhold MA, Bayer M, Reuter C (2020) Rapid relevance classification of social media posts in disasters and emergencies: a system and evaluation featuring active, incremental and online learning. Inf Process Manag 57(1):102132. https://doi.org/10.1016/j.ipm.2019.102132

Khwairakpam E, Khosa R, Gosain A, Nema A (2019) Monitoring and modelling water quality of Loktak Lake catchment. SN Appl Sci 1(5). https://doi.org/10.1007/s42452-019-0517-1

Kim K-S, Park J-H (2009) A survey of applications of artificial intelligence algorithms in eco-environmental modelling. Environ Eng Res 14(2):102–110. https://doi.org/10.4491/EER.2009.14.2.102

Kochkov D, Yuval J, Langmore I, Norgaard P, Smith J, Mooers G, Klöwer M, Lottes J, Rasp S, Düben P, Hatfield S, Battaglia P, Sanchez-Gonzalez A, Willson M, Brenner MP, Hoyer S (2024) Neural general circulation models for weather and climate. Nature 632(8027):1060–1066. https://doi.org/10.1038/s41586-024-07744-y

Kumar P, Dimri AP, Tandon SK (2021) Modeling of Indian monsoon extremes during 850-2000AD using the proxy-data from speleothems. Quat Int 599–600:117–127. https://doi.org/10.1016/j.quaint.2021.02.009

Kumar P, Barat A, Sarthi PP, Shankar A (2024) Evaluating future urban temperature over smart cities of the Gangetic plains using statistically downscaled CMIP6 projections. Theor Appl Climatol. https://doi.org/10.1007/S00704-024-04896-9

LaGrandeur K (2021) How safe is our reliance on AI, and should we regulate it? AI Ethics 1(2):93–99. https://doi.org/10.1007/S43681-020-00010-7/METRICS

Larbi I, Nyamekye C, Hountondji FCC, Okafor GC, Odoom PRE (2021) Climate change impact on climate extremes and adaptation strategies in the Vea catchment, Ghana. In: African handbook of climate change adaptation, pp 1–17. https://doi.org/10.1007/978-3-030-42091-8_95-1

Leal Filho W, Wall T, Rui Mucova SA, Nagy GJ, Balogun AL, Luetz JM, Ng AW, Kovaleva M, Safiul Azam FM, Alves F, Guevara Z, Matandirotya NR, Skouloudis A, Tzachor A, Malakar

K, Gandhi O (2022) Deploying artificial intelligence for climate change adaptation. Technol Forecast Soc Chang 180:121662. https://doi.org/10.1016/J.TECHFORE.2022.121662
Li X, Babovic V (2019) Multi-site multivariate downscaling of global climate model outputs: an integrated framework combining quantile mapping, stochastic weather generator and empirical copula approaches. Clim Dyn 52(9–10):5775–5799. https://doi.org/10.1007/S00382-018-4480-0
Liakos KG, Busato P, Moshou D, Pearson S, Bochtis D (2018) Machine learning in agriculture: a review. Sensors (Basel, Switzerland) 18:2674
Lin P, Jenkins R, Abney K (2017) Robot ethics 2.0: from autonomous cars to artificial intelligence. Oxford University Press
Liu Y, Cheng Z, Liu J, Yassin B, Nan Z, Luo J (2019) AI for earth: rainforest conservation by acoustic surveillance. http://arxiv.org/abs/1908.07517
Lu S, Christie GA, Nguyen TT, Freeman JD, Hsu EB (2022) Applications of artificial intelligence and machine learning in disasters and public health emergencies. Disaster Med Public Health Prep 16(4):1674–1681. https://doi.org/10.1017/DMP.2021.125
Maizland L (2023) Global climate agreements: successes and failures | Council on Foreign Relations. Council on Foreign Relations. https://www.cfr.org/backgrounder/paris-global-climate-change-agreements
Marrable D, Barker K, Tippaya S, Wyatt M, Bainbridge S, Stowar M, Larke J (2022) Accelerating species recognition and labelling of fish from underwater video with machine-assisted deep learning. Front Mar Sci 9. https://doi.org/10.3389/FMARS.2022.944582/FULL
McGovern A, Elmore KL, Gagne DJ, Haupt SE, Karstens CD, Lagerquist R, Smith T, Williams JK (2017) Using artificial intelligence to improve real-time decision-making for high-impact weather. Bull Am Meteorol Soc 98(10):2073–2090. https://doi.org/10.1175/BAMS-D-16-0123.1
Miller RL, Schmidt GA, Nazarenko LS, Tausnev N, Bauer SE, Delgenio AD, Kelley M, Lo KK, Ruedy R, Shindell DT, Aleinov I, Bauer M, Bleck R, Canuto V, Chen Y, Cheng Y, Clune TL, Faluvegi G, Hansen JE et al (2014) CMIP5 historical simulations (1850–2012) with GISS ModelE2. J Adv Model Earth Syst 6(2):441–477. https://doi.org/10.1002/2013ms000266
Mohan, A. (2017). Occasional paper 130 130 from Rio to Paris: India in global climate politics
Mohseni S, Zarei N, Ragan ED (2021) A multidisciplinary survey and framework for design and evaluation of explainable AI systems. ACM Trans Interact Intell Syst (TiiS) 11(3–4). https://doi.org/10.1145/3387166
Mujumdar PP, Kumar DN (n.d.) Hydrologic modeling for floods. Cambridge University Press, Cambridge, pp 5–42
Murshed M, Abbass K, Rashid S (2021) Modelling renewable energy adoption across south Asian economies: empirical evidence from Bangladesh, India, Pakistan and Sri Lanka. Int J Finan Econ 26(4):5425–5450. https://doi.org/10.1002/ijfe.2073
Nahvi B, Habibi J, Mohammadi K, Shamshirband S, Al Razgan OS (2016) Using self-adaptive evolutionary algorithm to improve the performance of an extreme learning machine for estimating soil temperature. Comput Electron Agric 124:150–160
Newton C, Singleton J, Copland C, Kitchen S, Hudack J (2021) Scalability in modeling and simulation systems for multi-agent, AI, and machine learning applications. 11746:534–552. https://doi.org/10.1117/12.2585723
Nost E, Colven E (2022) Earth for AI: a political ecology of data-driven climate initiatives. Geoforum 130:23–34. https://doi.org/10.1016/J.GEOFORUM.2022.01.016
Ntoutsi E, Fafalios P, Gadiraju U, Iosifidis V, Nejdl W, Vidal ME, Ruggieri S, Turini F, Papadopoulos S, Krasanakis E, Kompatsiaris I, Kinder-Kurlanda K, Wagner C, Karimi F, Fernandez M, Alani H, Berendt B, Kruegel T, Heinze C et al (2020) Bias in data-driven artificial intelligence systems—an introductory survey. Wiley Interdiscip Rev: Data Min Knowl Discov 10(3):e1356. https://doi.org/10.1002/WIDM.1356
Okkan U, Ersoy ZB, Ali Kumanlioglu A, Fistikoglu O (2021) Embedding machine learning techniques into a conceptual model to improve monthly runoff simulation: a nested hybrid rainfall-runoff modeling. J Hydrol 598. https://doi.org/10.1016/J.JHYDROL.2021.126433

Ouyang W, Lawson K, Feng D, Ye L, Zhang C, Shen C (2021) Continental-scale streamflow modeling of basins with reservoirs: towards a coherent deep-learning-based strategy. J Hydrol 599. https://doi.org/10.1016/J.JHYDROL.2021.126455

Owino VO, Mouratidou T (2019) Strengthened data systems to mitigate the double burden of malnutrition: the role of stable isotope technique-derived nutrition indicators. Ann Nutr Metab 75(2):119–122. https://doi.org/10.1159/000503669

Owino V, Kumwenda C, Ekesa B, Parker ME, Ewoldt L, Roos N, Lee WT, Tome D (2022) The impact of climate change on food systems, diet quality, nutrition, and health outcomes: a narrative review. Front Clim 4:941842. https://doi.org/10.3389/FCLIM.2022.941842/BIBTEX

Parfitt J, Barthel M, MacNaughton S (2010) Food waste within food supply chains: quantification and potential for change to 2050. Philos Trans R Soc B: Biol Sci 365(1554):3065–3081. https://doi.org/10.1098/RSTB.2010.0126

Park C, Min S-K, Lee D, Cha D-H, Suh M-S, Kang H-S, Hong S-Y, Lee D-K, Baek H-J, Boo K-O, Kwon W-T (n.d.) Evaluation of multiple regional climate models for summer climate extremes over East Asia. Clim Dyn 46(7–8):2469–2486

Patterson D, Gonzalez J, Le Q, Liang C, Munguia L-M, Rothchild D, So D, Texier M., Dean J (2021) Carbon emissions and large neural network training. http://arxiv.org/abs/2104.10350

Peters DPC, Rivers A, Hatfield JL, Lemay DG, Liu S, Basso B (2020) Harnessing AI to transform agriculture and inform agricultural research. IT Prof 22(3):16–21. https://doi.org/10.1109/MITP.2020.2986124

Piadeh F, Behzadian K, Alani AM (2022) A critical review of real-time modelling of flood forecasting in urban drainage systems. J Hydrol 607. https://doi.org/10.1016/J.JHYDROL.2022.127476

Pickering B (2021) Trust, but Verify: informed consent, AI technologies, and public health emergencies. Future Internet *13*(5):132. https://doi.org/10.3390/FI13050132

Racah E (2017) Extreme weather: a large-scale climate dataset for semi-supervised detection, localization, and understanding of extreme weather events. Adv Neural Inform Process Syst 30:3405–3416

Raghavan M, Barocas S, Kleinberg J, Levy K (2020) Mitigating bias in algorithmic hiring: evaluating claims and practices. In: FAT* 2020 – Proceedings of the 2020 conference on fairness, accountability, and transparency, pp 469–481. https://doi.org/10.1145/3351095.3372828

Raihan A (2023) Artificial intelligence and machine learning applications in forest management and biodiversity conservation. Nat Resour Conserv Res 6(2):3825. https://doi.org/10.24294/NRCR.V6I2.3825

Raza A, Shahid MA, Safdar M, Zaman M, Mehmood Sabir R (2023) The role of artificial intelligence in climate-smart agriculture: a review of recent advances and future directions. Biol Life Sci Forum. https://doi.org/10.3390/IOCAG2023-16877

Rodrigues M, De la Riva J (2014) An insight into machine-learning algorithms to model human-caused wildfire occurrence. Environ Model Softw 57:192–201. https://doi.org/10.1016/J.ENVSOFT.2014.03.003

Rolnick D, Donti PL, Kaack LH, Kochanski K, Lacoste A, Sankaran K, Ross AS, Milojevic-Dupont N, Jaques N, Waldman-Brown A, Luccioni A, Maharaj T, Sherwin ED, Mukkavilli SK, Kording KP, Gomes C, Ng AY, Hassabis D, Platt JC et al (2019) Tackling climate change with machine learning. ACM Comput Surv 55(2) http://arxiv.org/abs/1906.05433

Sajjad A, Lu J, Chen X, Chisenga C, Mazhar N (2023) Rapid assessment of riverine flood inundation in Chenab floodplain using remote sensing techniques. Geoenviron Disasters 10(1):1–18. https://doi.org/10.1186/S40677-023-00236-7/TABLES/6

Samek W, Wiegand T, Müller K-R (2017) Explainable artificial intelligence: understanding, visualizing and interpreting deep learning models. http://arxiv.org/abs/1708.08296

Sankarananth S, Karthiga M, Suganya E, Sountharrajan S, Bavirisetti DP (2023) AI-enabled metaheuristic optimization for predictive management of renewable energy production in smart grids. Energy Rep 10:1299–1312. https://doi.org/10.1016/J.EGYR.2023.08.005

Sarker MNI, Hossain B, Shi G, Firdaus RBR (2023) Promoting net-zero economy through climate-smart agriculture: transition towards sustainability. Sustain Sci 18(5):2107–2119. https://doi.org/10.1007/S11625-023-01379-0

Sayad YO, Mousannif H, Al Moatassime H (2019) Predictive modeling of wildfires: a new dataset and machine learning approach. Fire Saf J 104:130–146. https://doi.org/10.1016/J.FIRESAF.2019.01.006

Scher S, Jewson S, Messori G (2021) Robust worst-case scenarios from ensemble forecasts. Weather Forecast 36(4):1357–1373. https://doi.org/10.1175/WAF-D-20-0219.1

Serban AC, Lytras MD (2020) Artificial intelligence for smart renewable energy sector in europe – smart energy infrastructures for next generation smart cities. IEEE Access 8:77364–77377. https://doi.org/10.1109/ACCESS.2020.2990123

Shabandri B, Madara SR, Maheshwari P (2020) IoT-based smart tree management solution for green cities. Stud Big Data 67:181–199. https://doi.org/10.1007/978-981-15-0663-5_9

Shivanna KR (2020) The sixth mass extinction crisis and its impact on biodiversity and human welfare. Resonance 25(1):93–109. https://doi.org/10.1007/S12045-019-0924-Z

Shivanna KR (2022) Climate change and its impact on biodiversity and human welfare. Proc Indian Natl Sci Acad A Phys Sci 88(2):160. https://doi.org/10.1007/S43538-022-00073-6

Shivaprakash KN, Swami N, Mysorekar S, Arora R, Gangadharan A, Vohra K, Jadeyegowda M, Kiesecker JM (2022) Potential for artificial intelligence (AI) and machine learning (ML) applications in biodiversity conservation, managing forests, and related Services in India. Sustainability (Switzerland) 14(12):7154. https://doi.org/10.3390/SU14127154/S1

Slater LJ, Anderson B, Buechel M, Dadson S, Han S, Harrigan S, Kelder T, Kowal K, Lees T, Matthews T, Murphy C, Wilby RL (2021) Nonstationary weather and water extremes: a review of methods for their detection, attribution, and management. Hydrol Earth Syst Sci 25(7):3897–3935. https://doi.org/10.5194/HESS-25-3897-2021

Smart Cities | Free Full-Text | Artificial Intelligence Techniques in Smart Grid: A Survey (n.d.) Retrieved June 7, 2024, from https://www.mdpi.com/2624-6511/4/2/29

Smart Grid System Report, U.S. Department of Energy (2018)

Smith PJ, Pappenberger F, Wetterhall F, Thielen Del Pozo J, Krzeminski B, Salamon P, Muraro D, Kalas M, Baugh C (2016) On the operational implementation of the European flood awareness system (EFAS). In: Flood Forecasting: a Global Perspective, pp 313–348. https://doi.org/10.1016/B978-0-12-801884-2.00011-6

Tamburrini G (2022) The AI carbon footprint and responsibilities of AI scientists. Philosophies 7(1):4. https://doi.org/10.3390/PHILOSOPHIES7010004

UNDP (2024) What is climate change adaptation and why is it crucial? Climate Promise. https://climatepromise.undp.org/news-and-stories/what-climate-change-adaptation-and-why-it-crucial

UNFCCC (2020) *Introduction to mitigation*. https://unfccc.int/topics/introduction-to-mitigation

Wei Z (2021) Forecasting wind waves in the US Atlantic Coast using an artificial neural network model: towards an AI-based storm forecast system. Ocean Eng 237:109646. https://doi.org/10.1016/J.OCEANENG.2021.109646

Weitz K, Schiller D, Schlagowski R, Huber T, André E (2019) I "do you trust me?": increasing user-trust by integrating virtual agents in explainable AI interaction design. In: IVA 2019 – proceedings of the 19th ACM international conference on intelligent virtual agents, pp 7–9. https://doi.org/10.1145/3308532.3329441

Wildbook – Wild Me (n.d.). Retrieved April 28, 2025, from https://www.wildme.org/wildbook.html

Xu Y, Liu X, Cao X, Huang C, Liu E, Qian S, Liu X, Wu Y, Dong F, Qiu CW, Qiu J, Hua K, Su W, Wu J, Xu H, Han Y, Fu C, Yin Z, Liu M et al (2021) Artificial intelligence: a powerful paradigm for scientific research. Innov (Camb) 2(4). https://doi.org/10.1016/J.XINN.2021.100179

Yip SC, Wong KS, Hew WP, Gan MT, Phan RCW, Tan SW (2017) Detection of energy theft and defective smart meters in smart grids using linear regression. Int J Elec Power Energy Syst 91:230–240. https://doi.org/10.1016/J.IJEPES.2017.04.005

Yu H, Shen Z, Miao C, Leung C, Lesser VR, Yang Q (2018) Building ethics into artificial intelligence. In: *IJCAI international joint conference on artificial intelligence, 2018-July*, pp 5527–5533. https://doi.org/10.24963/ijcai.2018/779

Zhang D, Han X, Deng C (2018) Review on the research and practice of deep learning and reinforcement learning in smart grids. CSEE J Power Energy Syst 4(3):362–370. https://doi.org/10.17775/CSEEJPES.2018.00520

Zhang Y, Zhan C, Huang Q, Sornette D (2024) Seismically informed reference models enhance AI-based earthquake prediction systems. J Geophys Res: Solid Earth 129(3):e2023JB028037. https://doi.org/10.1029/2023JB028037

Zhao Z, Wang Y, Zang Z, Deng S, Lan T, Xie Z, Xiong G, Li J, Shen G (2020) Climate warming has changed phenology and compressed the climatically suitable habitat of Metasequoia glyptostroboides over the last half century. Glob Ecol Conserv 23. https://doi.org/10.1016/J.GECCO.2020.E01140

Zhao X, Lovreglio R, Kuligowski E, Nilsson D (2021) Using artificial intelligence for safe and effective wildfire evacuations. Fire Technol 57(2):483–485. https://doi.org/10.1007/S10694-020-00979-X

Zurek M, Hebinck A, Selomane O (2022) Climate change and the urgency to transform food systems. Science 376(6600):1416–1421. https://doi.org/10.1126/SCIENCE.ABO2364/ASSET/84400FAE-F20A-40B5-BE87-B34BCF7F9B69/ASSETS/IMAGES/LARGE/SCIENCE.ABO2364-F1.JPG

Artificial Intelligence and Environmental Sustainability: A Path Toward a Greener Future

8

Shikha Daga, Kiran Yadav, Pardeep Singh, and Sonal Thukral

Abstract

Climate change and environmental degradation are the pressing issues that require immediate action at the micro as well as the macro level. Modernization that includes digitalization, technical innovation, use of artificial intelligence, and machine learning all address the pertinent issues of resource depletion and environmental degradation to ensure that there is a paradigm shift toward a development that is resilient, green, and sustainable. The chapter is an attempt to address the contribution of artificial intelligence (AI) toward facilitating a paradigm shift toward ensuring a development that is resilient, green, and sustainable. The chapter further discusses proactive intervention and adaptive management strategies to achieve the same. The chapter provides an outline of the use of AI in promoting sustainable development, its growth, and trends over the years, along with providing a bibliometric analysis of the prominent authors, countries, institutions, and thematic evolution. Thus, the chapter contributes in understanding the role of AI in attaining the goals set in the 2030 Agenda of Sustainable Development Goals. AI not only helps in the identification of sources of extreme carbon footprints but can also assist in the development of energy-efficient practices and waste management, thereby helping in reducing carbon footprints and promoting green development. This chapter will therefore be of significant relevance to policymakers, academicians, industry leaders, and insti-

S. Daga (✉) · K. Yadav
Department of Commerce, PGDAV College, University of Delhi, New Delhi, India
e-mail: Shikha.menani@pgdav.du.ac.in

P. Singh
Department of Environmental Studies, PGDAV College, University of Delhi, New Delhi, India

S. Thukral
Department of Commerce, Delhi School of Economics, University of Delhi, New Delhi, India

B. Thokchom et al. (eds.), *Artificial Intelligence Enabled Real Time Environmental Monitoring*, https://doi.org/10.1007/978-981-95-3767-9_8

tutions seeking to contribute meaningfully to the advancement of a green economy.

Keywords

Artificial intelligence · Sustainable development · Bibliometric analysis · Carbon footprints · Waste management

8.1 Introduction

Rapid industrialization, growing population, and increase in energy demand worldwide have led to an exorbitant increase in carbon emissions globally, thereby disrupting the ecological and environmental balance (Zhu et al. 2017). Climate change today is an issue that requires immediate and strict actions to ensure that the Earth is not converted to another Mars and there is a reverse aging that the planet witnesses in the decades to come. Severe change in the environment has led to the initiation of policy making at the macro as well as the micro level. Be it the path breaking. Paris Agreement or the establishment of the Sustainable Development Goals at the global level and its acceptance at the country level. The severe changes in the global environment have prompted discussions on the formulation of policies aimed at addressing these challenges at both the macro and micro levels. The landmark Paris Agreement and the establishment of the Sustainable Development Goals (SDGs) are pivotal global efforts toward environmental sustainability, which have been further translated into country-specific targets. It highlights the importance of environmental protection by enhancing renewable energy use that leads to emission reduction and sustainable development (Guo et al. 2023). Rise in emissions backed by the unmindful use of natural resources, has led to the paradigm shift towards greener and cleaner sources of energy that not only assist in wealth generation but also ensures that it is green (Jiakui et al. 2023). The practice is equally supported by the international organisations that support the promotion of green financial sources like green bonds, green monetary measures like carbon tax among others. Policies taken at the macro level and its integration at the micro level also ensure that the SDGs are achieved in time and its benefit is equally spread across countries. The situation is not only supported by policymakers but is equally backed by academicians as is visible from the rise in literature related to growth and sustainability. One such area that has garnered the attention recently is the adoption and use of modern technology across different spheres. The most advanced and prominent tool that can make a paradigm shift toward sustainability attainment is artificial intelligence. Technical advancements, specifically the use of artificial intelligence and machine learning, have transformed the way businesses are conducted or policies are framed. The use of robotics and machines for analysis and deep learning has brought significant changes in various sectors and can surely bring a drastic impact on environmental sustainability as well (Goralski and Tan 2020). AI not only helps in the development of efficient tools for carbon footprint reduction but can also devise modular and advanced techniques for the efficient use of renewable energy sources that help in the promotion of the Sustainable Development Goals (Hannan et al. 2021). This is because AI helps in real time monitoring and hence assist in efficient decision making. The chapter focuses on highlighting the different perspectives of artificial intelligence and its role in promoting sustainable development. This study offers readers a comprehensive overview of the research conducted in the field of artificial intelligence (AI) and the potential developments that may be made to enhance its role in promoting environmental sustainability.

8.2 Challenges in Promoting Sustainable Development

The turn of the century has seen drastic changes in the environment that has led to a spike in the policies related to climate change and a step up towards green growth. The 2019 Pandemic has added to the fuel wherein there is mass movement towards net zero emissions across the countries. Be it the promotion of circular economy practices or the adoption of green sources of energy and finance there is a rapid rise in the policies that can act towards environmental protection. It has led to the initiation of steps that is conducive to growth as well as the environment (Di Vaio et al. 2020; Kartika et al. 2024). The policies promoting greener energy have found a place in the growth agendas of various countries whether it be the installation of solar panels, use of wind energy, promotion of hydrogen fuel, or subsidies for the usage of electronic vehicles. The set of 17 Sustainable Development Goals (SDGs) adopted by the United Nations in the year 2015 specifically focuses on attaining diverse goals that promote growth that is resilient and sustainable. The time frame set for attaining these goals was 2030; however, it still faces a lot of challenges, a few of which are the lack of adequate financial support, absence of necessary awareness, resistance to change, technological barriers and lack of global collaborative efforts. This is because making a shift towards sustainable practices be it the production or consumption needs adoption of sustainable practices across different levels of supply chain that needs backing of requisite resources. Absence or shortage of the same specially in case of not so affluent economies poses a burden and raises the dilemma of meeting the other social and economic goals. Making a shift toward sustainable practices in various sectors, viz., transportation, agriculture, consumption, production, and others, requires significant investments in technology, learning, research and development. It becomes difficult for low-income countries to undertake investments for adopting and implementing sustainable practices that are already under the economic pressure of meeting basic needs. Therefore, financial constraints act as a significant barrier that inhibits such countries from promoting and attaining sustainable growth. Lack of education and awareness among the general masses with regard to the technological advancements that can contribute toward sustainability is a major hindrance to achieving the target of sustainable development in various countries. Resistance to Change is another hurdle. The promotion of sustainable development is often hindered by both a lack of understanding and opposition to change. According to De Guinea & Markus, (2009) individuals often encounter challenges while attempting to transition from traditional habits to modern technologies and procedures, unless there is a compelling motivation to do so.

To ensure that there is a development that is sustainable across the countries, there must be international cooperation to handle the above obstacles. Countries should come forward to help others in terms of finances, technical advancements, and policy formulation as it is well accepted that the green development objectives attained in one country have a spillover effect in other parts of the world as well. Other than the above, it is also imperative to introduce AI and other technical advancements that can accelerate the progress of the countries toward the attainment of Sustainable Development Goals (Liengpunsakul 2021).

8.3 Leveraging Artificial Intelligence for Promoting Sustainable Development

The impact of artificial intelligence on the achievement of sustainable development objectives is significant as it prioritizes inclusive progress via its facilitation of efficient, expeditious, and dependable data collecting and analysis (Mhlanga 2021; Liengpunsakul 2021). Therefore, it facilitates the distribution of government responsibilities by fostering innovation, infrastructural advancement, and economic expansion. Literature on AI readiness index across Asian nations, revealed significant differences among countries that underlined the need for improvement to strengthen readiness. This is crucial for achieving global progress in the usage of AI and its contribution to the Sustainable Development Goals (SDGs) (Singh et al. 2024).

Utilizing digital technologies to create, use, send, or get electronic data for organizational activities may be employed to accomplish Sustainable Development Goals (SDGs) (Mondejar et al. 2021). A few ways in which AI can be utilized to achieve the SDG targets are having a mechanism that can monitor and assess its usage in real time. This is because efficient technology backed by use of AI and contemporary technology can assist in promoting green growth and emission mitigation through enhanced technology integration as is documented in "AI driven traffic management" (Bakioglu, 2025). AI is also documented to reduce agricultural waste that occurs during the entire supply chain along with providing an efficient AI enabled environmental control system that can promote resilient greenhouse farming (Lee et al., 2024). It is also documented to support food security management as AI helps in handling large amount of data and hence can assist in urban healthcare (Koppula et al., 2024). This is because climate change has a negative impact on food security that ultimately impact the healthcare system, AI can assist in handling the same by reducing its negative impact and assiting in managing food security that foster healthcare. AI is thus documented to have an influence across diverse fields such as poverty mitigation, healthcare, agriculture, resource management and other issues that arise because of climate change (Neuerburg, 2020).

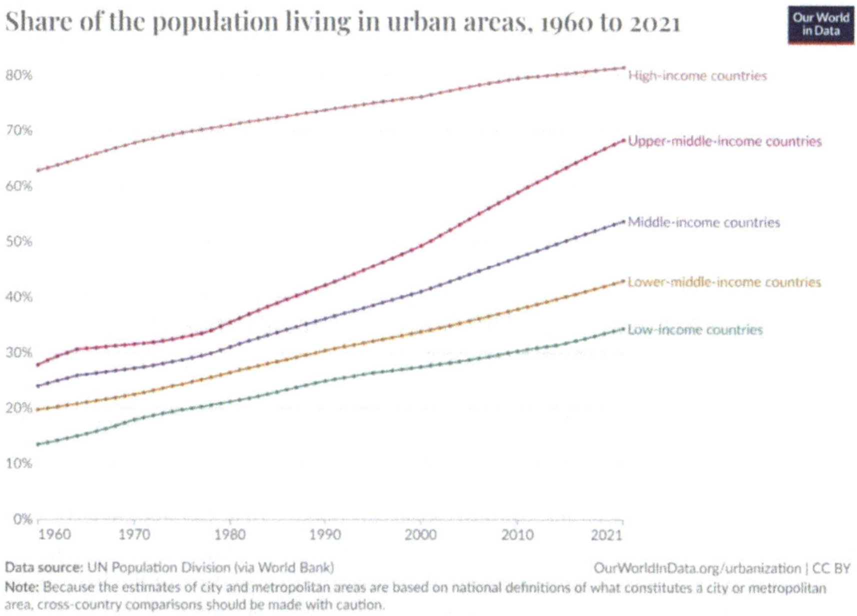

Fig. 8.1 Rising urbanization in the world: 1960–2021. (Source: ourworldindata.org)

As an economy progresses, there is a significant shift in the relative status of the area under the modern cities and urban areas, along with a rapid increase in transportation and hence greenhouse gas emissions. AI and machine learning can thus provide support to promote energy-efficient transportation systems and sustainable urban planning, thereby targeting on the attainment of SDG 11 that focuses on smart cities with reduced emissions and enhanced efficiency (Al-Raeei, 2024). Figure 8.1 illustrates the upward trajectory of urbanization in various regions worldwide, thereby emphasizing the significance of investigating the potential impact of artificial intelligence (AI) in this context.

AI not only lead to a healthier planet but also satisfies the SDG 3 target of good health and well-being. There has been a significant improvement and modernization in the healthcare sector, and it can be further accelerated by introducing AI in disease detection, its prognosis, management, medicine development, and testing (Zaman, 2024). AI also help in access to education, and it has been proven worldwide, during the Covid 2019 pandemic. This time has demonstrated how education can be provided efficiently and effectively through digital means. The use of AI would further ensure that there is cost-effective education and it is accessible to all, thereby targeting the attainment of SDG 4 Quality Education. AI can also help in the integration of education across global schools and universities while focusing on inclusive education (Pagliara et al., 2024). It can also help in targeting SDG 1, which is poverty reduction by generating avenues for income generation (Mhlanga, 2021). The use of AI can help in leveraging the advantage of a rising population, increased education, and demographic dividend by promoting innovation, creating entrepreneurial opportunities and hence generating employment. Therefore, it is crucial to have a rapid comprehension of the several Sustainable Development Goals (SDGs) and the role of AI in achieving them to facilitate a transition toward sustainable development. This necessitates endeavours at both the individual and societal levels.

8.4 Promoting Artificial Intelligence for Advancing Sustainable Development

Artificial intelligence is a contemporary tool that has the potential to assist in achieving Sustainable Development Goals, as well as improving health, education, transportation, infrastructure, climate, and other associated objectives. While the advantages of applying AI and digitalization are greater than the associated costs, there are still certain issues that must be tackled. This section highlights many initiatives that may be implemented to ensure the equitable distribution of the benefits of artificial intelligence (AI) within different socioeconomic groups.

To facilitate the adoption of AI and machine learning for addressing prevailing challenges and fostering resilient development, it is imperative to conduct thorough research in the field to align it with the specific requirements of the sector and economy in which it is intended to be implemented. Figure 8.2

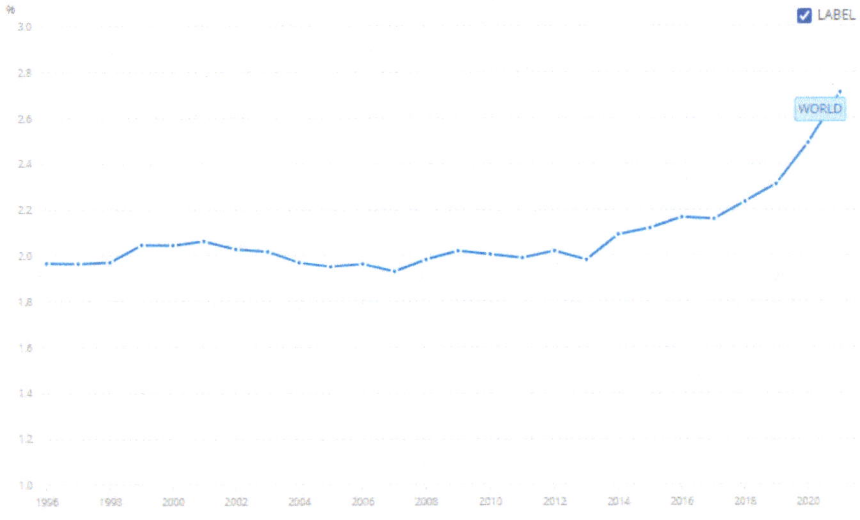

Fig. 8.2 Research and development expenditure (% of GDP): world data from 1996 to 2020. (Source: data.worldbank.org)

presents the comprehensive global research and development spending. Nevertheless, it is crucial to consider its impact on emerging sectors such as artificial intelligence (AI) to achieve various objectives that have the potential to enhance global well-being.

Awareness also plays an important role in promoting AI among shareholders. Advanced understanding of its use, advantages, and outcomes helps to achieve the desired outcomes via the provision of access and sources for the utilization and execution of artificial intelligence (AI). The importance of AI literacy and digital empowerment cannot be underestimated to effectively contribute to the advancement of sustainable development. It is essential to promote digital literacy among stakeholders for the necessary adaptability of AI (Baskara, 2025). To facilitate the advancement of sustainable development via the use of artificial intelligence (AI) tools such as satellite imaging sensors, trackers, and similar technologies, it is essential to establish a population that has the necessary adaptability. Internet penetration serves as a metric for comprehending the extent of digital literacy. Government also plays an important role as governmental support in the form of grants and subsidies is required for the initiation of any new technology To further the development of artificial intelligence (AI), it must get the requisite support and promotion from governmental bodies and international organizations. The prevalent rules and regulations need to make it simple for everyone to make use of AI at different levels and for diverse uses to ensure that the penetration of AI is not limited to a small group but its benefit is leveraged by the majority. The government also need to ensure that there is literacy among the masses regarding the use of AI, along with providing security against any misuse of the contemporary tool. To ensure that the use of AI gets promoted, there is a need to provide benefits in the form of subsidies that can promote its use vis-à-vis the competing polluting technique (Ran, 2021). However, digital education, awareness, and government grants cannot generate desired results until and unless there are technical incubators that can provide guidance and support to the naïve partners to develop models based on AI that have the potential to deliver sustainable results. Technical incubators help in providing the necessary assistance and support to the idea that is at its initial stage (Thottoli et al., 2025). AI, being highly innovative, risky, and difficult to implement, requires expertise of the technically advanced professionals who can maximize its potential while mitigating the associated risk.

The realization of the favorable influence that artificial intelligence (AI) can have on sustainable development should be a priority for policymakers and governmental bodies. This can be accomplished by successfully addressing the challenges that are linked with AI and ensuring that the aforementioned elements are managed in an appropriate manner.

8.5 Ethical and Policy Considerations in Artificial Intelligence and Sustainable Development

The use of artificial intelligence (AI) has the potential to generate significant changes in several sectors, including agriculture, tourism, education, and infrastructure, to name a few. Moreover, artificial intelligence (AI) has the potential to make significant contributions toward the attainment of Sustainable Development Goals (Pigola et al. 2021). However, it is important to ensure that ethical and policy issues are taken into account while adopting the function of AI to achieve sustainable development initiatives. It is necessary to include concepts of openness, justice, and accountability in the use of AI-driven technology. Impacts, results, and mechanisms of AI can be quite confidential, hidden, and dark. To ensure that the benefits of AI outperform its drawbacks, it is necessary to ensure that there is transparency in its functioning, algorithm development, and execution. Transparency in the usage of AI will thus help in building confidence that will further aggravate its use and hence will promote attaining the goal of sustainable development. Transparency in AI development and execution will also ensure that there is accountability that helps in building the trust of the diverse group of stakeholders, which can then work in tandem to attain the goal of sustainable development. Many times, in AI usage, there are malpractices and the absence of accountability that leads to a lack of trust and hence underutilization. Thus, the presence of transparency and hence accountability will ensure that there is a multidimensional use of AI techniques that would further foster resilient growth. Fairness is also essential with transparency and accountability. The use of artificial intelligence ought to guarantee that there is a fair playing field, in which no one group has an edge over the other. To guarantee that there is cooperation across a wide range of sectors, organizations, and economies that are capable of working together to achieve the objective of sustainable development, fairness in the use of artificial intelligence will also be necessary. Another concern is data protection as AI makes use of a large quantity of data that often raises doubts about data privacy (Manheim & Kaplan, 2019). The use of AI for solving issues like that of sustainable development should be backed by strict rules and regulations regarding data privacy. It is essential for building trust to ensure that the use of AI can go a long way in attaining the Sustainable Development Goals. However, access to digital literacy and, hence use of AI for problem-solving is not equally distributed and has a wide disparity within and across countries. The policies should ensure that there is balanced access and literacy of digital use across diverse segments, thereby promoting the usage at a global scale. Ethical considerations thus should take into account the diverse impact that AI can have on the varied sectors of education, health, employment, infrastructure, and others. By taking into account the ethical and policy considerations of AI, the economies at large can ensure that the goal of sustainable development is attained by harnessing the full potential of AI.

8.6 Bibliometric Analysis of Studies on Artificial Intelligence and Sustainable Development

A sound, unbiased literature review is essential to get an in-depth analysis of the research conducted in a particular domain. It helps in identifying the most prominent authors, countries, and areas of research, along with providing the research gaps that can be carried out in the future. One of the scientific approaches to conduct literature analysis is the bibliometric analysis that is a scientific method and excludes any author bias in selection of the literature. The biggest advantage of bibliometric analysis is that it produces results that are unbiased, scientific, and replicable. To extract the literature in social science, there are two databases: SCOPUS and Web of Science. This chapter has made use of Elsevier SCOPUS as it provides a more extensive list of articles over a long time. The second step here is to identify the research based on selected keywords. The strings need to be carefully assessed as it helps in tracing the right articles for further analysis. It is one of the most important steps as any missed information in this step can lead to faulty data extraction for bibliometric analysis, thereby making the entire activity futile. The study made use of two keywords artificial intelligence and sustainability the string used for the study was searched in the title of the paper, abstract, and keywords. The specific string used was (TITLE-ABS-KEY ("artificial Intelligence" AND "sustain*"). It resulted in a corpus of 13,448 papers spanning from 1984 to 2024. However, the search results provided research papers, conference proceedings, book reviews, and others, but for the analysis, only articles are selected that too in English and full papers. It gave rise to 5159 papers, which is the final lot on which bibliometrix R package biblioshiny is then used to get an overview of the research growth in a particular domain (Aria & Cuccurullo, 2017). Bibliometric analysis provides a detailed review of the growth in the research in a particular domain by highlighting the number of documents over the years, along with identifying the most influential authors, journals, most frequently used keywords, and trends in the research in a particular field over the years (Fig. 8.3).

8.6.1 Research Trends Over the Years on Artificial Intelligence and Sustainable Development

Research in the field of artificial intelligence was almost negligible until the global financial crisis of 2008, after which research in the domain started picking up; however, the area of interest was still not appealing until 2020, after which it spiked and crossed the 1250 mark by the end of 2023 (Fig. 8.4). The reason for rising interest in the area could be because of the government, industries, and interest in the area of AI and its role in promoting sustainable growth. The research in the area is however still not diverse and is limited to a few countries as elaborated in the next section.

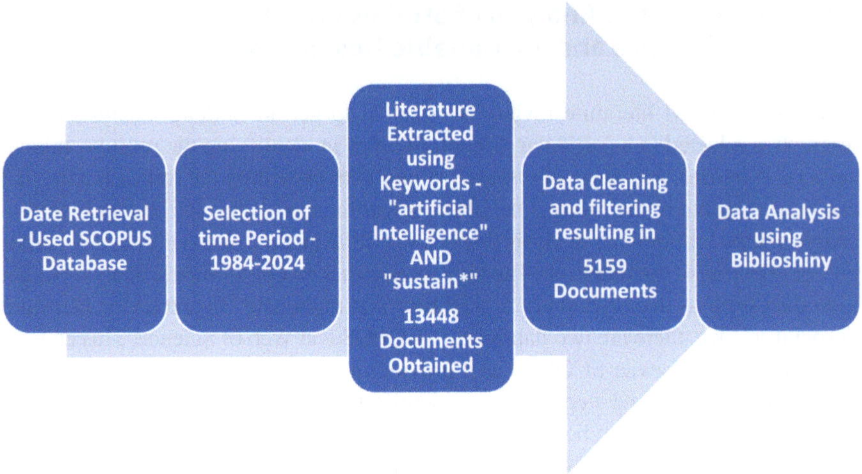

Fig. 8.3 Steps in bibliometric analysis. (Source: author's compilation)

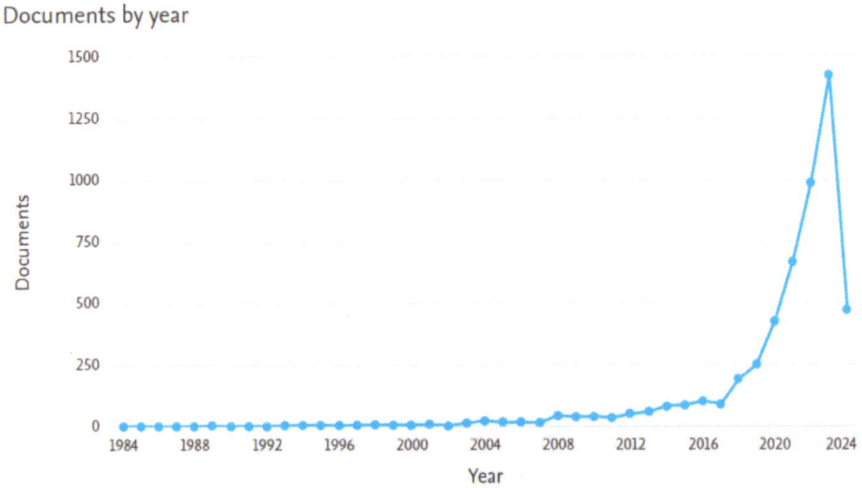

Fig. 8.4 Publications per year on AI and sustainable development. (Source: SCOPUS)

8.6.2 Publications Across Countries on Artificial Intelligence and Sustainable Development

Figure 8.5 displays the data publications in the domain of artificial intelligence and sustainable development, categorized by countries. The chart identifies China, the United States, India, the United Kingdom, and Italy as the leading countries in this sector. While China and the United States top the chart in terms of research in the area of AI and sustainable development. The emerging economy of India is also not

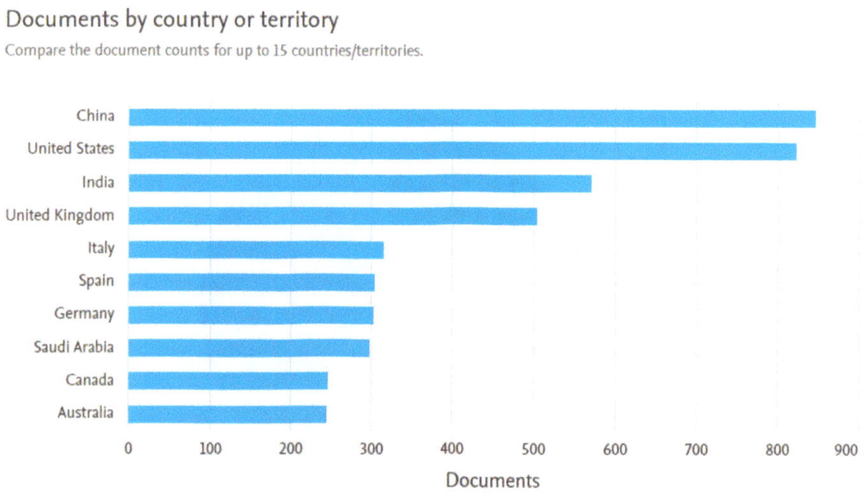

Fig. 8.5 Publications countrywise on AI and sustainable development. (Source: SCOPUS)

far behind in depicting the use of contemporary techniques in countries other than the developed ones.

8.6.3 Publications Across Different Subject Areas on Artificial Intelligence and Sustainable Development

Most research in the area of AI and sustainable development is in the field of computer science followed by engineering, environmental science, social science, and energy (Fig. 8.6). A vast distribution of the research across different subject areas depicts its multidisciplinary nature. It shows how the use of AI is not limited to computer science and informational technology, but it can be of immense use for sorting out issues in agriculture, transportation, food security, urbanization, and others.

8.6.4 Three Field Plot Depicting Prominent Countries, Authors, and Keywords for Research in AI and Sustainable Development

An integrative chart depicting prominent countries, authors, and main keywords in the research of AI and sustainable development is provided in Fig. 8.7, which reiterates China as the most prominent country conducting research in the domain of artificial intelligence, machine learning, deep learning, and sustainable development. Three field plot provides a comprehensive analysis of the frequently used keywords, authors, and countries working in a specific field of research.

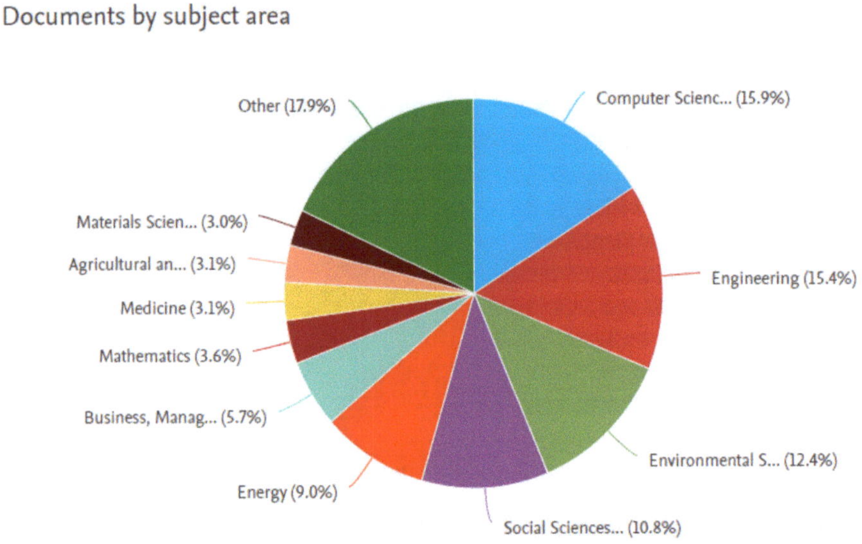

Fig. 8.6 Publications subject wise on AI and sustainable development. (Source: SCOPUS)

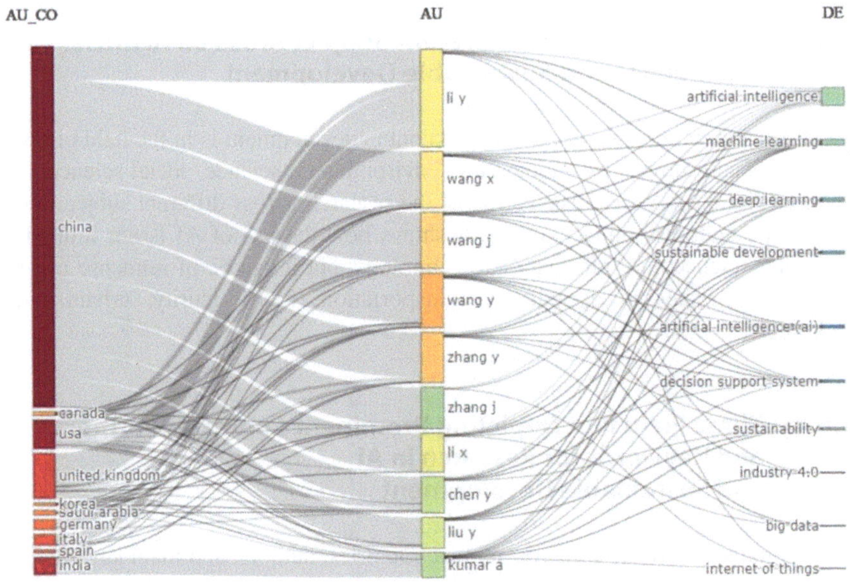

Fig. 8.7 Three field plot based on countries, authors, and keywords. (Source: Biblioshiny)

8.6.5 Prominent Sources in the Field of Artificial Intelligence and Sustainable Development

To get an idea of the most cited, reliable, and important documents related to the field of AI and sustainable development finding out the most prominent sources/journals is provided in this section (Fig. 8.8). *Sustainability, Journal of Cleaner Production, IEEE Access,* and *Energies* are a few of the top-rated journals in terms of the number of documents being published. However, an important thing to notice here is that the first source is way ahead in terms of publication compared to the other prominent journals.

8.6.6 Prominent Keywords in the Field of Artificial Intelligence and Sustainable Development

Authors' keywords are an important indicator for identifying the evolution of the research over a period of time. Figure 8.9 highlights the authors' keywords that have found the most number of occurrences in the relevant documents. Artificial intelligence, machine learning, sustainability, deep learning, and sustainable development

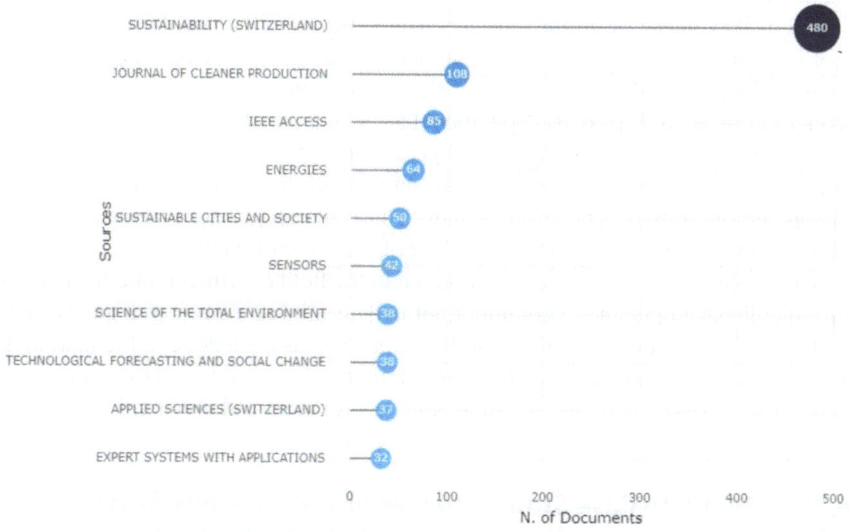

Fig. 8.8 Most relevant sources in the domain of artificial intelligence and sustainable development. (Source: Biblioshiny)

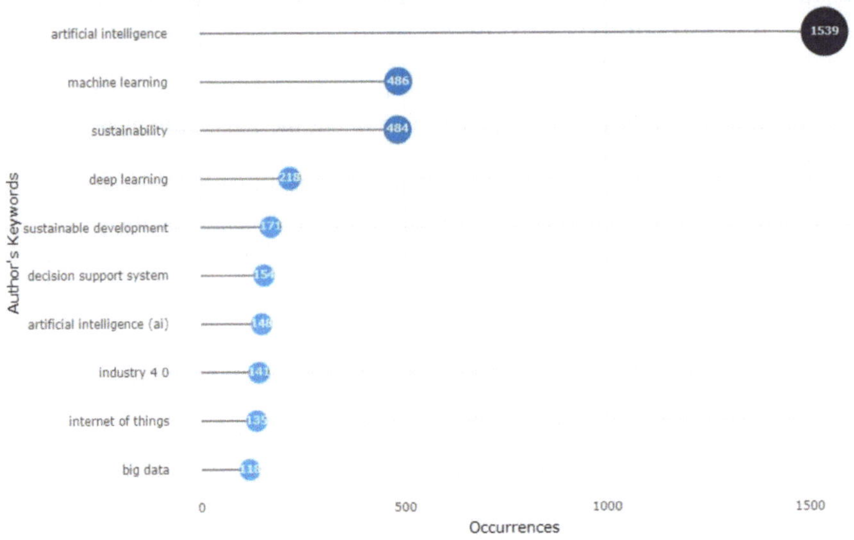

Fig. 8.9 Most relevant keywords in the domain of artificial intelligence and sustainable development. (Source: Biblioshiny)

are the top 5 keywords in terms of the number of occurrences, where artificial intelligence is way ahead of all the other keywords depicting the popularity of the concept in the research domain. Another important factor that needs to be assessed to identify the popularity and the shift of the research interests over the years is presented in Fig. 8.10. It provides how there has been a paradigm shift in the research areas from water supply, ambient intelligence, and developing countries in 2008 to fuzzy logic, data mining a decade later, and finally to deep learning, Internet of Things, among others. The shifting interest in research provides advancements in research and maturity in the understanding of the topic of interest.

This bibliometric analysis of the research in the field of artificial intelligence and sustainable development is very important in determining the research gap because it offers a comprehensive analysis of the work that has been done in the field up to this point as well as the areas that are still underdeveloped and need to be addressed in order to achieve the target of sustainable development and green growth.

8.7 Real-Life Case Studies to Elaborate the Role of Artificial Intelligence in Promoting Sustainable Development

Artificial intelligence has been utilized to identify prominent lifestyle issues affecting the contemporary world, such as detecting difficulties in drinkable water, recognizing climatic threats, and scenarios like floods and water shortages. This section highlights a few real-life scenarios where AI has been instrumental in providing assistance.

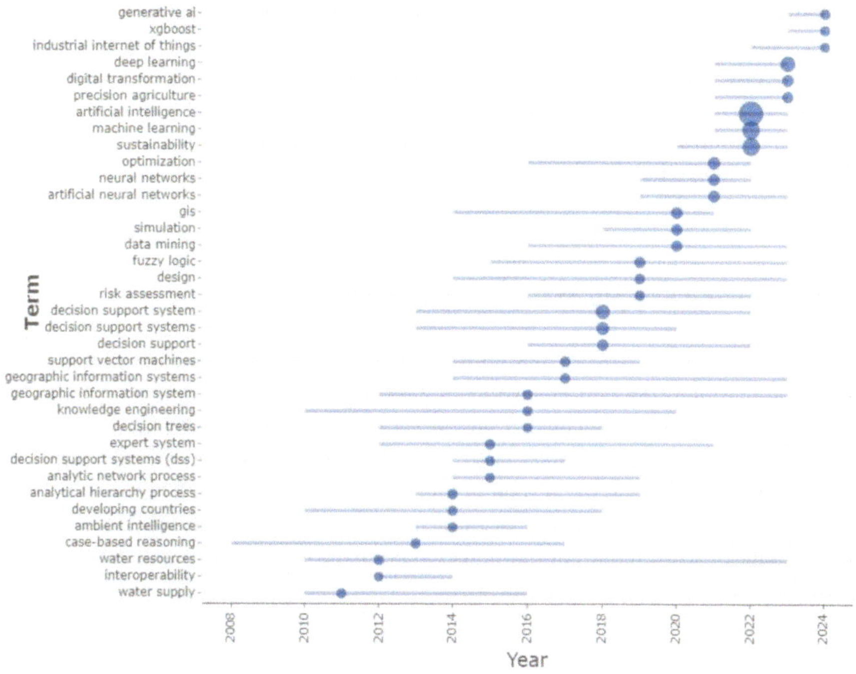

Fig. 8.10 Trend topics in the field of artificial intelligence and sustainable development. (Source: Biblioshiny)

8.7.1 Artificial Intelligence to Detect Arsenic Poisoning

AI technology has been used to identify specific regions inside the Ganga Delta and categorize them into different zones based on the levels of arsenic presence. The researchers have used AI algorithms to analyze environmental, geological, and human usage aspects. The research is beneficial for determining the contamination of groundwater by using modern techniques to assure the reliability and validity of the findings (Chakraborty et al., 2020) and ("AI-based system to detect arsenic pollution," Press Trust of India, 2020).

8.7.2 Geo AI and its Role in Targeting Brick Kiln Hotspots of Air Pollution

India is amongst the one of the largest emitters of the world. Brick manufacturing is one of the significant contributors to the total pollution level in the city, in addition to the many other causes that contribute to the emissions. The use of artificial intelligence was essential in determining the precise sites of brick kiln manufacturing and their contribution to the degree of pollution. This information was then communicated to the authorities so that they could take the appropriate measures. It was shown by the GeoAI platform that it is possible to discover instances of environmental noncompliance via the use of artificial intelligence, specifically through the utilization of data from space. As a result, the vast amount of work and financial resources that are required to monitor every industrial plant in India are significantly reduced (Kolluri, Singh, & Raghavan, 2020; www.undp.org). The use of AI in identifying and hence helping in tackling the issue of pollution emission thus provides how it saves on the resources of time, energy, as well as cost.

8.7.3 AI and Supply of Rural Drinking Water

In an effort to make use of the advantages offered by artificial intelligence and machine learning, the Union Jal Shakti Ministry of the Government of India has implemented sensory devices that will assist in monitoring the drinking water supply and the issues that are associated with it in lakhs of Indian villages. Artificial intelligence has come to the rescue in this endeavor, which is an attempt to achieve the Sustainable Development Goal of clean drinking water. Not only does it lower the amount of work required, but it also maximizes the accuracy obtained (indiaai.gov.in).

8.7.4 AI and Prediction of Floods

Artificial intelligence is not only useful for making people's lives simpler and more comfortable, but it also contributes to the reduction of a variety of hazards. Floods and the aftermath of floods are one kind of natural calamity that, on an annual basis, compels millions of people to leave their homes. Google and the International Federation of Red Cross and Red Crescent Societies (IFRC) have launched an initiative in India that sends alerts in advance to people in a variety of local languages. The purpose of this initiative is to ensure that individuals are transferred to a safer location at the appropriate time, thereby saving lives. This initiative was launched to both anticipate potential risks and make appropriate preparations in advance by informing those who are concerned about the situation. "*In 2018, Google's flood forecasting programme was first tested in the Patna area of Bihar. Since then, it has been expanded to cover the whole of India by the year 2020, including more than 250,000 square kilometres and covering a population of 200 million people. The technology developed by Google is being used to enhance the targeting of each and every warning that the government sends out, thereby reducing the risks and the associated losses*" (indiaai.gov.in).

Each of the above-mentioned real-world case studies contributes to the provision of a look into the many applications of artificial intelligence (AI), and it may be used in other countries as well. Consequently, the case studies contribute to the generation of ideas and the resolution of problems.

8.8 Conclusion

As a result of climate change, environmental degradation, and imbalances in flora and fauna, the global economy has been disrupted, and governments and organizations all over the globe have been convinced to shift toward sources of growth that are more robust, clean, and sustainable. A revolution that has altered how the situation can be better in the future is the use of digital technology, which is artificial intelligence that can provide new and modernized methods to promote a development that is not only clean, green, and resilient but is also sustainable (Nahar 2024). This revolution has changed how the situation can be improved in the future. The global economy continues to struggle with the problem of digital literacy and increased digital penetration throughout the world, both of which are necessary for the revolution to be successful in producing the outcomes that are sought. On the other hand, artificial intelligence is a double-edged sword that may have both good and bad consequences, and if it is balanced well, it can generate exponential benefits in terms of achieving sustainable development objectives (Mondejar et al. 2021). The employment of artificial intelligence should, however, makes certain that ethics and rules in its application are of the highest significance. This will not only assist in the establishment of trust and the promotion of data privacy, but it will also go a long way toward resolving modern difficulties that call for sustainable growth. In certain areas, the application of artificial intelligence to the advancement of the various goals of the United Nations' seventeen Sustainable Development Goals (SDGs) is quite widespread, while in other areas, there is still a major need for progress (Nasir et al. 2023). According to the findings of a study conducted by Vinuesa et al. in 2020, the advantages of artificial intelligence are not without any negative consequences. The study reveals that out of the 17 Sustainable Development Goals and 169 targets that need to be achieved by the year 2030, the utilization of AI can assist in the achievement of 134 targets, but it can also cause a delay in the achievement of 59 targets. To prevent any instances of cross-cannibalization, it is necessary to closely supervise the use of artificial intelligence.

References

Al-Raeei M (2024) Artificial intelligence for climate resilience: Advancing sustainable goals in SDGs 11 and 13 and its relationship to pandemics. In Discover Sustainability (Vol. 5, Issue 1). Springer Nature. https://doi.org/10.1007/s43621-024-00775-5

Aria M & Cuccurullo C (2017) bibliometrix: An R-tool for comprehensive science mapping analysis. Journal of Informetrics, 11(4), 959–975. https://doi.org/10.1016/j.joi.2017.08.007

Baskara FR (2025) Conceptualizing digital literacy for the AI era: A framework for preparing students in an AI-driven world. Data and Metadata, 4(530), 1–13

Bakioglu G (2025) Assessing the role of digital transition in achieving net zero carbon transport: A hybrid Q-Rung orthopair fuzzy approach. Environment, Development and Sustainability, 27(9), 23081–23131. Scopus. https://doi.org/10.1007/s10668-025-06664-9

Chakraborty M, Sarkar S, Mukherjee A, Shamsudduha M, Ahmed KM, Bhattacharya A, & Mitra A (2020) Modeling regional-scale groundwater arsenic hazard in the transboundary Ganges River Delta, India and Bangladesh: Infusing physically-based model with machine learning. Science of The Total Environment, 748, 141107. https://doi.org/10.1016/j.scitotenv.2020.141107

De Guinea AO, & Markus ML (2009) Why break the habit of a lifetime? Rethinking the roles of intention, habit, and emotion in continuing information technology use. MIS Quarterly, 433–444

Di Vaio A et al (2020) Artificial intelligence and business models in the sustainable development goals perspective: a systematic literature review. J Bus Res 121(August):283–314. https://doi.org/10.1016/j.jbusres.2020.08.019

Goralski MA, Tan TK (2020) Artificial intelligence and sustainable development. Int J Manag Educ 18(1). https://doi.org/10.1016/j.ijme.2019.100330

Guo X et al (2023) Renewable energy for balancing carbon emissions and reducing carbon transfer under global value chains: a way forward. Sustainability (Switzerland) 15(1). https://doi.org/10.3390/su15010234

Hannan MA et al (2021) Impact of renewable energy utilization and artificial intelligence in achieving sustainable development goals. Energy Rep 7:5359–5373. https://doi.org/10.1016/j.egyr.2021.08.172

https://data.worldbank.org/indicator/GB.XPD.RSDV.GD.ZS?end=2021&start=1996&type=shaded&view=chart

India AI. (2021). IoT devices to monitor rural drinking water supply. Retrieved fromhttps://indiaai.gov.in/case-study/iot-devices-to-monitor-rural-drinking-water-supply

India AI. (2020). Using AI to predict floods and save lives. Retrieved fromhttps://indiaai.gov.in/case-study/using-ai-to-predict-floods-and-save-lives

Lee M.- H, Yao M.-H, Kow P.-Y, Kuo B.-J, & Chang F.-J (2024) An Artificial Intelligence-Powered Environmental Control System for Resilient and Efficient Greenhouse Farming. Sustainability (Switzerland), 16(24). Scopus. https://doi.org/10.3390/su162410958

Pagliara SM, Bonavolonta G, Pia M, Falchi S, Zurru AL, Fenu G, & Mura A (2024) The integration of artificial intelligence in inclusive education: A scoping review. Information, 15(12), 774

Pigola A, Da Costa PR, Carvalho LC, Silva L FD, Kniess C. T, & Maccari EA (2021) Artificial Intelligence-Driven Digital Technologies to the Implementation of the Sustainable Development Goals: A Perspective from Brazil and Portugal. Sustainability, 13(24), 13669. https://doi.org/10.3390/su132413669

Press Trust of India (2020, September 16). IIT researchers develop AI-based system to detect arsenic pollution. The Week. Retrieved fromhttps://www.theweek.in/news/health/2020/09/16/IIT-researchers-develop-AI-based-system-to-detect-arsenic-pollution.html

Kolluri, S., Singh, R., & Raghavan, K. S. (2020, November 19). How a GeoAI platform is helping target brick kiln hotspots of air pollution. United Nations Development Programme. fromhttps://www.undp.org/india/blog/how-geoai-platform-helping-target-brick-kiln-hotspots-air-pollution

Koppula VK, Birundadevi M, Ramprathap K, Parthasarathi P, & Nageswaran A (2024) Climate Change and Food Security Management–based Urban Health Care Systems Using Artificial Intelligence Techniques. Remote Sensing in Earth Systems Sciences, 7(4), 328–336. Scopus. https://doi.org/10.1007/s41976-024-00125-z

Jiakui C et al (2023) Green technological innovation, green finance, and financial development and their role in green total factor productivity: empirical insights from China. J Clean Prod 382(October 2022):135131. https://doi.org/10.1016/j.jclepro.2022.135131

Liengpunsakul S (2021) Artificial intelligence and sustainable development in China. Chin Econ 54(4):235–248. https://doi.org/10.1080/10971475.2020.1857062

Manheim K, & Kaplan L (2019) Artificial intelligence: Risks to privacy and democracy. Yale JL & Tech., 21, 106

Mhlanga D (2021) Artificial intelligence in the industry 4.0, and its impact on poverty, innovation, infrastructure development, and the sustainable development goals: Lessons from emerging economies? Sustainability, 13(11), 5788

Mondejar ME et al (2021) Digitalization to achieve sustainable development goals: steps towards a smart green planet. Sci Total Environ 794(June). https://doi.org/10.1016/j.scitotenv.2021.148539

Nahar S (2024) Modeling the effects of artificial intelligence (AI)-based innovation on sustainable development goals (SDGs): applying a system dynamics perspective in a cross-coun-

try setting. Technol Forecast Soc Change 201(January):123203. https://doi.org/10.1016/j.techfore.2023.123203
Nasir O et al (2023) artificial intelligence and sustainable development goals nexus via four vantage points. Technol Soc 72(November 2022):102171. https://doi.org/10.1016/j.techsoc.2022.102171
Neuerburg KM (2020) Applications of Artificial Intelligence in Achieving Development Goals. In International Journal of Technology, Knowledge and Society (Vol. 18, Issue 1, pp. 35–47). Common Ground Research Networks. https://doi.org/10.18848/1832-3669/CGP/v18i01/35-47
Ran L (2021) Influence of government subsidy on high-tech enterprise investment based on artificial intelligence and fuzzy neural network. Journal of Intelligent & Fuzzy Systems, 40(2), 2553–2563. https://doi.org/10.3233/JIFS-189248
Singh A et al (2024) Artificial intelligence for sustainable development goals: bibliometric patterns and concept evolution trajectories. Sustain Dev 32(1):724–754. https://doi.org/10.1002/sd.2706
Thottoli MM, Cruz ME, & Al Abri SSS (2025) The incubation revolution: Transforming entrepreneurial education with artificial intelligence. Asia Pacific Journal of Innovation and Entrepreneurship, 19(1), 2–23
Vinuesa R et al (2020) The role of artificial intelligence in achieving the sustainable development goals. Nat Commun 11(1):1–10. https://doi.org/10.1038/s41467-019-14108-y
Zaman Q (2024) The role of artificial intelligence in early disease detection: Transforming diagnostics and treatment. Multidisciplinary Journal of Healthcare (MJH), 1(2), 43–54
Zhu Z et al (2017) CO2 emissions from the industrialization and urbanization processes in the manufacturing center Tianjin in China. J Clean Prod 168:867–875. https://doi.org/10.1016/j.jclepro.2017.08.245

The Role of Artificial Intelligence in Achieving Seafood Sustainability

9

Seda Yıldırım and Durmus Cagrı Yıldırım

Abstract

Food security is seen as a serious problem in both developed and developing countries. Hunger and food shortages continue in underdeveloped countries, and different problems related to food have become a threat to developed and developing countries. As a result of climate change, changing seasons and ecological systems have begun to deplete food resources. At this point, there are also different problems for the seafood resources provided by marine and ocean ecosystems. Seafood is an important resource in terms of both health and food. The subjects of seafood security and seafood sustainability will be examined in the study. It is planned to review the relationship between the sustainability of seafood and marine ecosystems with the effect of technology, digitalization, and artificial intelligence.

Keywords

Seafood security · Seafood sustainability · Food security · Artificial intelligence · Marine ecosystem

9.1 Introduction

Achieving seafood sustainability has become a vital issue for food security in the twenty-first century. Historically, humanity has always struggled with issues such as hunger, poverty, and inequality (Käferstein and Abdussalam 1999; Fung et al. 2018; Yıldırım and Yıldırım 2021; Yıldırım et al. 2024). Unfortunately, it is seen that some societies are tested by hunger and their primary problems are food

S. Yıldırım (✉) · D. C. Yıldırım
Tekirdag Namık Kemal University, Tekirdag, Turkey

B. Thokchom et al. (eds.), *Artificial Intelligence Enabled Real Time Environmental Monitoring*, https://doi.org/10.1007/978-981-95-3767-9_9

insufficiency and lack of food. Especially, people in the African and Middle East countries have had to live with water and food deprivation for a long time (Hadipour et al. 2025). On the other hand, food security has been threatened in developed and developing countries for different reasons. Due to reasons such as climate change, global warming, changes in seawater temperatures, and the continuation of nonenvironmentally friendly production systems, the problem of finding food in sufficient quantity, variety, and at affordable prices has emerged in societies (Kaplan and Yildirim 2023). While ensuring food security is difficult even under normal conditions, unexpected situations such as the Russia–Ukraine crisis have also caused conditions that trigger food crises (Leal Filho et al. 2023; Yildirim et al. 2022b). During the COVID-19 period, many problems were encountered, especially food supply, and it was observed that some countries did not have sufficient food stocks (FAO 2020; Laborde et al. 2020; Kakaei et al. 2022; Word Food Programme 2020).

Globally, water and food are basic resources, and the sustainability of life must be ensured first and foremost. However, depending on geographical conditions and settlement opportunities, the food resources of each country differ from each other and the missing food must be supplied from other countries. In this context, terrestrial foods and water-based foods differ from country to country. Seafood sustainability and seafood security have an important place in food resources for countries with sea or ocean coasts and island countries. In island countries, aquaculture has a positive impact both in terms of economy and food and social welfare. In European countries, aquaculture has an important place as a healthy food source. In many countries with a coast, fishing is a sector that contributes to national income and creates employment (Yildirim and Kaplan 2022a, b; Kaplan et al. 2022).

Seafood is an important source of protein that is recognized worldwide. Seafood supply chain management is also important as it has a large trade network. The seafood market differs between the Global South and the Global North depending on consumption factors. While the sustainability theme is prominent in seafood marketing in the Global Northside, the themes of quality and safety are prominent in the Global South (Malcorps et al. 2021). The seafood market value was recorded as $165.44 billion in 2023. By 2032, the seafood market value is expected to reach $201.49 billion. As consumers transition to a healthier and more sustainable lifestyle, seafood has emerged as a source of healthy nutrition. The seafood market is also increasing with the change in consumer lifestyle (Chaudhary 2025). The seafood market has different values across main regions in the world. For example, the Asia-Pacific region had 41 percent of the seafood market; North America had 28 percent of the seafood market; Europe region had 28 percent of the seafood market; the Latin America region had 5 percent of the seafood market; and the Middle East and Africa region had 3 percent of the seafood market by 2023 (Kakroo 2024).

According to recent literature, it is seen that there is a close link between digitalization and sustainability in the long term. Green innovation, green technology, artificial intelligence, robots, etc., are all smart technologies that support sustainable development goals positively (Mondejar et al. 2021; Bhutani and Paliwal 2015; Kondratenko et al. 2022). With the help of digital technology, sustainability standards can be monitored more successfully in institutions and businesses, and environmental and directly nature-related situations such as monitoring changes in nature and climate and monitoring biodiversity can be realized more effectively.

The issue of seafood sustainability and security addresses all three fundamental dimensions (economic, social, and environmental) of sustainable development. Therefore, all kinds of policies and strategic approaches to ensure sustainability in seafood are important. In this study, a compilation of the effects of artificial intelligence on seafood sustainability is planned. The study will focus on how AI is used for seafood security and the marine ecosystem and the benefits of using AI.

9.2 Challenges in Seafood Security and Sustainability

Seventeen main goals of the 2030 SDGs are expected to be achieved by 2030. Since 2015, developed and developing countries have begun to adopt sustainable policies and strategies and to adopt new legal regulations and economic models suitable for the green transition model to achieve the goals (Yildirim 2022). Although progress has been made in some goals, it has become clear that a sustainable future scenario will not be easily achieved in the face of COVID-19, the Russia–Ukraine war, and the accelerating effects of climate change (Allam et al. 2022; Lass-Hennemann et al. 2024).

Food and water security, which require the greatest effort to achieve, is a very challenging goal globally. Food resources, seas, oceans, and freshwater resources, which are directly and most affected by climate change, are considered as an area waiting for a solution to global climate change (UNDP 2024; Hoegh-Guldberg et al. 2023; Lincoln et al. 2022).

Food security can be defined as the ability to access food in sufficient quantity and variety that is within people's purchasing power. At this point, "food quantity, food variety, and food prices" can be considered as important components. While the primary problem of most societies that have difficulty accessing food is food quantity and accessible food variety, concerns about food quality and food prices are also increasing in developed and developing societies (World Bank Group n.d.; Barrett 2002; Pinstrup-Andersen 2009; Yıldırım et al. 2024). Therefore, strategies and policies aimed at ensuring food security require different approaches to be followed among societies and countries.

The concept of seafood security, like food security, refers to the fact that "seafood is available in sufficient quantity, variety and price" and accessible to people (Yıldırım and Yıldırım 2021). While seafood is the primary protein source for coastal and island countries, seafood stands out as a healthy protein source in developed countries. It is known that seafood is an important source of protein as a source of nutrition and with its rich content. In terms of a sustainable lifestyle, eating seafood in the nutritional model also supports a healthy life (Boyd et al. 2022; Shalders et al. 2022; Viana et al. 2023). In this context, identifying the factors that prevent the sustainability of seafood and producing solutions for this will be a meaningful goal globally.

There are many hindering factors and challenges to seafood security and seafood sustainability, especially climate change. These factors include human overfishing, environmental pollution, excessive waste, and ineffective treatment systems. These factors and challenges are explained separately here for a better understanding of the hindering factors.

9.2.1 Challenges Related to Climate Change

As demand for fish and seafood increases, the growth of the aquaculture sector has become more important. However, accelerating climate change, droughts, ocean acidification, excessive salinity, and sea level rise threaten the sustainability of the seafood market. Due to climate change, water shortages are occurring in many parts of the world and aquaculture shortages are emerging (Ahmed and Thompson 2017).

The first negative impact on the sustainability of seafood products comes from the factors created by climate change. Due to climate change, there are changes in the temperature and content of sea and ocean water (Earth Observatory, NASA 2023). Since changes in seawater temperatures primarily change the habitat of local species, biodiversity decreases, and they are under extinction stress (Venegas et al. 2023; Freitas et al. 2021; Antão et al. 2020; Hodapp et al. 2023).

The change in seawater temperature causes the migration of marine and oceanic creatures. Migratory species are not always harmless. At this point, invasive marine alien species have begun to migrate between seas (Yildirim and Kaplan 2022b). The warming of seawater has caused many species of jellyfish and fish, such as pufferfish, lionfish, and compass jellyfish, which normally live in warm and temperate waters, to migrate to local and closed seas. This effect of climate change has created a primary struggle area in terms of seafood sustainability. Sustainable policies are being developed to combat invasive species and cleanse these species from local seas or reduce their population. For example, to combat the pufferfish migrating to the Mediterranean in Turkey, incentive programs have been organized for fishermen to catch these fish. With the help of this incentive, 28 million 12 thousand 50 puffer fish were prevented from entering the ecosystem (T.C. Tarim Ve Orman Bakanliği 2025).

The migration of invasive fish species, threatening the future of the local population, is a significant threat to seafood sustainability. Many fisheries and fish industries have been damaged due to the spread of invasive fish species (Bax et al. 2003; Haubrock et al. 2022).

The change in oxygen levels in sea and ocean waters and the change in other minerals also pose a risk to the lives of marine life. The change in water temperatures of seas and oceans and the differences in the amount and content of minerals pose a sustainability threat to local species. Therefore, sea and ocean oxidation and content change also pose a serious threat to the seafood market (Garcia-Soto et al. 2021; Branch et al. 2013; World Economic Forum 2020; EPA 2024).

9.2.2 Challenges Related to Human Activity

The problems in the fishing sector became visible in the twentieth century. In the fishing sector, the harmful situations toward fishermen and fishing activities started to be examined in this period. The decrease in seafood, seafood shortage, and the shrinkage of fisheries can be considered as the prominent negative effects in the fishing sector (Vázquez-Rowe 2020). Economic activities such as fishing, aquaculture, and seafood processing produce greenhouse gases. First, the greenhouse gas effect, which is caused by humans, also accelerates climate change. Carbon emissions from fishing activities contribute to ocean acidification (Xu et al. 2024).

Overfishing and unsustainable fishing activities both damage the marine ecosystem and negatively affect the seafood sector and the lives of many vulnerable low-income people. Illegal fishing, unregulated fishing, and employment of people in working conditions that violate human rights are unsustainable activities. Fishing is negatively affected both by human-based and climate change. Aquaculture may be a better solution for the continuity of natural marine ecosystems and biodiversity. Sustainable aquaculture is among the economic practices that are encouraged (Our Shared Seas 2019). Mielgo et al. (2024) studied how wealthy countries affect the seafood markets of developing countries by fishing in distant waters. Illegal fishing, by-catch, and overfishing activities particularly affect coastal countries in the Global South. Unfortunately, the excessive damage done by developed and wealthy countries to marine biodiversity in the Global South region negatively affects local people there (Mielgo et al. 2024). Alsaleh and Abdul-Rahim (2024) examined the relationship between the fishing sector and seawater pollution in 27 European countries in their study. Because oil consumption, waste fishing, etc., pollute the seas and can poison the ecosystem. In the study, the evolution of fishing activities and seawater pollution increased together. In other words, sea krill increases with increasing fishing. Therefore, the reorganization of fishing activities with sustainable policies will be a vital step for the sustainability of the seafood market in the long term (Alsaleh and Abdul-Rahim 2024).

9.3 Digitalization, Artificial Intelligence, and Seafood Sustainability

It can be said that human activities have become more efficient in the twenty-first century with digitalization and digital transformation. Sustainability and digital technologies have begun to be mentioned together (Yildirim et al. 2022a, b). Developments in digital technologies and the development of artificial intelligence can provide positive support for people to produce more sustainable policies and strategies. At this point, artificial intelligence and digital technologies play a greater role in following developments related to nature in natural sciences and in researching climate change.

Marine ecology is a very difficult field to understand. Since the marine ecological environment has a very complex structure, it was also a difficult process to follow ecological changes and biodiversity changes. At this point, it is expected that more efficient identification, classification, and prediction models will be developed with the artificial intelligence system. For example, a deep convolutional neural network was revealed by tracking from some points in the North Pacific in the identification of humpback whale songs. A new 3D modeling approach can be used for seabed examination with artificial intelligence. Artificial intelligence can also be used to follow seawater temperature and level changes. In summary, artificial intelligence is seen as an effective tool in areas such as marine meteorology monitoring, sea temperature monitoring, seabed examination, biodiversity monitoring, and examination. More accurate analysis of the marine ecosystem and accurate monitoring of climate change effects can provide a more sustainable future for the seafood market in the long term (Jiang and Zhu 2022).

Artificial intelligence (AI) approaches assist in many ocean and marine events prediction studies such as maritime transportation, data collection/analysis, and ocean ecosystem monitoring, and even become the primary tool (Kavitha et al. 2023). A large part of the degradation in the marine ecosystem is caused by pollution and includes large amounts of marine pollutant effects based on human activities. It is important to examine the marine ecosystem with effective monitoring systems to maintain the balance in the marine ecosystem (Rodrigues et al. 2021). Zare et al. (2024) studied the usage of AI tools in marine ecology, and they determined that AI was mostly used to make better predictions on marine pollution levels, make various positioning, and monitor and control the marine ecosystem. Rodrigues et al. (2021) achieved very high accuracy in some culture studies with 2D convolutional neural networks in their study and showed the benefit of combining machine learning and deep learning algorithm systems. Bakker (2022) argues that more effective results can be obtained for tracking and monitoring marine ecosystems and marine biodiversity with vehicles configured with artificial intelligence systems. For example, the MMPA system mobile marine protection vehicles are geographically mobile structures that can track endangered species that change location in the ocean. It is a fact that digital transformation is also seen in environmental control and management (Bakker 2022). Yang et al. (2024) suggest that effective environmental monitoring models can be established for marine ecosystem protection with the use of artificial intelligence. With AI systems, richer content data can be examined and analyzed more quickly. Although the ocean ecosystem is considered one of the most difficult areas to study, its study and research are also essential for the sustainability of the marine ecosystem and seafood products. With the use of AI systems, low-cost monitoring and observation models can be obtained (Katija et al. 2022). Chen et al. (2024) modeled an AI-based marine conservation model. They found that this system is useful for monitoring marine biodiversity and marine movements.

Machine learning is important in understanding marine ecology. Marine ecology data analysis, data collection, and data tracking can be done more effectively and quickly with machine AI tools (Rubbens et al. 2023). Rubbens et al. (2023) give a detailed examination of the dimension of AI in Fig. 9.1.

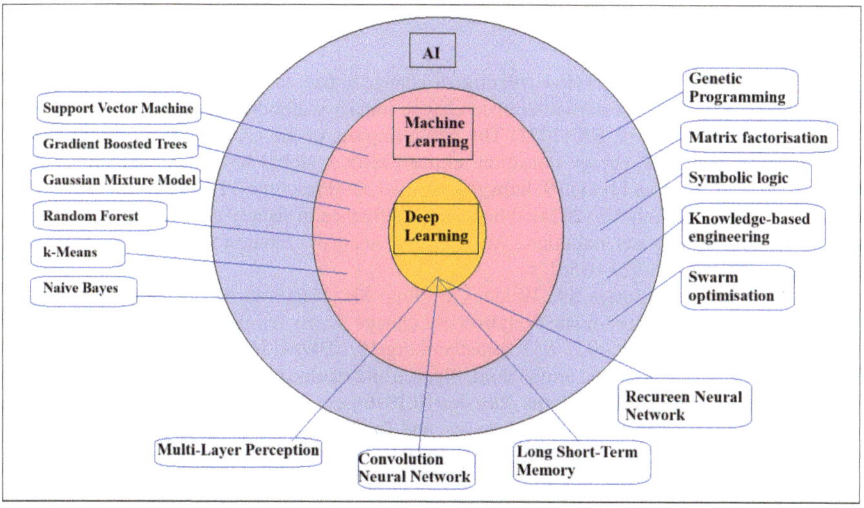

Fig. 9.1 Dimensions for AI-related machine learning and deep learning. (Source: adapted from Rubbens et al. (2023), then created by the authors)

9.4 Conclusion

Monitoring physical changes in the oceans has always involved some difficulties. Monitoring new changes due to climate change and making predictions have become more important. Artificial intelligence helps obtain more efficient results in ocean monitoring and prediction studies. Elements such as smart sensor systems, smart monitoring, underwater robots, and digitalization in observation technologies show the relationship between AI and ocean research (Song et al. 2023). AI applications stand out in the field of combating marine pollution. AI applications in combating three main pollution types—oil pollution, water quality monitoring, and plastic waste pollution—are very good examples of digitalization in the marine field (Ning et al. 2024). On the other hand, AI applications for effective management of the seafood supply chain system have recently come to the fore. The uncertainty in the seafood market and demand forecasting makes forecasting studies with AI more important (Ijadi et al. 2024).

The increasing use of digitalization and AI applications in seafood also promises hope for the realization of the 2030 SDGs targets. The processing of seafood also provides a useful step for food safety. Thanks to AI systems, it has become possible to extend the shelf life of seafood and increase its quality (Hassoun et al. 2022).

The sustainability of seafood is an important issue in ensuring food safety. Apart from land-based food products, seafood is an important source of protein and a healthy food source. For the sustainability of seafood, healthy monitoring and control of marine organisms, marine biodiversity, and marine ecosystems should be protected. Afterward, necessary measures should be taken for product diversity and quality in the marine market. AI is an important intersection point for both marine biodiversity sustainability and the sustainability of the seafood market.

References

Ahmed N, Thompson S (2017) How will climate change impact seafood production? https://www.globalseafood.org/advocate/will-climate-change-impact-seafood-production/

Allam Z, Bibri SE, Sharpe SA (2022) The rising impacts of the COVID-19 pandemic and the Russia–Ukraine war: energy transition, climate justice, global inequality, and supply chain disruption. Resources 11(11):99. https://doi.org/10.3390/resources11110099

Alsaleh M, Abdul-Rahim AS (2024) What are the influence of fishery activities and their implication on marine water pollution? An empirical analysis. Environ Dev Sustain. https://doi.org/10.1007/s10668-024-05561-x

Antão LH, Bates AE, Blowes SA, Waldock C, Supp SR, Magurran AE, Dornelas M, Schipper AM (2020) Temperature-related biodiversity change across temperate marine and terrestrial systems. Nat Ecol Evol 4:927–933. https://doi.org/10.1038/s41559-020-1185-7

Bakker K (2022) Smart oceans: artificial intelligence and marine protected area governance. Earth Syst Governance 13:100141. https://doi.org/10.1016/j.esg.2022.100141

Barrett CB (2002) Chapter 40: Food security and food assistance programs. In: Handbook of agricultural economics, vol 2, Part B. Elsevier, pp 2103–2190. https://doi.org/10.1016/S1574-0072(02)10027-2

Bax N, Williamson A, Aguero M, Gonzalez E, Geeves W (2003) Marine invasive alien species: a threat to global biodiversity. Mar Policy 27(4):313–323. https://doi.org/10.1016/S0308-597X(03)00041-1

Bhutani S, Paliwal Y (2015) Digitalization: a step towards sustainable development. OIDA Int J Sustain Dev 08(12):11–24. Available at SSRN: https://ssrn.com/abstract=2713132

Boyd CE, McNevin AA, Davis RP (2022) The contribution of fisheries and aquaculture to the global protein supply. Food Sec 14:805–827. https://doi.org/10.1007/s12571-021-01246-9

Branch TA, DeJoseph BM, Ray LJ, Wagner CA (2013) Impacts of ocean acidification on marine seafood. Trends Ecol Evol 28(3):178–186. https://doi.org/10.1016/j.tree.2012.10.001

Chaudhary T (2025) Global seafood market overview. https://www.marketresearchfuture.com/reports/seafood-market-1971?utm_term=&utm_campaign=&utm_source=adwords&utm_medium=ppc&hsa_acc=2893753364&hsa_cam=19912237177&hsa_grp=148712481999&hsa_ad=659589502539&hsa_src=g&hsa_tgt=dsa-2081392597831&hsa_kw=&hsa_mt=&hsa_net=adwords&hsa_ver=3&gad_source=1

Chen VY, Lu D-J, Han Y-S (2024) Hybrid intelligence for marine biodiversity: integrating citizen science with AI for enhanced intertidal conservation efforts at Cape Santiago, Taiwan. Sustainability 16(1):454. https://doi.org/10.3390/su16010454

Earth Observatory, NASA (2023) The ocean has a fever. https://earthobservatory.nasa.gov/images/151743/the-ocean-has-a-fever

EPA (2024) Effects of ocean and coastal acidification on marine life. https://www.epa.gov/ocean-acidification/effects-ocean-and-coastal-acidification-marine-life#:~:text=Laboratory%20studies%20suggest%20changing%20ocean,feed%20on%20these%20sensitive%20organisms

FAO (2020) Coronavirus disease 2019 (COVID-19) | Addressing the impacts of COVID-19 in food crises: April–December 2020, May update, FAO'S component of the global COVID-19 Humanitarian Response Plan. Rome. https://doi.org/10.4060/ca9192en

Freitas C, Villegas-Ríos D, Moland E, Olsen EM (2021) Sea temperature effects on depth use and habitat selection in a marine fish community. J Anim Ecol 90:1787–1800. https://doi.org/10.1111/1365-2656.13497

Fung F, Wang HS, Menon S (2018) Food safety in the 21st century. Biomed J 41(2):88–95. https://doi.org/10.1016/j.bj.2018.03.003

Garcia-Soto C, Cheng L, Caesar L, Schmidtko S, Jewett EB, Cheripka A, Rigor I, Caballero A, Chiba S, Báez JC, Zielinski T, Abraham JP (2021) An overview of ocean climate change indicators: sea surface temperature, ocean heat content, ocean pH, dissolved oxygen concentration, Arctic Sea ice extent, thickness and volume, sea level and strength of the AMOC (Atlantic

Meridional Overturning Circulation). Front Mar Sci 8:642372. https://doi.org/10.3389/fmars.2021.642372

Hadipour A, Ahmadi SH, Sepaskhah AR (2025) Consequences of water and land scarcity and unhealthy diets on the productive capacity of the Middle East and North Africa. Environ Sustain Indic 25:100594. https://doi.org/10.1016/j.indic.2025.100594

Hassoun A, Siddiqui SA, Smaoui S, Ucak İ, Arshad RN, Garcia-Oliveira P, Prieto MA, Aït-Kaddour A, Perestrelo R, Câmara JS, Bono G (2022) Seafood processing, preservation, and analytical techniques in the age of Industry 4.0. Appl Sci 12(3):1703. https://doi.org/10.3390/app12031703

Haubrock PJ, Bernery C, Cuthbert RN, Liu C, Kourantidou M, Leroy B, Turbelin AJ, Kramer AM, Verbrugge LNH, Diagne C, Courchamp F, Gozlan RE (2022) Knowledge gaps in economic costs of invasive alien fish worldwide. Sci Total Environ 803(2022):149875. https://doi.org/10.1016/j.scitotenv.2021.149875

Hodapp D, Roca IT, Fiorentino D, Garilao C, Kaschner K, Kesner-Reyes K, Schneider B, Segschneider J, Kocsis ÁT, Kiessling W, Brey T, Froese R (2023) Climate change disrupts core habitats of marine species. Glob Chang Biol 29:3304–3317. https://doi.org/10.1111/gcb.16612

Hoegh-Guldberg O, Northrop E et al (2023) The ocean as a solution to climate change: updated opportunities for action. Special report. World Resources Institute, Washington, DC. Available online at https://oceanpanel.org/publication/ocean-solutions-to-climate-change

Ijadi B, Wingreen SC, Reyes PM (2024) Artificial intelligence meets seafood supply chain management. In: CONFIRM 2024 proceedings. 16. https://aisel.aisnet.org/confirm2024/16

Jiang M, Zhu Z (2022) The role of artificial intelligence algorithms in marine scientific research. Front Mar Sci 9:920994. https://doi.org/10.3389/fmars.2022.920994

Käferstein F, Abdussalam M (1999) Food safety in the 21st century. Bull World Health Organ 77(4):347–351. PMID: 10327714; PMCID: PMC2557642

Kakaei H, Nourmoradi H, Bakhtiyari S, Jalilian M, Mirzaei A (2022) Effect of COVID-19 on food security, hunger, and food crisis. COVID-19 Sustain Dev Goals:3–29. https://doi.org/10.1016/B978-0-323-91307-2.00005-5. Epub 2022 Jul 29. PMCID: PMC9335023

Kakroo V (2024) Seafood market (by type: fish, crustaceans, mollusca, cephalopods, groundfish, salmonids, pelagics, tuna, others; by application: retails, food service, institutions; by nature: organic, conventional; by source: farm raised, wild caught; by end use: food, pharmaceutical, cosmetics, industrial, biotechnology, household retail; by category; by distribution channel) – global industry analysis, size, share, growth, trends, regional outlook, and forecast 2024–2034. https://www.precedenceresearch.com/seafood-market

Kaplan M, Yildirim S (2023) The sustainability of seafood market under the attack of invasive alien species: a review of Turkish marines. In: Ben Salem A, Rhazi L, Karmaoui A (eds) Climatic and environmental significance of wetlands: case studies from Eurasia and North Africa. IGI Global Scientific Publishing, pp 74–95. https://doi.org/10.4018/978-1-7998-9289-2.ch005

Kaplan M, Yildirim S, Yildirim DC (2022) Pufferfish versus lionfish: comparing risks for Turkish marine economics. Mar Econ Manag 5(2):173–187. https://doi.org/10.1108/MAEM-09-2022-0008

Katija K, Orenstein E, Schlining B, Lundsten L, Barnard K, Sainz G, Boulais O, Cromwell M, Erin Butler E, Woodward B, Bell KLC (2022) FathomNet: a global image database for enabling artificial intelligence in the ocean. Sci Rep 12:15914. https://doi.org/10.1038/s41598-022-19939-2

Kavitha MS, Jackson A, Manikandarajan V, Mariappan M, Paul A, Venmathi Maran BA (2023) Artificial intelligence methods in marine biotechnology. In: Shah MD, Ransangan J, Venmathi Maran BA (eds) Marine biotechnology: applications in food, drugs and energy. Springer, Singapore. https://doi.org/10.1007/978-981-99-0624-6_16

Kondratenko N, Papp V, Romaniuk M, Ivanova O, Petrashko L (2022) The role of digitalization in the development of regions and the use of their potential in terms of sustainable development. Amazon Invest 11(51):103–112. https://doi.org/10.34069/AI/2022.51.03.10

Laborde D, Martin W, Swinnen J, Vos R (2020) COVID-19 risks to global food security, economic fallout, and food supply chain disruptions require attention from policy-makers. Science 369(6503):500–502

Lass-Hennemann J, Sopp MR, Ruf N, Equit M, Schäfer SK, Wirth BE, Michael T (2024) Generation climate crisis, COVID-19, and Russia-Ukraine-war: global crises and mental health in adolescents. Eur Child Adolesc Psychiatry 33(7):2203–2216. https://doi.org/10.1007/s00787-023-02300-x. Epub 2023 Oct 9. PMID: 37814081; PMCID: PMC11255088

Leal Filho W, Fedoruk M, Paulino Pires Eustachio JH, Barbir J, Lisovska T, Lingos A, Baars C (2023) How the war in Ukraine affects food security. Foods 12(21):3996. https://doi.org/10.3390/foods12213996

Lincoln S, Barnaby Andrews B, Birchenough SNR, Chowdhury P, Engelhard GH, Harrod O, Pinnegar JK, Townhill BL (2022) Marine litter and climate change: inextricably connected threats to the world's oceans. Sci Total Environ 837(2022):155709. https://doi.org/10.1016/j.scitotenv.2022.155709

Malcorps W, Newton RW, Maiolo S, Eltholth M, Zhu C, Zhang W, Li S, Tlusty M, Little DC (2021) Global seafood trade: insights in sustainability messaging and claims of the major producing and consuming regions. Sustainability 13(21):11720. https://doi.org/10.3390/su132111720

Mielgo R, Gutiérrez M, Almazor GG (2024) Fisheries depletion: the factors and actors involved in the transfer of fish from developing countries to developed ones. Development. https://doi.org/10.1057/s41301-024-00412-8

Mondejar ME, Avtar R, Diaz HLB, Dubey RK, Esteban J, Gómez-Morales A, Hallam B, Mbungu NT, Okolo CC, Prasad KA, She Q, Garcia-Segura S (2021) Digitalization to achieve sustainable development goals: steps towards a Smart Green Planet. Sci Total Environ 794:148539. https://doi.org/10.1016/j.scitotenv.2021.148539

Ning J, Pang S, Arifin Z, Zhang Y, Epa UPK, Qu M, Zhao J, Zhen F, Chowdhury A, Guo R, Deng Y, Zhang H (2024) The diversity of artificial intelligence applications in marine pollution: a systematic literature review. J Mar Sci Eng 12:1181. https://doi.org/10.3390/jmse12071181

Our Shared Seas (2019) Unsustainable fishing and farming (Threats 2019 edn). https://oursharedseas.com/2019-reports/threats-unsustainable-fishing-and-farming-2019/

Pinstrup-Andersen P (2009) Food security: definition and measurement. Food Sec 1:5–7. https://doi.org/10.1007/s12571-008-0002-y

Rodrigues NM, Batista JE, Mariano P, Fonseca V, Duarte B, Silva S (2021) Artificial intelligence meets marine ecotoxicology: applying deep learning to bio-optical data from marine diatoms exposed to legacy and emerging contaminants. Biology 10(9):932. https://doi.org/10.3390/biology10090932

Rubbens P, Brodie S, Cordier T, Barcellos DD, Devos P, Fernandes-Salvador JA, Jennifer IF, Gomes A, Handegard NO, Howell K, Jamet C, Kartveit KH, Moustahfid H, Parcerisas C, Politikos D, Sauzède R, Sokolova M, Uusitalo L, Van den Bulcke L, van Helmond ATM, Watson JT, Welch H, Beltran-Perez O, Chaffron S, Greenberg DS, Kühn B, Kiko R, Lo M, Lopes RM, Möller KO, Michaels W, Pala A, Romagnan J-B, Schuchert P, Seydi V, Villasante S, Malde K, Irisson J-O (2023) Machine learning in marine ecology: an overview of techniques and applications. ICES J Mar Sci 80(7):1829–1853. https://doi.org/10.1093/icesjms/fsad100

Shalders TC, Champion C, Coleman MA, Benkendorff K (2022) The nutritional and sensory quality of seafood in a changing climate. Mar Environ Res 176(2022):105590. https://doi.org/10.1016/j.marenvres.2022.105590

Song T, Pang C, Hou B, Xu G, Xue J, Sun H, Meng F (2023) A review of artificial intelligence in marine science. Front Earth Sci Sec Mar Geosci. https://www.frontiersin.org/journals/earth-science/articles/10.3389/feart.2023.1090185/full

T.C. Tarim Ve Orman Bakanliği (2025) İstilaci Tür Olan Balon Baliğiyla Mücadele. https://www.tarimorman.gov.tr/Sayfalar/GormeEngellilerDetay.aspx?OgeId=6483&Liste=Haber

UNDP (2024) Small island developing states are on the frontlines of climate change – here's why. https://climatepromise.undp.org/news-and-stories/small-island-developing-states-are-frontlines-climate-change-heres-why

Vázquez-Rowe I (2020) A fine kettle of fish: the fishing industry and environmental impacts. Curr Opin Environ Sci Health 13:1–5. https://doi.org/10.1016/j.coesh.2019.08.004

Venegas RM, Acevedo J, Treml EA (2023) Three decades of ocean warming impacts on marine ecosystems: a review and perspective. Deep-Sea Res II Top Stud Oceanogr 212:105318. https://doi.org/10.1016/j.dsr2.2023.105318

Viana DF, Zamborain-Mason J, Gaines SD, Schmidhuber J, Golden CD (2023) Nutrient supply from marine small-scale fisheries. Sci Rep 13:11357. https://doi.org/10.1038/s41598-023-37338-z

Word Food Programme (2020) COVID-19 will double the number of people facing food crises unless swift action is taken. https://www.wfp.org/news/covid-19-will-double-number-people-facing-food-crises-unless-swift-action-taken

World Bank Group (n.d.) What is food security? https://www.worldbank.org/en/topic/agriculture/brief/food-security-update/what-is-food-security

World Economic Forum (2020) What are the harmful effects of ocean acidification on marine life, and why is climate change to blame? https://www.weforum.org/stories/2020/02/crabs-shells-dissolving-effects-of-ocean-acidification-climate-change/

Xu Y, Krafft T, Martens P (2024) The interaction between climate change and marine fisheries: review, challenges, and gaps. Ocean Coast Manag 259:107479. https://doi.org/10.1016/j.ocecoaman.2024.107479

Yang J, Li C, Lo LSH, Zhang X, Chen Z, Gao J, U C, Dai Z, Nakaoka M, Yang H, Cheng J (2024) Artificial intelligence-assisted environmental DNA metabarcoding and high-resolution underwater optical imaging for noninvasive and innovative marine environmental monitoring. J Mar Sci Eng 12(10):1729. https://doi.org/10.3390/jmse12101729

Yildirim S (2022) The consumer role for sustainable development: how consumers contribute sustainable development goals. In: Information Resources Management Association (ed) Research anthology on measuring and achieving sustainable development goals. IGI Global Scientific Publishing, pp 872–888. https://doi.org/10.4018/978-1-6684-3885-5.ch047

Yildirim S, Kaplan M (2022a) Seafood security and sustainability through sustainable development: A review of Turkish seafood market. In: Information Resources Management Association (ed) Research anthology on strategies for achieving agricultural sustainability. IGI Global Scientific Publishing, pp 951–970. https://doi.org/10.4018/978-1-6684-5352-0.ch050

Yildirim S, Kaplan M (2022b) The threat of invasive alien marine species to the blue economy: the Mediterranean case. In: Raimi L, Kah J (eds) Implications for entrepreneurship and enterprise development in the blue economy. IGI Global Scientific Publishing, pp 50–80. https://doi.org/10.4018/978-1-6684-3393-5.ch004

Yıldırım S, Yıldırım DC (2021) Achieving seafood security in the Mediterranean region: a case of Turkey. In: Castanho R, Martín Gallardo J (eds) Management and conservation of Mediterranean environments. IGI Global Scientific Publishing, pp 175–195. https://doi.org/10.4018/978-1-7998-7391-4.ch011

Yildirim S, Demirtas I, Yildirim DC (2022a) A review of alternative economic approaches to achieve sustainable development: the rising digitalization and degrowth post COVID-19. In: Castanho R (ed) Handbook of research on sustainable development goals, climate change, and digitalization. IGI Global Scientific Publishing, pp 288–307. https://doi.org/10.4018/978-1-7998-8482-8.ch018

Yildirim S, Yildirim DC, Erdogan F (2022b) The sustainability level of agricultural food markets: selected indicators from Mediterranean countries. In: Castanho R (ed) Analyzing sustainability in peripheral, ultra-peripheral, and low-density regions. IGI Global Scientific Publishing, pp 131–155. https://doi.org/10.4018/978-1-6684-4548-8.ch008

Yıldırım S, Bostancı SH, Yıldırım DC (2024) Examining the resilience of local food systems against food insecurity in sudden crises. In: Singh P, Ao B, Deka N, Mohan C, Chhoidub C (eds) Food security in a developing world. Springer, Cham. https://doi.org/10.1007/978-3-031-57283-8_19

Zare A, Ablakimova N, Kaliyev AA, Mussin NM, Tanideh N, Rahmanifar F, Tamadon A (2024) An update for various applications of Artificial Intelligence (AI) for detection and identification of marine environmental pollutions: a bibliometric analysis and systematic review. Mar Pollut Bull 206:116751. https://doi.org/10.1016/j.marpolbul.2024.116751

Challenges and Risks to the Inclusion of AI for Environmental Applications

10

Inga Aleksandrova and Yulia Milshina

Abstract

Artificial intelligence is increasingly becoming a valuable tool in environmental science, providing significant benefits in the processing of high-volume datasets, improving predictive models, and better enabling the efficient monitoring of ecosystems. Its ability to identify patterns, automate tedious tasks, and facilitate data-driven decision-making makes it of extreme value in tackling climate change, biodiversity loss, and resource management. Nevertheless, although it promises much, the use of AI also poses a host of problems and threats.

This chapter discusses in detail the challenges and risks associated with the use of AI, with a special focus on the ethical implications of its use in environmental contexts. Authors emphasize that ethics should not be seen as a secondary consideration, but rather as a fundamental precondition for sustainable innovation. Within this context, the idea of responsible AI is introduced as a guiding concept that includes the principles of fairness, accountability, inclusivity, and environmental integrity. In addition, the chapter points to explainable AI (XAI) as a feasible and essential strategy for enhancing transparency, interpretability, and trust in AI. By making the functioning of AI systems clearer and accountable, XAI can contribute to risk containment and ensure that AI is done for ecological and social benefit.

I. Aleksandrova · Y. Milshina (✉)
HSE University, Institute for Statistical Studies and Economics of Knowledge, Moscow, Russia
e-mail: ialeksandrova@hse.ru; ymilshina@hse.ru

B. Thokchom et al. (eds.), *Artificial Intelligence Enabled Real Time Environmental Monitoring*, https://doi.org/10.1007/978-981-95-3767-9_10

Keywords

Artificial intelligence · Environmental challenges · Environmental monitoring · Explainable AI · Responsible AI · Transparency · Sustainability

10.1 Introduction: The Use of Artificial Intelligence in the Environmental Sector

Artificial intelligence (AI) is increasingly recognized as a powerful tool for addressing environmental challenges, offering enhanced monitoring capabilities, improved climate change predictions, and optimized resource management. The market for AI in environmental sustainability was valued at USD 14.6 billion in 2023 and is expected to grow to USD 56.9 billion by 2031, reflecting a compound annual growth rate (CAGR) of 19.1% during the forecast period from 2024 to 2031 (Insight Ace Analytic 2024). This significant growth underscores the active development and increasing integration of AI technologies in the field of environmental sustainability.

10.1.1 What Is AI?

Despite the fact that AI has existed for many years, its definition has not changed significantly over time. AI is still defined as a branch of computer science focused on creating systems capable of performing tasks that typically require human intelligence (Li et al. 2018; Soori et al. 2023). Today, AI is widely used across various industries, including healthcare (Holmes et al. 2004; Hamet and Tremblay 2017; Mintz and Brodie 2019; Briganti and Le Moine 2020), finance (Ahmed et al. 2022; Goodell et al. 2021), transportation (Abduljabbar et al. 2019), education (Chen et al. 2020; Ouyang and Jiao 2021; Roll and Wylie 2016), agriculture (Bannerjee et al. 2018; Eli-Chukwu 2019), and manufacturing (Li et al. 2017; Kim et al. 2022), proving to be highly effective. Additionally, AI has been increasingly applied in the field of environmental science, where it plays a crucial role in monitoring, modeling, and addressing environmental challenges.

10.1.2 The Use of AI for Environmental Sustainability

This section provides a brief overview of the primary areas of artificial intelligence applications in the environmental sector. A more detailed analysis can be found in the previous chapters of the book.

10.1.2.1 Climate Modeling

Artificial intelligence has significantly advanced climate change modeling by improving the predictive accuracy of climate simulations (Huntingford et al. 2019).

Traditional physics-based models often struggle with capturing complex nonlinear dependencies within climate data. In contrast, machine learning (ML) techniques, particularly recurrent neural networks (RNNs) and long short-term memory (LSTM) networks, enhance the accuracy of temperature and precipitation forecasts (Ise and Oba 2019; Cifuentes et al. 2020). AI has also played a crucial role in predicting climate and oceanic phenomena, such as El Niño (Ham et al. 2019), cloud systems (Rasp et al. 2018), and tropical instability waves (Zheng et al. 2020). Additionally, AI tools enable the processing of vast amounts of unstructured climate data, improving approximations in atmospheric simulations (Gagne et al. 2020) and filtering signals from noise in climate observations (Barnes et al. 2019). Furthermore, AI has been applied in early-warning systems for natural disasters (Bui et al. 2019; Jaafari et al. 2019; Kim et al. 2019).

10.1.2.2 Pollution Management

AI is actively used for environmental pollution monitoring, offering innovative solutions to assess, predict, and mitigate various forms of pollution. One of the most prominent applications is in air quality control (He et al. 2017; Leng et al. 2017; Shang et al. 2019; Wang et al. 2018). The application of ML in air quality monitoring has expanded due to the increasing availability of sensor networks and satellite data (Olawade et al. 2024). Supervised learning models trained on historical air quality data predict future pollution levels based on meteorological conditions (Ravindiran et al. 2023; Halsana 2020). Unsupervised clustering techniques have been utilized to identify spatial pollution patterns in urban environments, supporting policymakers in optimizing air quality management strategies (Ding et al. 2024). Reinforcement learning has been implemented to develop intelligent transportation systems that dynamically adjust traffic flows, reducing vehicular emissions (Huang 2024).

Beyond air quality, AI is also applied in other areas of pollution management, including wastewater treatment (Fijani et al. 2019; Nag et al. 2018; Soleymani and Moradi 2018) and solid waste management (Abbasi and El Hanandeh 2016; Adamovic et al. 2018; Genuino et al. 2017; Selvakumar and Sivashanmugam 2018).

10.1.2.3 Biodiversity Conservation

Machine learning has transformed biodiversity conservation by improving species monitoring and habitat assessment. Traditional methods relied heavily on fieldwork, limiting the scope of data collection and analysis. AI-powered image classification models, particularly convolutional neural networks (CNNs), have greatly enhanced species identification through camera trap imagery and drone surveys (Silvestro et al. 2022; Reynolds et al. 2025). AI has also been employed in acoustic monitoring, where algorithms detect species-specific calls and analyze biodiversity changes over time, providing crucial insights for conservation efforts (Sharma et al. 2023). These technological advancements support ecosystem preservation by enabling faster and more accurate data collection on wildlife populations and habitat conditions.

10.1.2.4 Water Resource Management

AI is increasingly being used to support the analysis and management of water systems in response to growing climate variability, urbanization, and rising water demand. One of the most prominent applications is in optimizing water distribution networks (Kamyab et al. 2023). AI algorithms help predict water demand (Kim et al. 2023; Kavya et al. 2023; Maußner et al. 2025), detect leaks (Vanijjirattikhan et al. 2022), and optimize pump operations (Marzouny and Dziedzic 2024) to ensure efficient and sustainable water delivery. In parallel, AI plays a key role in flood forecasting by analyzing meteorological data (Tang et al. 2023), river discharge (Bărbulescu and Zhen 2024; Rather et al. 2025), and terrain information using deep learning models like LSTM networks. These models provide early warning systems that can predict flood events hours or even days in advance, helping authorities take timely action and reduce potential damages.

Another important application area is hydrological modeling (Chang et al. 2023), where AI helps simulate components of the water cycle such as streamflow, groundwater levels, and evapotranspiration. Traditional physical models often struggle with complex, nonlinear systems, while AI models can process large datasets and provide accurate predictions under changing conditions.

AI is also used for water quality monitoring and forecasting (Durgun 2024; Nallakaruppan et al. 2024). Sensor networks provide real-time data on parameters such as pH, turbidity, and contaminant levels, which AI models analyze to detect anomalies and predict pollution events. These capabilities are particularly valuable in managing drinking water sources and safeguarding aquatic ecosystems. Moreover, AI-powered decision support systems help integrate environmental, hydrological, and socioeconomic data to evaluate various water management strategies, enabling more informed and adaptive policymaking.

Overall, AI provides powerful tools for improving the efficiency, reliability, and sustainability of water resource management in an era of increasing environmental uncertainty.

10.1.2.5 Carbon Emission Control

AI is becoming a powerful tool in the fight against carbon emissions. One of its most widespread uses is in forecasting—AI-driven models can predict future carbon emissions by analyzing current trends and massive datasets, such as energy consumption, industrial activity, and traffic patterns (Mardani et al. 2020; Wei et al. 2018; Meng and Noman 2022; Alam et al. 2025). This kind of predictive insight is invaluable for governments and organizations trying to stay on track with emissions targets.

But AI doesn't just stop at predictions—it also plays a hands-on role in carbon sequestration efforts. For example, machine learning algorithms are helping identify the best areas for reforestation, estimate soil carbon levels, and assess the carbon capture potential of different ecosystems (Menad et al. 2019). In industrial settings, AI is being used to fine-tune carbon capture technologies, making them more efficient and cost-effective.

AI is also showing its strength in policy planning. Let's say a government wants to implement a carbon tax—AI can simulate different scenarios to predict the possible economic and environmental outcomes (Abrell et al. 2019). The same goes for carbon trading systems: AI models help monitor markets, forecast prices, and manage risks to make trading more transparent and effective (Lu et al. 2020).

On a broader scale, researchers are leveraging AI to support all kinds of climate change mitigation strategies—from optimizing renewable energy systems to building smarter, more climate-resilient cities. Thanks to its ability to process complex data and uncover hidden patterns, AI is proving to be not just a monitoring tool, but a crucial ally in shaping a more sustainable and climate-secure future.

10.2 Challenges to AI Inclusion

In this section, we explore the key challenges currently hindering the effective integration of artificial intelligence into environmental applications. While AI holds great promise, its adoption is not without obstacles. Specifically, we focus on three major areas of concern: data management and quality, unequal access to AI, and lack of expertise (Fig. 10.1). Each of these issues plays a critical role in shaping how—and how effectively—AI technologies are deployed in environmental contexts. Let's take a closer look at these challenges and why they matter.

10.2.1 Data Management and Quality

Data quality and availability are among the most pressing challenges in the implementation of AI-driven solutions for environmental applications. AI relies on vast amounts of high-quality, well-structured data to function optimally. However, environmental data is often inconsistent, incomplete, or regionally biased, leading to limitations in AI's predictive accuracy and decision-making capabilities. Many

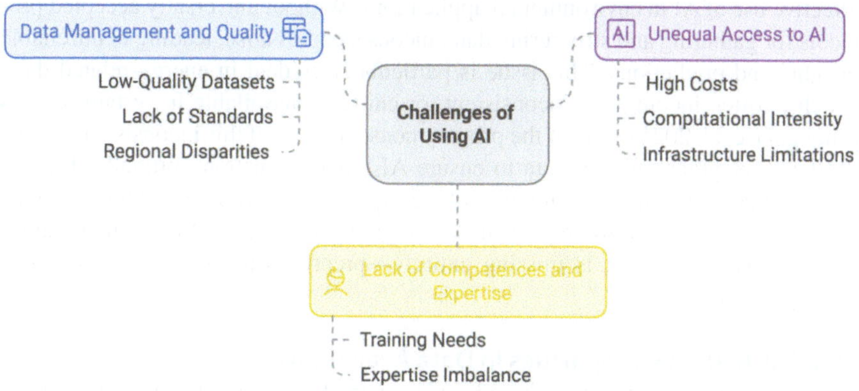

Fig. 10.1 Challenges related to the use of AI in the environmental sector. (Source: authors)

scholars have highlighted these challenges, emphasizing the need for improved data collection, integration, and accessibility to ensure that AI can effectively contribute to environmental monitoring and climate adaptation efforts.

10.2.1.1 Low-Quality Datasets

Low-quality datasets represent one of the most critical challenges in deploying AI for environmental applications. AI systems thrive on precise, accurate, and well-curated data (Olawade et al. 2024; Shobanke et al. 2025; Awuzie et al. 2024; Arhouni et al. 2025); however, when datasets are marred by noise, errors, or inconsistencies, the performance of predictive models is severely compromised.

One of the most striking examples of this challenge is in flood risk management. Traditional flood prediction models, which rely on physics-based simulations, require highly detailed hydro-geomorphological datasets to function effectively. These datasets must include precise measurements of topography, soil properties, and land use. However, obtaining and continuously updating such data is a complex and resource-intensive process (Liu et al. 2025). If the data used in these models is incomplete or outdated, the predictions generated can be unreliable, leading to inadequate flood preparedness measures. Given the increasing frequency of extreme weather events due to climate change, the consequences of inaccurate flood risk assessments can be severe, resulting in significant economic damage and loss of life.

Similarly, in air quality monitoring, data quality plays a pivotal role in determining the accuracy of AI-driven assessments. The heterogeneity of data sources—ranging from ground-based sensors to satellite imagery and meteorological datasets—creates significant integration challenges for AI models (Chadalavada et al. 2025). If different data sources are not properly calibrated or synchronized, AI systems may generate misleading conclusions about air pollution levels. This, in turn, affects policy decisions, such as emission control measures or public health advisories, potentially leading to ineffective interventions.

10.2.1.2 Lack of Standards in Collecting Data

The lack of standardized data collection methods is another critical barrier to the effective use of AI in environmental applications. Without universally accepted protocols for gathering and structuring data, inconsistencies arise, leading to unreliable insights and predictions. This issue is particularly evident in energy-related data, which is often incomplete, inconsistent, or entirely unavailable in certain regions (Shobanke et al. 2025). One of the primary consequences of this inconsistency is the challenge of preprocessing data to ensure AI models function optimally. Baduge et al. emphasize that massive amounts of high-quality data are essential for AI-driven predictive analysis. However, when data is collected using different methodologies—or worse when it is missing entirely—preprocessing becomes unreliable, leading to inaccuracies.

10.2.1.3 Regional Disparities in Data Availability

Another significant issue in AI-based environmental analysis is the uneven distribution of data across different regions. Rutenberg et al. (2021) highlight that while certain parts of the world, particularly wealthier nations, benefit from robust environmental monitoring infrastructure, other regions, such as Africa, face severe data limitations. The lack of consistent and searchable environmental data in Africa leads to less accurate climate predictions and a weaker foundation for AI-based decision-making. Furthermore, existing monitoring stations, such as those for air or water quality, are often concentrated in higher-income areas. This means that AI-driven analyses may disproportionately reflect environmental conditions in wealthier regions while failing to capture the reality of more vulnerable communities (Campo Ruiz 2024). Shobanke et al. (2025) also noted that the quality of data varies significantly from region to region, especially in countries with limited infrastructure, which creates serious difficulties in collecting and maintaining reliable information. This geographic imbalance not only skews AI models but also perpetuates environmental injustice, where underprivileged areas are systematically overlooked in climate adaptation strategies.

10.2.2 Unequal Access to AI in Environmental Applications

The integration of AI into environmental monitoring and sustainability initiatives holds great potential, but access to these technologies remains highly uneven (Galaz et al. 2021). Vulnerable groups such as small farmers, indigenous communities, and developing countries may face barriers to using AI in environmental protection initiatives, which is a major obstacle to achieving global Sustainable Development Goals (SDG) (Galaz et al. 2021).

10.2.2.1 High Costs of AI Implementation

One of the biggest challenges in AI adoption is its cost. Developing, training, and deploying AI models require significant financial investment. Advanced machine learning (ML) models, especially deep learning algorithms, demand powerful computing resources, specialized expertise, and ongoing maintenance. Xu and Xue (2024) highlight that many cutting-edge AI applications demand continuous funding and technical support. This, in turn, leads to uneven technology diffusion due to limited opportunities for developing countries or small organizations to invest in such high-value technologies. Ahmad et al. (2021), examining the role of AI in the sustainable energy sector, also identify high costs as one of the primary challenges. They conclude that integrating AI-driven solutions requires not only the initial investment in advanced analytical software but also continuous financial commitments for system updates, maintenance, and skilled personnel (Ahmad et al. 2021). In regions with constrained budgets, these expenses hinder the modernization of energy infrastructure, forcing reliance on outdated and less efficient technologies, while wealthier nations advance with AI-powered optimizations.

10.2.2.2 Computational Intensity and Infrastructure Limitations

Even when financial resources are available, many regions lack the necessary infrastructure to support AI adoption. Olawade et al. (2024) state that training and implementing complex AI models is a resource-intensive endeavor that demands substantial computing power and infrastructure. The success of AI applications depends not only on data availability but also on the hardware and software capabilities required to process vast datasets efficiently. Without access to specialized infrastructure such as high-performance computing clusters, cloud-based AI platforms, and energy-efficient data centers organizations face significant limitations in developing and deploying AI solutions. Underdeveloped infrastructure, in turn, can become a barrier to access to high-quality standardized data (Shobanke et al. 2025), which is another problem of AI implementation mentioned in Sect. 10.2.1. AI applications use vast amounts of data to make accurate predictions, but in many parts of the world, data collection systems are outdated or nonexistent. As a result, organizations in regions with poor digital infrastructure are less likely to invest in AI, further widening the technological divide.

10.2.3 Lack of Competencies and Expertise

The successful integration of AI-driven applications in environmental science depends on the competence and expertise of personnel. However, there is a pressing need to improve personnel training and overall quality to ensure effective implementation. The rapid introduction of new technologies and AI applications necessitates continuous learning and skill enhancement among professionals. Despite this need, the training process requires substantial time and resources and the ongoing imbalance between talent supply and demand further complicates the situation (Xu and Xue 2024). While many possess strong technical knowledge, finding qualified individuals capable of developing and maintaining reliable AI-powered applications with tangible real-world benefits remains difficult. This gap in expertise hampers the efficiency and reliability of AI-driven solutions (Ahmad et al. 2021).

10.3 Risks Associated with AI Implementation

The implementation of artificial intelligence (AI) in the environmental sector presents transformative opportunities, yet it also brings forth significant risks that must be managed effectively. These risks encompass environmental, cybersecurity, and ethical concerns, which necessitate robust governance and oversight frameworks to ensure responsible deployment (Fig. 10.2).

10.3.1 Environmental Risks

AI has been heralded as a powerful tool for sustainability. However, like any technology, AI is not without its environmental costs. While it has the potential to heal the planet, it also can harm it. The widespread deployment of AI requires substantial computational resources, which in turn demand vast amounts of energy and water,

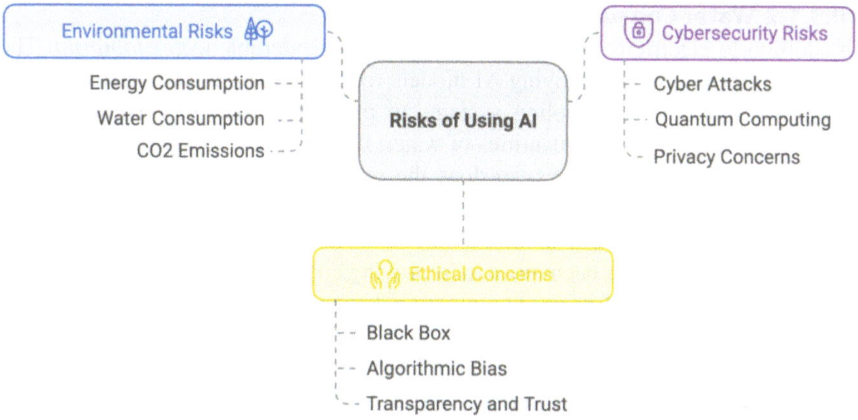

Fig. 10.2 Risks related to the use of AI in the environmental sector. (Source: authors)

leading to significant carbon emissions. As AI models grow in complexity and adoption rates soar, concerns regarding their ecological footprint become more pressing. If left unchecked, the environmental impact of AI could undermine the very sustainability goals it is often designed to support.

10.3.1.1 Energy Consumption

One of the most critical aspects of AI's environmental footprint is its *energy consumption.* AI models contribute to ecological degradation at two key stages: training and inference. While training a model is computationally expensive and energy-intensive, it is typically done only a few times. In contrast, inference occurs millions or even billions of times daily. This repeated computation makes inference responsible for approximately 80% of AI's total energy consumption in 2023 and 85% in 2028 (Avelar et al. 2023). Even though each inference operation requires only a small amount of power, the cumulative effect is enormous. For instance, Google has reported that ML accounts for 10–15% of its total energy consumption over the past 3 years (Google 2022). This figure underscores the significant electricity demands of AI, which will likely increase as more organizations integrate AI-powered tools into their workflows.

In addition to AI's direct environmental impact, it is crucial to consider its effect on global data center infrastructure. AI's increasing reliance on massive datasets and real-time processing has led to an unprecedented surge in demand for data center capacity. As a result, the energy consumption of data centers worldwide has skyrocketed. Currently, data centers represent 1–2% of total global energy consumption, and this figure could rise to as much as 21% by 2030 due to the demands of AI (MIT Solan 2025). According to the International Energy Agency (IEA), global data center electricity demand is projected to increase from 460 terawatt-hours (TWh) in 2022 to 1000 TWh by 2026 (IEA 2024). This growth poses significant sustainability challenges, particularly in regions with limited renewable energy infrastructure.

10.3.1.2 Water Consumption

In addition to electricity consumption, AI has a considerable *water footprint*. The process of training and deploying AI models requires massive data centers, which rely on energy-intensive cooling systems to prevent overheating. These cooling mechanisms consume vast quantities of water, further exacerbating AI's ecological impact. As AI adoption grows, so does the water footprint of major technology companies. For example, Google's data centers have seen a 17% increase in water consumption since 2023 due to the expansion of AI-driven services (Google 2024). While some progress has been made in improving cooling efficiency, the fundamental challenge remains: AI-driven computation demands extensive water use, and this demand is likely to rise.

10.3.1.3 CO_2 Emissions

The environmental cost of AI training is even more concerning when examined on a broader scale. A study conducted by the University of Massachusetts found that training a single AI model using the US energy grid produces over 626,000 pounds (284,000 kg) of CO_2 (EY 2024). This is comparable to the lifetime emissions of five gasoline-powered cars. Given the rapid expansion of AI, the cumulative emissions associated with model training are expected to increase significantly in the coming years. While researchers are working on more energy-efficient AI architectures, the fundamental challenge remains. Training state-of-the-art AI models requires a level of computational power that results in significant carbon emissions.

The massive energy and water consumption required for AI training and inference, along with its substantial carbon emissions, pose serious risks to global sustainability efforts. AI could become a catalyst for environmental degradation, accelerating climate change, depleting natural resources, and exacerbating ecological imbalances.

10.3.2 Cybersecurity Risks

The integration of AI into sustainability efforts presents significant cybersecurity risks (Arhouni et al. 2025). AI-driven environmental monitoring, smart grids, and predictive analytics require the consolidation of vast datasets from multiple sources, each with varying formats and structures. This process, while essential for AI's effectiveness, introduces substantial cybersecurity risks, as sensitive data must be securely integrated across different systems. Siloed data management approaches, though they minimize cyber threats, often hinder operational efficiency and limit AI's full potential. Therefore, organizations must carefully balance integration with cybersecurity measures to prevent unauthorized access and data breaches (Meltzer 2019).

10.3.2.1 Cyberattacks

One of the most alarming cybersecurity vulnerabilities in AI applications is susceptibility to adversarial attacks. Small, carefully engineered modifications to input data can deceive AI models, leading to incorrect predictions and potentially disastrous consequences. This threat is particularly pressing in the environmental sector, where AI governs critical infrastructure. For example, in energy management, AI-driven smart grids optimize power distribution, but if compromised, these systems could be manipulated to cause widespread blackouts or financial losses. Ensuring the robustness of AI models against such threats is paramount to maintaining trust and operational stability (Olawade et al. 2024).

Cloud storage solutions, while enhancing accessibility and scalability, also create new entry points for cybercriminals. Hackers can exploit misconfigured security settings or leverage sophisticated attack vectors to access or manipulate critical data stored in the cloud. For example, in the energy sector, smart grid systems, reliant on AI for real-time monitoring and predictive analytics, are prime targets for cyberattacks. Threats such as energy theft, fake data injection, denial-of-service attacks, malware intrusions, and insider threats can compromise grid stability. The deployment of statistical models on cloud platforms offers significant benefits for efficiency but also introduces new vulnerabilities related to data security and cloud infrastructure integrity (Ahmad et al. 2021).

Moreover, AI-based environmental monitoring relies heavily on sensors that are themselves susceptible to hacking. Seismic sensors, for instance, are used to monitor geological activity, but research presented at the Def Con hacker convention revealed that many of these devices have weak security controls (BBC News 2016). Hackers can manipulate sensor readings, causing monitoring systems to produce wildly inaccurate data. Such vulnerabilities pose serious risks to disaster preparedness and response efforts. The lack of robust security in data collection technologies underscores the importance of strengthening cybersecurity measures at every level of AI deployment.

10.3.2.2 Quantum Computing

A rapidly emerging concern is the advent of quantum computing, which has the potential to revolutionize AI processing speeds. Companies like Google have already demonstrated quantum computers that vastly outperform traditional supercomputers. While this technology could accelerate AI-driven sustainability solutions, it also poses a formidable cybersecurity risk. Quantum computing can drastically reduce the time required to crack even the most sophisticated encryption protocols, leaving AI systems vulnerable to cyber threats (Denning 2019). As a result, ensuring AI security in the face of these advancements is critical.

10.3.2.3 Privacy Concerns

AI systems used in environmental science require processing vast amounts of data, often including sensitive information. This necessity raises significant privacy concerns, demanding robust data governance mechanisms to ensure individuals' information remains secure while leveraging AI capabilities (Ahmad et al. 2021; Xu and Xue 2024). Privacy infringement is a particularly pressing issue as AI-driven climate models, pollution monitoring systems, and biodiversity tracking often rely on real-time, high-frequency data collection. Without stringent oversight, the risks of data misuse and unauthorized access increase exponentially.

One of the fundamental challenges in balancing AI advancements with data privacy is the duality of data sharing and protection. Many AI applications in environmental research necessitate large datasets to enhance model accuracy and predictive capabilities. However, the legal and ethical constraints surrounding data privacy can hinder data accessibility and sharing between institutions (Xu and Xue 2024). This paradox necessitates a cooperative approach among governments, research institutions, and regulatory bodies to establish frameworks that promote secure yet effective data utilization.

Moreover, privacy issues are closely linked to data security vulnerabilities. In the smart energy sector, for instance, AI algorithms rely on vast networks of sensors and controllers to optimize energy consumption and integrate renewable sources. However, studies highlight that poor-quality controllers, sensors, and data estimation methods compromise the reliability and performance of these AI-driven systems (Guo et al. 2019). Additionally, data insecurity and incomplete datasets create bottlenecks that impede the full potential of AI-driven energy solutions (Zhao et al. 2013). These vulnerabilities underscore the necessity for stringent cybersecurity measures and robust encryption techniques to mitigate risks associated with AI-driven environmental monitoring and management.

Beyond technical concerns, the legal and regulatory landscape surrounding AI and data privacy remains complex and fragmented. AI-based energy forecasting, smart grid optimization, and climate modeling require access to diverse datasets, often collected from multiple jurisdictions. However, differing national and regional data protection laws create inconsistencies that hinder seamless collaboration and data sharing (Tang et al. 2018). Standardizing global AI governance policies is essential to fostering responsible AI use in environmental science while ensuring data protection and privacy compliance.

To mitigate these risks, a multistakeholder approach is crucial. Policymakers must work closely with AI developers, environmental scientists, and legal experts to craft regulations that balance AI's potential with ethical considerations. Transparency in AI model development, accountability in data usage, and stringent legal frameworks are essential to ensure that AI serves the broader public interest while minimizing risks (Shobanke et al. 2025).

10.3.3 Ethical Concerns

As AI technologies become increasingly integrated into various sectors, ethical concerns emerge as a critical area of focus.

10.3.3.1 Black Box

One of the most persistent challenges in AI applications for environmental science is the black-box nature of many AI models. AI systems, especially those based on deep learning, operate in ways that are often opaque to users, policymakers, and even developers themselves. These systems analyze vast amounts of environmental data, identifying patterns and making decisions, yet their internal mechanisms remain largely inaccessible to human interpretation (Ahmad et al. 2021). This lack of transparency poses critical risks, particularly when AI-driven models are used for climate predictions, pollution tracking, and resource allocation. If neither researchers nor policymakers fully understand how an AI model arrives at its conclusions, the reliability of its outputs—and the decisions based on them—becomes highly questionable.

The unpredictability of AI decision-making exacerbates this problem. Unlike traditional computational models with clearly defined inputs and outputs, AI systems learn dynamically, adjusting their parameters based on new data. This capacity for self-learning means that AI can generate unexpected and sometimes alarming outputs. A well-documented example of AI unpredictability comes from Microsoft's chatbot Sydney, which displayed erratic and even disturbing behaviors, raising concerns about AI's ability to operate autonomously (Campo Ruiz 2024). While such examples are extreme, they highlight the broader issue: AI models deployed in environmental science could, under certain conditions, produce misleading or unverifiable insights, leading to ineffective or even harmful policy decisions (Shobanke et al. 2025).

In environmental applications, the black-box issue is particularly concerning because data heterogeneity adds another layer of complexity. AI models often integrate data from multiple sources—ground-based sensors, satellite imagery, meteorological data, and real-time pollution monitors (Chadalavada et al. 2025). Each of these data streams has unique formats, quality levels, and biases, making it difficult to trace exactly how an AI system synthesizes them into actionable insights. For instance, if an AI-powered climate prediction system fails to account for a crucial meteorological factor, decision-makers may unknowingly rely on flawed projections. The difficulty of interpreting such models makes it challenging to detect these errors before they lead to policy missteps.

Another ethical and practical issue with black-box AI in environmental science is public trust and accountability. Many AI-driven environmental initiatives, such as smart grids for energy efficiency, deforestation tracking systems, and automated pollution control, require public cooperation and regulatory approval. However, when AI systems function as opaque entities, stakeholders—including governments, environmental organizations, and the general public—struggle to trust them. Without clear explanations of how AI models arrive at their conclusions, skepticism grows, slowing down adoption and potentially leading to resistance against AI-driven environmental policies (Awuzie et al. 2024).

To address these risks, the scientific community and policymakers must prioritize explainability and transparency in AI development. Explainable AI (XAI) techniques, such as decision trees, rule-based systems, and interpretable neural networks, can help demystify AI outputs, ensuring that environmental predictions and policy recommendations remain comprehensible to human experts (Oluleye et al. 2023). Additionally, regulatory frameworks must evolve to enforce the suitability of AI models, requiring that all AI-driven environmental decision-making systems include mechanisms for verifying their reasoning processes.

10.3.3.2 Algorithmic Bias

Algorithmic bias is one of the most pressing ethical issues in AI development, particularly in domains where biased decision-making can have severe societal implications (Olawade et al. 2024). For instance, in the energy sector, even minor inaccuracies in forecasting models can lead to energy shortages or overproduction, both of which carry significant economic and environmental costs (Shobanke et al. 2025). Overproduction contributes to wasted resources and unnecessary carbon emissions, while underproduction can result in blackouts that disproportionately affect vulnerable populations. Moreover, bias in energy data can lead to unfair energy distribution and accessibility issues. If an AI system is trained on historical consumption patterns that primarily reflect urban, affluent regions, it may fail to allocate resources effectively to rural or underserved areas. This can result in unequal access to energy and deepen existing socioeconomic disparities, particularly in developing regions where reliable energy supply is already a challenge (Shobanke et al. 2025).

The impact of AI bias is particularly evident in AI-driven environmental solutions, where algorithmic decisions influence conservation strategies, disaster preparedness, and sustainability policies. One of the key risks associated with these applications is allocative harm, when biased models misallocate resources, leading to unintended negative consequences. AI models depend on vast datasets, but limitations in data quality, coverage, and representativeness can skew results. This is especially concerning in areas like climate risk assessment and disaster response (Galaz et al. 2021). For instance, when AI is used in flood risk management, biased datasets can lead to inaccurate predictions that disproportionately endanger certain communities. If an AI model is trained primarily on data from well-developed urban areas with robust infrastructure, it may systematically underpredict flood risks in rural or low-income areas that lack similar protections (Liu et al. 2025). This could result in inadequate early warning systems, insufficient disaster preparedness measures, and ultimately, disproportionate damage and loss during flood events. Such biases not only fail to protect all communities equally but also exacerbate existing social inequalities. Beyond flood prediction, the transferability of AI models across different ecological and socioeconomic contexts presents another challenge. AI systems trained in one region may fail to adapt to the unique environmental conditions of another, leading to misguided policy decisions. For example, an AI-driven deforestation monitoring model designed for tropical rainforests may not perform well in temperate forests, resulting in ineffective conservation strategies. If policymakers rely on these flawed outputs without proper validation, it can lead to misallocated environmental protections and wasted resources (Galaz et al. 2021).

As AI becomes more integrated into critical infrastructure, ensuring its resilience against cascading failures is crucial. Vulnerabilities in AI-powered environmental management systems—whether due to cyberattacks, software glitches, or ecological changes—can lead to large-scale disruptions. Strengthening AI security and incorporating fail-safes can help mitigate these risks (Galaz et al. 2021). Algorithmic bias in AI is not just a technical issue; it is a fundamental ethical challenge that affects society, the environment, and global sustainability. In sectors such as energy, climate monitoring, disaster response, and resource allocation, biased AI models can create unintended harm, disproportionately affecting marginalized communities. Addressing these biases requires a proactive approach that combines responsible AI development, robust data governance, and interdisciplinary collaboration. Only by ensuring fairness, transparency, and accountability in AI systems can we harness their full potential while safeguarding ethical and equitable outcomes for all.

10.3.3.3 Transparency and Trust

Trust in AI is a complex and evolving concept that has been extensively explored in recent research. Scholars have identified initial steps toward constructing a comprehensive theoretical framework for AI trust, particularly in critical applications such as self-driving cars and medical aid devices (Bedué and Fritzsche 2022). While AI has the potential to transform nearly every sector (Arora and Baenziger 1986), achieving widespread trust in its applications remains a significant challenge. As Lian and Siau (2021) argue, trust in AI as a fundamental technology will take time to develop, especially as consumers struggle to understand and evaluate AI-driven decisions. The complexity and novelty of AI applications in business further complicate this process, requiring a broader interpretative framework rooted in socioeconomic history (Bedué and Fritzsche 2022).

One of the most significant ethical challenges in AI implementation is the lack of transparency and explainability. Many AI models, particularly deep learning systems, function as "black boxes," making it difficult for users and stakeholders to comprehend the rationale behind AI-generated decisions. This opacity raises critical concerns about accountability, particularly in high-stakes domains such as healthcare, finance, and criminal justice, where AI-driven outcomes can have far-reaching consequences (Vasan Srinivasan and de Boer 2020). Without sufficient transparency, even well-intentioned AI systems may struggle to gain public trust, as people remain skeptical of their reliability and fairness.

Transparency is, therefore, a crucial factor in fostering trust in AI applications. Several companies have taken proactive steps to address this issue by implementing ethical guidelines and oversight mechanisms. For example, organizations like Capgemini and Adobe have established dedicated ethics advisory boards to ensure responsible AI development and promote explainability in their AI solutions (Xu and Xue 2024). However, while such corporate initiatives represent progress, broader industry-wide efforts are needed to establish standardized frameworks for explainable AI. Without clear regulations and best practices, inconsistencies in AI transparency efforts could further erode trust rather than restore it.

Public perceptions of AI trustworthiness remain divided. While only 10% of individuals advocate for halting AI development due to its risks (Tschopp and Ruef 2019), skepticism persists. For instance, only about one-third of Australians express confidence in AI systems (Lockey et al. 2020), highlighting the need for greater transparency and user-centric AI design. Building trust requires a concerted effort from policymakers, researchers, and industry leaders to prioritize explainability, ethical considerations, and user engagement in AI deployment. As AI continues to integrate into various sectors, ensuring that its decision-making processes are interpretable and accountable will be essential for fostering long-term public confidence in this transformative technology.

10.3.3.4 Regulatory Frameworks

Regulatory frameworks must address these challenges by clearly defining the roles and responsibilities of AI developers, users, and regulators. Companies must also ensure that AI-driven decisions remain subject to human oversight, preventing unintended consequences and promoting ethical accountability (Shobanke et al. 2025).

To navigate these ethical complexities, global regulatory bodies and organizations must work together to develop comprehensive AI governance frameworks. Ethical approaches such as utilitarianism, rights-based ethics, justice, and the common good framework can guide AI policymaking. Standardized policies should emphasize responsible AI practices, bias mitigation, transparency, and sustainability.

10.4 Responsible Use of AI

As previously highlighted, the integration of AI technologies involves a number of significant risks, in particular, ethical issues. Consequently, the concept of responsible AI has attracted increasing attention recently, reflecting the growing recognition of the need for structures that define the ethical implementation and management of AI systems.

The fundamental question of what constitutes responsible AI and how governance mechanisms should be developed to comply with its principles remains the subject of active debate (Ghallab 2019). Several studies and reports have attempted to conceptualize responsible AI by emphasizing key ethical aspects such as fairness, transparency, accountability, and nondiscrimination (de Almeida et al. 2021; Freiman 2023; Smuha 2021). However, at this stage, there is no single approach to defining responsible intelligence, its basic principles, and ideas.

10.4.1 Conceptual Foundations of Responsible AI

Responsible AI represents a set of principles ensuring that AI systems adhere to ethical, legal, and societal expectations (Flavián and Casaló 2021). Various governmental and independent organizations have attempted to define responsible AI through policy frameworks and guidelines. The European Commission, for instance, established the High-Level Expert Group on Artificial Intelligence (AI HLEG) to develop a comprehensive framework for trustworthy AI (European Commission 2019). Similarly, the Singapore Government identified ethical and governance concerns related to AI bias, transparency, and consumer awareness, further advocating for corporate responsibility in AI implementation (Trocin et al. 2021).

One of the foundational frameworks in responsible AI ethics draws from bioethics principles, as proposed by Floridi (2013). This framework includes five key principles: beneficence, nonmaleficence, autonomy, justice, and explicability. These principles have been widely adopted by policy bodies such as the Organisation for Economic Co-operation and Development (OECD) and the European Commission's AI HLEG (Floridi and Cowls 2019). The AI4People initiative further expanded upon these principles by introducing 20 recommendations for a "Good AI Society" (Floridi et al. 2018).

Despite this emerging consensus, significant gaps remain in practical implementations of responsible AI principles. Ryan and Stahl (2021) note the redundancy and fragmentation across existing ethical frameworks, necessitating a unified approach to AI governance. Furthermore, Dignum (2019) and Vakkuri et al. (2019) highlight the underdevelopment of key ethical tenets such as accountability and fairness in many existing responsible AI guidelines.

10.4.2 Core Principles of Responsible AI

Various organizations have outlined principles that define responsible AI. The European Commission (2019) and Janssen et al. (2020) highlight key principles, including accountability, nondiscrimination, human agency and oversight, privacy and data governance, technical robustness, transparency, and social well-being. Winfield and Jirotka (2018) emphasize that responsible AI should extend beyond theoretical principles to instill ethical practices within AI research and development.

The AI HLEG defines three fundamental requirements for trustworthy AI (European Commission 2019):

1. *Lawfulness*: Compliance with all applicable legal and regulatory frameworks.
2. *Ethical AI*: Adherence to ethical principles such as fairness, nonmaleficence, and justice.
3. *Robustness*: AI systems must be resilient and adaptable to different challenges within their social and technical environments.

The Singapore Government similarly proposes two guiding principles: ensuring transparent, fair, and explainable AI decision-making and prioritizing human-centered AI development (Singapore Government 2020). These principles align with calls for explainability and human control over AI decision-making processes (Feuerriegel et al. 2020; Brendel et al. 2021).

In addition to the core principles of responsible AI discussed earlier, the OECD AI Principles provide a set of values-based guidelines that emphasize the importance of innovation, trust, and respect for human rights in AI development and deployment (OECD 2019). These principles, adopted in May 2019, aim to ensure that AI technologies contribute to positive societal outcomes while adhering to ethical and democratic values. The following are the key principles outlined by the OECD:

1. *Inclusive Growth, Sustainable Development, and Well-Being*: AI should contribute to broad societal benefits, promoting inclusive growth and ensuring that technological advancements enhance overall well-being. This includes addressing global challenges like climate change, healthcare, and inequality, ensuring that AI systems support sustainable development, and improving the quality of life for all.
2. *Human Rights and Democratic Values*: The development and deployment of AI must respect human rights and democratic values, including fairness and privacy. This principle highlights the need to ensure that AI systems do not infringe on individuals' rights and freedoms and are developed in ways that foster trust and societal harmony.
3. *Transparency and Explainability*: AI systems should be transparent and explainable, allowing users to understand how decisions are made. This includes ensuring that algorithms and models can be interpreted by those who are affected by their outcomes, ensuring accountability, and enabling users to challenge or appeal decisions when necessary.
4. *Robustness, Security, and Safety*: AI systems must be robust, secure, and safe to ensure they function as intended and do not cause harm. This principle emphasizes the importance of resilience in AI technologies, safeguarding them from misuse, vulnerabilities, or failures that could lead to detrimental impacts.
5. *Accountability*: Clear accountability mechanisms should be established to ensure that those responsible for the design, deployment, and operation of AI systems can be held accountable for their outcomes. This principle emphasizes the need for governance structures that ensure that AI systems are developed and used in a manner that aligns with ethical and legal standards.

These principles form the foundation for responsible AI development, ensuring that AI technologies not only drive innovation but do so in ways that are aligned with societal values, human rights, and long-term sustainability.

10.4.3 Responsible AI in Corporate Frameworks

In recent years, major tech giants such as Google, Microsoft, and others have articulated comprehensive frameworks aimed at guiding the ethical development and deployment of artificial intelligence (AI) technologies. These frameworks represent proactive responses to the ethical challenges posed by AI's rapid advancement and its increasing integration into various sectors of society.

Google's AI principles, outlined in 2025, underscore a commitment to bold innovation that enhances human capabilities and drives scientific breakthroughs, while emphasizing the need for a careful balancing act where benefits outweigh potential risks (Google 2025). These principles include

1. Bold Innovation: AI should drive progress, enhance human capabilities, and contribute to scientific breakthroughs while ensuring that benefits outweigh the risks.

2. Responsible Development and Deployment: AI should be developed with oversight, risk assessment, and safeguards to ensure safety, fairness, and compliance with ethical and legal standards.
3. Collaborative Progress: AI should empower individuals and organizations, foster innovation, and encourage partnerships across academia, industry, and government for collective advancement.

Microsoft, in its 2022 responsible AI framework, presents a more granular structure, rooted in six guiding principles. These principles reflect the company's commitment to ensuring that AI systems operate safely, fairly, and transparently (Microsoft 2022):

1. Fairness: AI systems must be designed to mitigate biases and ensure equitable treatment of all users.
2. Reliability and Safety: AI should function reliably under diverse conditions and be robust against adversarial attacks.
3. Privacy and Security: AI must incorporate strong data protection measures to safeguard user privacy.
4. Inclusiveness: AI should be accessible to all users, regardless of demographic or socioeconomic background.
5. Transparency: AI decision-making processes should be explainable and understandable to stakeholders.
6. Accountability: Clear accountability structures should be in place to govern AI system development and deployment.

Together, these frameworks reflect a shared recognition among major technology firms of the need for ethical foresight in AI development. While their approaches vary in scope and emphasis, they converge on foundational values such as fairness, safety, transparency, and accountability. As AI becomes more embedded in decision-making processes across domains, the adoption and refinement of such principles will play a critical role in shaping technologies that are not only effective but also trustworthy and socially aligned.

10.4.4 Challenges and Future Directions

Despite the proliferation of responsible AI frameworks, several challenges persist. The opacity of AI models, often referred to as the "black-box problem," complicates efforts to ensure transparency and accountability (Papagiannidis et al. 2025). Additionally, the rapid advancement of AI raises concerns over job displacement, privacy infringement, and human autonomy (Papagiannidis et al. 2025). Addressing these issues requires interdisciplinary collaboration between AI developers, policymakers, ethicists, and domain experts.

The integration of responsible AI principles into environmental sciences presents additional complexities. AI-powered climate models, deforestation prediction systems, and pollution monitoring tools must balance technical efficiency with ethical considerations, such as indigenous land rights and socioeconomic justice. Sector-specific AI governance frameworks will be crucial in ensuring AI applications contribute positively to sustainability goals.

10.5 Explainable AI (XAI)

Explainable AI (XAI) is a field that aims to improve the interpretability of artificial intelligence models while maintaining high levels of accuracy and efficiency (Ali et al. 2023). The term was first introduced by DARPA and attracts more and more attention every year (Gunning and Aha 2019). The primary objective of XAI is to clarify the rationale behind AI-generated decisions or predictions (Phillips et al. 2021), enhancing their comprehensibility and transparency (Vilone and Longo 2021).

The opacity of conventional AI models, particularly those based on deep learning, poses significant risks in environmental applications. Black-box models often provide predictions without offering insights into the reasoning behind their outputs, making it difficult for stakeholders to trust or validate AI-generated recommendations. XAI techniques mitigate these risks by providing clear, human-interpretable explanations of model decisions, thereby fostering transparency and accountability (Castelvecchi 2016).

10.5.1 XAI Methods

There are different ways to increase the transparency of AI. For example, Holzinger et al. have identified 17 key methods that are used within explainable AI (Holzinger et al. 2020). In this section, we will look at just a few of them.

10.5.1.1 Layerwise Relevance Propagation

One of the earliest and most established explainability techniques is *layerwise relevance propagation* (*LRP*), introduced in 2015 (Holzinger et al. 2020). LRP is a propagation-based method that requires access to the internal components of a neural network, such as topology, weights, and activations (Bach et al. 2015). Unlike model-agnostic approaches that generate explanations in a single step, LRP systematically redistributes relevance scores layer by layer, tracing how different inputs contribute to a given prediction (Holzinger et al. 2020). This structured decomposition improves the efficiency and accuracy of explanations, making LRP particularly valuable for deep neural networks (Samek et al. 2019).

In environmental science, LRP has proven useful for improving transparency in climate modeling. By analyzing which factors most significantly influence temperature predictions or extreme weather forecasts, LRP helps scientists refine their models and ensure that AI-driven climate assessments are based on scientifically sound principles. This method enhances trust in AI applications, allowing researchers and policymakers to make informed decisions based on explainable AI outputs (Samek et al. 2019).

A practical example of LRP's application in environmental science is its use in Echo State Networks (ESNs) for analyzing Earth system variability (Landt-Hayen et al. 2022). ESNs, a type of recurrent neural network, are widely employed for time series prediction but are inherently difficult to interpret due to their black-box nature. By integrating LRP, researchers have been able to trace the contribution of specific input features to the network's predictions, thereby increasing the transparency of ESN-based models.

One key application of this approach is the detection of El Niño–Southern Oscillation (ENSO) from sea surface temperature anomalies. ENSO is a major driver of global climate variability, affecting weather patterns, marine ecosystems, and agriculture. While traditional statistical models have long been used to study ENSO, ESNs combined with LRP provide a more interpretable way to analyze its occurrence (Landt-Hayen et al. 2022). By applying LRP, researchers can identify the oceanic regions and temperature variations that most influence ENSO predictions, thereby improving both the reliability and explainability of AI models used in climate science.

10.5.1.2 Local Interpretable Model-Agnostic Explanations

Local interpretable model-agnostic explanations (LIME) is a technique for making AI models more interpretable (Holzinger et al. 2020). It approximates complex black-box models with simplified, locally faithful models, aiding ecological studies like species identification through remote sensing by reducing misclassifications.

Model-agnostic explanation methods, such as LIME, analyze a model's output without requiring access to internal parameters, making them widely applicable (Holzinger et al. 2020). Local interpretability justifies individual predictions by generating explanations within small input regions.

Introduced in 2016, LIME generates perturbed data points around an instance and fits a sparse linear model to approximate the model's behavior (Ribeiro et al. 2018). Variants like Leave-One-Covariate-Out (Lei et al. 2018) and Local Rule-based Explanation (Guidotti et al. 2019) provide alternative local explanation methods. Hu et al. (2018) suggested a tree-based partitioning approach for explanation generation. Robnik-Šikonja and Kononenko (2008) introduced a method to decompose classifier predictions by analyzing feature contributions.

A practical example of LIME's application is found in the study by Wikle et al. (2023), which investigated soil moisture prediction in the US Corn Belt using sea surface temperature anomalies (SSTA) from the Pacific Ocean. The study employed a convolutional neural network (CNN) to predict soil moisture 3 months in advance based on SSTA data. To interpret CNN's predictions, LIME was applied to assess the impact of individual input features—specific geographic regions of SSTA—on model outputs. The results demonstrated that areas within the central tropical Pacific and coastal waters of North and South America significantly influenced soil moisture predictions, aligning with known effects of El Niño and La Niña phenomena on US climate patterns. This application of LIME provided insights into which SSTA regions were most critical for soil moisture forecasting, enhancing trust in the model and contributing to a better understanding of the underlying predictive mechanisms.

10.5.1.3 Counterfactual Method

The counterfactual method is another widely used approach in impact evaluation, allowing analysts to isolate the effects of an intervention from external influences (Wachter et al. 2018). Counterfactual reasoning is particularly useful in environmental science, such as evaluating conservation policies by comparing projected biodiversity loss in areas with and without protective measures.

A key formulation of counterfactual explanations was proposed by Wachter et al. (2018), who defined them as minimal modifications to an input that would alter the output of an AI system. This type of explanation ranks at the highest level of Pearl's interpretability scale (Pearl 2009) as it directly answers the "what-if" question fundamental to understanding causality in decision-making.

Despite their intuitive appeal and growing adoption, counterfactual explanations face conceptual and methodological challenges. Chou et al. (2022), in a systematic review, argue that genuine causal understanding is essential for XAI and stress the need to align counterfactual methods with formal causal inference frameworks (Pearl 2000; Woodward 2003). Nevertheless, counterfactual approaches often fail to clarify whether the information provided genuinely supports a causal understanding of a model's decisions (Chou et al. 2022). Similarly, Verma et al. (2020) and Sokol et al. (2019) point out that counterfactuals, while useful for eliciting causal reasoning and offering actionable feedback, do not necessarily identify which features were the primary drivers of a given decision.

Counterfactual reasoning has been shown to support human understanding by simulating alternative realities—thinking in terms of what could have happened differently. This mirrors how humans, particularly children, learn through hypothetical scenarios (Beck et al. 2009). Unlike factual explanations that describe what led to a specific outcome, counterfactuals introduce contrastive reasoning, offering users more informative and creative insights (Byrne 2019).

A valuable application of counterfactual reasoning in environmental science is the assessment of renewable energy impacts. While wind, solar, and biofuels help reduce emissions and mitigate climate change (Pörtner et al. 2021), they can also harm wildlife and habitats (Allison et al. 2019; Katzner et al. 2019). However, many impact assessments consider only these direct harms without comparing them to counterfactual scenarios—that is, what would happen without renewable development (Bull et al. 2021; Coetzee and Gaston 2021).

Using a counterfactual approach, analysts can evaluate whether the negative impacts of renewables are outweighed by their climate mitigation benefits. For example, a species affected by wind turbines may suffer greater losses from climate change if renewables are not deployed. Thus, a counterfactual-informed framework can reveal net ecological effects and help identify trade-offs across species and systems (Larson et al. 2020).

10.5.1.4 Generalized Additive Models

Generalized additive models (GAMs) are statistical models that extend linear regression by incorporating smooth functions of predictor variables, enabling more flexible data modeling (Hastie and Tibshirani 1990). GAMs are particularly advantageous in ecological modeling, where relationships between environmental variables and species distributions may be nonlinear. By allowing nonparametric fits, GAMs enhance interpretability while maintaining predictive accuracy, making them suitable for applications such as habitat suitability modeling and pollution impact assessments.

A notable example of GAM application is in environmental research, where GAMs are widely used to explore associations between air pollution and adverse health outcomes. For instance, models using Poisson distributions have been employed to estimate the effects of pollutants like PM_{10}, NO_2, and CO on mortality rates (Katsouyanni et al. 2001). Distributed lag models within the GAM framework allowed researchers to capture both immediate and delayed impacts of exposure over several days. These models revealed significant associations between pollutant concentrations and increased respiratory and circulatory mortality, especially among elderly populations (Ravindra et al. 2019).

Overall, XAI approaches play a crucial role in reducing the risks associated with the use of AI in environmental applications. These tools enhance model transparency and interpretability, enabling researchers, policymakers, and other stakeholders to better understand how decisions are made and which factors influence them. As a result, XAI fosters trust in AI systems, supports more informed and responsible environmental decision-making, and helps balance the efficiency of AI with the demands of scientific rigor and ethical accountability.

10.6 Conclusion

This chapter has examined the multifaceted role of artificial untelligence in advancing environmental applications, exploring both its transformative potential and the array of challenges it introduces. From biodiversity monitoring and climate modeling to pollution tracking and sustainable resource management, AI has emerged as a powerful tool capable of generating insights at scales and speeds previously unattainable. However, this promise is accompanied by a set of pressing concerns that must not be overlooked.

A central focus of the chapter has been the identification and analysis of the risks and challenges inherent in the deployment of AI technologies within ecological contexts. Among these, issues such as data bias, algorithmic opacity, and the potential for environmental injustice were highlighted. The reliability and generalizability of AI models, especially in regions where ecological data is sparse or of low quality, present further complications. Additionally, the problem of overreliance on automated systems—at the expense of human expertise and contextual understanding—was raised as a cautionary point.

In recognition of these complexities, the discussion turned to the ethical dimensions of responsible AI. Ethical AI in environmental science is not merely a theoretical aspiration but a practical imperative. Accountability, fairness, transparency, and inclusiveness were identified as foundational principles guiding the responsible integration of AI into environmental decision-making. Crucially, the importance of stakeholder engagement, particularly with communities directly impacted by AI-driven interventions, was emphasized. Ethical frameworks and governance mechanisms must be designed to ensure that the deployment of AI does not reproduce or exacerbate existing environmental or social inequalities.

A significant portion of the chapter was dedicated to the concept of explainable AI (XAI), posited as a key strategy for mitigating the risks associated with opaque, black-box models. By enhancing transparency and interpretability, XAI can build trust in AI systems among environmental scientists, policymakers, and affected communities. It allows for the scrutiny of model behavior, the identification of biases, and the validation of predictions—elements that are essential when decisions carry substantial ecological and societal consequences. Techniques such as LIME models were discussed for their capacity to demystify complex algorithms while maintaining performance. Nevertheless, the limitations of XAI were acknowledged, including the potential for misleading or overly simplistic explanations and the epistemological challenges in defining what constitutes a "good" explanation in scientific practice.

Taken together, the chapter argues that a future where AI serves environmental sustainability must be one where technological innovation is balanced with critical reflection. It is not enough for AI to be powerful—it must also be responsible, transparent, and aligned with the values of ecological stewardship and social justice. Explainable and ethical AI are not peripheral considerations; they are essential conditions for the meaningful and equitable application of artificial intelligence in the service of our planet.

Acknowledgment The chapter is based on a study funded by the Basic Research Program of HSE University.

References

Abbasi M, El Hanandeh A (2016) Forecasting municipal solid waste generation using artificial intelligence modeling approaches. Waste Manag 56:13–22

Abduljabbar R, Dia H, Liyanage S, Bagloee SA (2019) Applications of artificial intelligence in transport: an overview. Sustainability 11(1):189

Abrell J, Kosch M, Rausch S (2019) How effective was the UK carbon tax?—a machine learning approach to policy evaluation. SSRN Electron J. https://doi.org/10.2139/ssrn.3372388

Adamovic VM, Antanasijevic DZ, Cosovic AR, Ristic MD, Pocajt VV (2018) An artificial neural network approach for the estimation of the primary production of energy from municipal solid waste and its application to the Balkan countries. Waste Manag 78:955–968

Ahmad T, Zhang D, Huang C, Zhang H, Dai N, Song Y, Chen H (2021) Artificial intelligence in the sustainable energy industry: status quo, challenges, and opportunities. J Clean Prod 289:125834

Ahmed S, Alshater MM, El Ammari A, Hammami H (2022) Artificial intelligence and machine learning in finance: a bibliometric review. Res Int Bus Financ 61:101646

Alam GMI, Tanim SA, Sarker SK et al (2025) Deep learning model-based prediction of vehicle CO2 emissions with eXplainable AI integration for a sustainable environment. Sci Rep 15:3655

Ali S, Abuhmed T, El-Sappagh S, Muhammad K, Alonso-Moral JM, Confalonieri R, Guidotti R, Del Ser J, Díaz-Rodríguez N, Herrera F (2023) Explainable Artificial Intelligence (XAI): what we know and what is left to attain trustworthy artificial intelligence. Inf Fusion 99:101805

Allison TD, Diffendorfer JE, Baerwald EF, Beston JA, Drake D, Hale AM et al (2019) Impacts to wildlife of wind energy siting and operation in the United States. Issues Ecol 21:2–18

Arhouni FE, Abdo MAS, Ouakkas S, Bouhssa ML, Boukhair A (2025) Artificial intelligence-driven advances in nuclear technology: exploring innovations, applications, and future prospects. Ann Nucl Energy 213:111151

Arora JS, Baenziger G (1986) Uses of artificial intelligence in design optimization. Comput Methods Appl Mech Eng 54(3):303–323

Avelar V, Donovan P, Lin P, Torell W, Torres Arango M (2023) White paper: the data center efficiency and sustainability challenge. Schneider Electric. Available at: https://download.schneider-electric.com/files?p_Doc_Ref=SPD_WP110_EN. Accessed 1 Apr 2025

Awuzie B, Ngowi A, Aghimien D (2024) Towards built environment decarbonization: a review of the role of Artificial Intelligence in improving energy and materials' circularity performance. Energy Build 319:114491

Bach S, Binder A, Montavon G, Klauschen F, Müller KR, Samek W (2015) On pixel-wise explanations for non-linear classifier decisions by layer-wise relevance propagation. PLoS One 10(7):e0130140

Bannerjee G, Sarkar U, Das S, Ghosh I (2018) Artificial intelligence in agriculture: a literature survey. Int J Sci Res Comput Sci Appl Manag Stud 7(3):1–6

Bărbulescu A, Zhen L (2024) Forecasting the river water discharge by artificial intelligence methods. Water 16(9):1248

Barnes EA, Hurrell JW, Ebert-Uphoff I, Anderson C, Anderson D (2019) Viewing forced climate patterns through an AI lens. Geophys Res Lett 46(22):13389–13398

BBC News (2016) Security of seismic sensor grid probed. Available at: https://www.bbc.com/news/technology-37001301. Accessed 1 Apr 2025

Beck SR, Riggs KJ, Gorniak SL (2009) Relating developments in children's counterfactual thinking and executive functions. Think Reason 15(4):337–354

Bedué P, Fritzsche A (2022) Can we trust AI? An empirical investigation of trust requirements and guide to successful AI adoption. J Enterp Inf Manag 35(2):530–549

Brendel AB, Mirbabaie M, Lembcke TB, Hofeditz L (2021) Ethical management of artificial intelligence. Sustainability 13:1974

Briganti G, Le Moine O (2020) Artificial intelligence in medicine: today and tomorrow. Front Med 7:509744

Bui DT, Tsangaratos P, Ngo PTT, Pham TD, Pham BT (2019) Flash flood susceptibility modeling using an optimized fuzzy rule-based feature selection technique and tree-based ensemble methods. Sci Total Environ 668:1038–1054

Bull JW, Strange N, Smith RJ, Gordon A (2021) Reconciling multiple counterfactuals when evaluating biodiversity conservation impact in social-ecological systems. Conserv Biol 35:510–521

Byrne RMJ (2019) Counterfactuals in explainable artificial intelligence (XAI): evidence from human reasoning. In: Kraus S (ed) Proceedings of the 28th international joint conference on artificial intelligence (IJCAI 2019), Macao, China, August 10–16, 2019, pp 6276–6282. ijcai.org

Campo Ruiz I (2024) Controlling the environment with Artificial Intelligence risks intensifying social inequalities and colonization. Open Res Eur 4:16

Castelvecchi D (2016) Can we open the black box of AI? Nature 538(7623):20–23

Chadalavada S, Faust O, Salvi M, Seoni S, Raj N, Raghavendra U, Gudigar A, Barua PD, Molinari F, Acharya R (2025) Application of artificial intelligence in air pollution monitoring and forecasting: a systematic review. Environ Model Softw 185:106312

Chang FJ, Chang LC, Chen JF (2023) Artificial intelligence techniques in hydrology and water resources management. Water 15(10):1846

Chen L, Chen P, Lin Z (2020) Artificial intelligence in education: a review. IEEE Access 8:75264–75278

Chou YL, Moreira C, Bruza P, Ouyang C, Jorge J (2022) Counterfactuals and causability in explainable artificial intelligence: theory, algorithms and applications. Inf Fusion 81:59–83
Cifuentes J, Marulanda G, Bello A, Reneses J (2020) Air temperature forecasting using machine learning techniques: a review. Energies 13(16):4215
Coetzee BW, Gaston KJ (2021) An appeal for more rigorous use of counterfactual thinking in biological conservation. Conserv Sci Pract 3:e409
de Almeida PGR, dos Santos CD, Farias JS (2021) Artificial intelligence regulation: a framework for governance. Ethics Inf Technol 23:505–525
Denning DE (2019) Is quantum computing a cybersecurity threat? Am Sci 107(2):83–85
Dignum V (2019) Responsible artificial intelligence: how to develop and use AI in a responsible way. Springer Nature
Ding J, Ren C, Wang J, Feng Z, Cao SJ (2024) Spatial and temporal urban air pollution patterns based on limited data of monitoring stations. J Clean Prod 434:140359
Durgun Y (2024) Real-time water quality monitoring using AI-enabled sensors: detection of contaminants and UV disinfection analysis in smart urban water systems. J King Saud Univ Sci 36(9):103409
Eli-Chukwu NC (2019) Applications of artificial intelligence in agriculture: a review. Eng Technol Appl Sci Res 9(4):4377–4383
European Commission (2019) Ethics guidelines for trustworthy AI. Available at: https://digital-strategy.ec.europa.eu/en/library/ethics-guidelines-trustworthy-ai. Accessed 4 Apr 2025
EY (2024) AI and sustainability: opportunities, challenges, and impact. Available at: https://www.ey.com/en_nl/insights/climate-change-sustainability-services/ai-and-sustainability-opportunities-challenges-and-impact. Accessed 1 Apr 2025
Feuerriegel S, Dolata M, Schwabe G (2020) Fair AI: challenges and opportunities. Bus Inf Syst Eng 62:379–384
Fijani E, Barzegar R, Deo R, Tziritis E, Konstantinos S (2019) Design and implementation of a hybrid model based on a two-layer decomposition method coupled with extreme learning machines to support real-time environmental monitoring of water quality parameters. Sci Total Environ 648:839–853
Flavián C, Casaló LV (2021) Artificial intelligence in services: current trends, benefits and challenges. Serv Ind J 41:853–859
Floridi L (2013) The ethics of information. Oxford University Press
Floridi L, Cowls J (2019) A unified framework of five principles for AI in society. SSRN Electron J. Available at: https://doi.org/10.2139/SSRN.3831321. Accessed 4 Apr 2025
Floridi L, Cowls J, Beltrametti M, Chatila R, Chazerand P, Dignum V, Luetge C, Madelin R, Pagallo U, Rossi F, Schafer B, Valcke P, Vayena E (2018) AI4People—an ethical framework for a good AI society: opportunities, risks, principles, and recommendations. Mind Mach 28(4):689–707
Freiman O (2023) Making sense of the conceptual nonsense 'trustworthy AI'. AI Ethics 3:1351–1360
Gagne DJ, Christensen HM, Subramanian AC, Monahan AH (2020) Machine learning for stochastic parameterization: generative adversarial networks in the Lorenz '96 model. J Adv Model Earth Syst 12:e2019MS001896
Galaz V, Centeno MA, Callahan PW, Causevic A, Patterson T, Brass I, Baum S, Farber D, Fischer J, Garcia D, McPhearson T, Jimenez D, King B, Larcey P, Levy K (2021) Artificial intelligence, systemic risks, and sustainability. Technol Soc 67:101741
Genuino DAD, Bataller BG, Capareda SC, de Luna MDG (2017) Application of artificial neural network in the modeling and optimization of humic acid extraction from municipal solid waste biochar. J Environ Chem Eng 5:4101–4107
Ghallab M (2019) Responsible AI: Requirements and challenges. AI Perspect 1:1–7
Goodell JW, Kumar S, Lim WM, Pattnaik D (2021) Artificial intelligence and machine learning in finance: identifying foundations, themes, and research clusters from bibliometric analysis. J Behav Exp Financ 32:100577

Google (2022) Good news about the carbon footprint of machine learning training. Available at: https://research.google/blog/good-news-about-the-carbon-footprint-of-machine-learning-training/. Accessed 1 Apr 2025
Google (2024) Environmental report 2024. Available at: https://www.gstatic.com/gumdrop/sustainability/google-2024-environmental-report.pdf. Accessed 1 Apr 2025
Google (2025) AI responsibility update – February 2025. Available at: https://ai.google/static/documents/ai-responsibility-update-published-february-2025.pdf. Accessed 4 Apr 2025
Guidotti R, Monreale A, Giannotti F, Pedreschi D, Ruggieri S, Turini F (2019) Factual and counterfactual explanations for black box decision making. IEEE Intell Syst 34:14–23
Gunning D, Aha DW (2019) DARPA's explainable artificial intelligence program. AI Mag 40(2):44–58
Guo Y, Wang J, Chen H, Li G, Huang R, Yuan Y, Ahmad T, Sun S (2019) An expert rule-based fault diagnosis strategy for variable refrigerant flow air conditioning systems. Appl Therm Eng 149:1223–1235
Halsana S (2020) Air quality prediction model using supervised machine learning algorithms. Int J Sci Res Comput Sci Eng Inf Technol:190–201
Ham YG, Kim JH, Luo JJ (2019) Deep learning for multi-year ENSO forecasts. Nature 573(7775):568–572
Hamet P, Tremblay J (2017) Artificial intelligence in medicine. Metabolism 69:S36–S40
Hastie T, Tibshirani R (1990) Generalized additive models. Wiley, New York
He JJ, Gong SL, Yu Y, Yu LJ, Wu L, Mao HJ, Song CB, Zhao SP, Liu HL, Li XY, Li RP (2017) Air pollution characteristics and their relation to meteorological conditions during 2014–2015 in major Chinese cities. Environ Pollut 223:484–496
Holmes J, Sacchi L, Bellazzi R (2004) Artificial intelligence in medicine. Ann R Coll Surg Engl 86:334–338
Holzinger A, Saranti A, Molnar C, Biecek P, Samek W (2020) Explainable AI methods – a brief overview. In: Holzinger A, Goebel R, Fong R, Moon T, Müller KR, Samek W (eds) xxAI – Beyond explainable AI. xxAI 2020. Lecture notes in computer science, vol 13200. Springer, Cham
Hu L, Chen J, Nair VN, Sudjianto A (2018) Locally interpretable models and effects based on supervised partitioning (LIME-SUP). arXiv preprint, arXiv:1806.00663
Huang Z (2024) Reinforcement learning-based adaptive control method for traffic lights in intelligent transportation. Alex Eng J 106:381–391
Huntingford C, Jeffers ES, Bonsall MB, Christensen HM, Lees T, Yang H (2019) Machine learning and artificial intelligence to aid climate change research and preparedness. Environ Res Lett 14(12):124007
Insight Ace Analytic (2024) AI in environmental sustainability market. Available at: https://www.insightaceanalytic.com/report/ai-in-environmental-sustainability-market/2753. Accessed 8 Apr 2025
International Energy Agency (2024) Electricity 2024 – executive summary. Available at: https://www.iea.org/reports/electricity-2024/executive-summary. Accessed 1 Apr 2025
Ise T, Oba Y (2019) Forecasting climatic trends using neural networks: an experimental study using global historical data. Front Robot AI 6:446979
Jaafari A, Zenner EK, Panahi M, Shahabi H (2019) Hybrid artificial intelligence models based on a neuro-fuzzy system and metaheuristic optimization algorithms for spatial prediction of wildfire probability. Agric For Meteorol 266–267:198–207
Janssen M, Brous P, Estevez E, Barbosa LS, Janowski T (2020) Data governance: organizing data for trustworthy artificial intelligence. Gov Inf Q 37:101493
Kamyab H, Khademi T, Chelliapan S, SaberiKamarposhti M, Rezania S, Yusuf M, Farajnezhad M, Abbas M, Jeon BH, Ahn Y (2023) The latest innovative avenues for the utilization of artificial intelligence and big data analytics in water resource management. Results Eng 20:101566
Katsouyanni K, Touloumi G, Samoli E, Gryparis A, Le Tertre A, Monopolis Y et al (2001) Confounding and effect modification in the short-term effects of ambient particles on total

mortality: results from 29 European cities within the APHEA2 project. Epidemiology 12(5):521–531

Katzner TE, Nelson DM, Diffendorfer JE, Duerr AE, Campbell CJ, Leslie D et al (2019) Wind energy: an ecological challenge. Science 366:1206–1207

Kavya M, Mathew A, Raja Shekar P, Sarwesh P (2023) Short-term water demand forecast modeling using artificial intelligence for smart water management. Sustain Cities Soc 95:104610

Kim S, Pan S, Mase H (2019) Artificial neural network-based storm surge forecast model: practical application to Sakai Minato, Japan. Appl Ocean Res 91:101871

Kim SW, Kong JH, Lee SW, Lee S (2022) Recent advances of artificial intelligence in manufacturing industrial sectors: a review. Int J Precis Eng Manuf:1–19

Kim D, Choi S, Kang S, Noh H (2023) A study on developing an AI-based water demand prediction and classification model for Gurye intake station. Water 15(23):4160

Landt-Hayen M, Kröger P, Claus M, Rath W (2022) Layer-wise relevance propagation for echo state networks applied to earth system variability. Signal Image Process Embed Syst Trends:115–130

Larson E, Greig C, Jenkins J, Mayefield E, Pascale A, Zhang C et al (2020) Net-zero America: potential pathways, infrastructure, and impacts, interim report. Princeton University, Princeton

Lei J, G'Sell M, Rinaldo A, Tibshirani RJ, Wasserman L (2018) Distribution-free predictive inference for regression. J Am Stat Assoc 113:1094–1111

Leng XZ, Wang JH, Ji HB, Wang QG, Li HM, Qian X, Li FY, Yang M (2017) Prediction of size-fractionated airborne particle-bound metals using MLR, BP-ANN and SVM analyses. Chemosphere 180:513–522

Li BH, Hou BC, Yu WT, Lu XB, Yang CW (2017) Applications of artificial intelligence in intelligent manufacturing: a review. Front Inf Technol Electron Eng 18(1):86–96

Li J, Sun C, Zhou H, Zhao Y, Bao X, Zhao Q (2018) A survey of development and application of artificial intelligence in smart grid. IOP Conf Ser Earth Environ Sci 186:012066

Lian R, Siau K (2021) Trust in AI in autonomous driving vehicles. In: MWAIS 2021 proceedings

Liu Z, Coleman N, Patrascu FI, Yin K, Li X, Mostafavi A (2025) Artificial intelligence for flood risk management: a comprehensive state-of-the-art review and future directions. Int J Disaster Risk Reduct 117:105110

Lockey S, Gillespie N, Curtis C (2020) Trust in artificial intelligence: Australian insights. The University of Queensland and KPMG Australia

Lu H, Ma X, Huang K, Azimi M (2020) Carbon trading volume and price forecasting in China using multiple machine learning models. J Clean Prod 249:119386

Mardani A, Liao H, Nilashi M, Alrasheedi M, Cavallaro F (2020) A multi-stage method to predict carbon dioxide emissions using dimensionality reduction, clustering, and machine learning techniques. J Clean Prod 275:122942

Marzouny H, Dziedzic R (2024) AI-assisted pump operation for energy-efficient water distribution systems. Eng Proc 69(1):3

Maußner C, Oberascher M, Autengruber A, Kahl A, Sitzenfrei R (2025) Explainable artificial intelligence for reliable water demand forecasting to increase trust in predictions. Water Res 268(Part B):122779

Meltzer JP (2019) Governing digital trade. World Trade Rev 18(S1):S23–S48

Menad NA, Hemmati-Sarapardeh A, Varamesh A, Shamshirband S (2019) Predicting solubility of CO2 in brine by advanced machine learning systems: application to carbon capture and sequestration. J CO_2 Util 33:83–95

Meng Y, Noman H (2022) Predicting CO2 emission footprint using AI through machine learning. Atmosphere 13(11):1871

Microsoft (2022) Responsible AI standard v2: general requirements. Available at: https://blogs.microsoft.com/wp-content/uploads/prod/sites/5/2022/06/Microsoft-Responsible-AI-Standard-v2-General-Requirements-3.pdf. Accessed 4 Apr 2025

Mintz Y, Brodie R (2019) Introduction to artificial intelligence in medicine. Minim Invasive Ther Allied Technol 28(2):73–81

MIT Sloan (2025) AI has high data center energy costs. There are solutions. Available at: https://mitsloan.mit.edu/ideas-made-to-matter/ai-has-high-data-center-energy-costs-there-are-solutions. Accessed 1 Apr 2025

Nag S, Mondal A, Roy DN, Bar N, Das SK (2018) Sustainable bioremediation of Cd (II) from aqueous solution using natural waste materials: kinetics, equilibrium, thermodynamics, toxicity studies and GA-ANN hybrid modelling. Environ Technol Innov 11:83–104

Nallakaruppan MK, Gangadevi E, Shri ML et al (2024) Reliable water quality prediction and parametric analysis using explainable AI models. Sci Rep 14:7520

OECD (2019) AI principles. Available at: https://www.oecd.org/en/topics/ai-principles.html. Accessed 4 Apr 2025

Olawade DB, Wada OZ, Ige AO, Egbewole BI, Olojo A, Oladapo BI (2024) Artificial intelligence in environmental monitoring: advancements, challenges, and future directions. Hyg Environ Health Adv 12:100114

Oluleye BI, Chan DW, Antwi-Afari P (2023) Adopting artificial intelligence for enhancing the implementation of systemic circularity in the construction industry: a critical review. Sustain Prod Consum 35:509–524

Ouyang F, Jiao P (2021) Artificial intelligence in education: the three paradigms. Comput Educ Artif Intell 2:100020

Papagiannidis E, Mikalef P, Conboy K (2025) Responsible artificial intelligence governance: a review and research framework. J Strateg Inf Syst 34(2):101885

Pearl J (2000) Causality: models, reasoning, and inference. Cambridge University Press, Cambridge

Pearl J (2009) Causal inference in statistics: an overview. Stat Surv 3:96–146

Phillips PJ, Hahn CA, Fontana PC, Yates AN, Greene K, Broniatowski DA, Przybocki MA (2021) Four principles of explainable artificial intelligence. NIST Interagency/Internal Report (NIST IR) 8312

Pörtner HO, Scholes RJ, Agard J, Archer E, Arneth A, Bai X et al (2021) IPBES-IPCC co-sponsored workshop report on biodiversity and climate change. IPBES and IPCC, Bonn

Rasp S, Pritchard MS, Gentine P (2018) Deep learning to represent subgrid processes in climate models. Proc Natl Acad Sci 115(39):9684–9689

Rather SA, Patel M, Kapoor K (2025) AI-driven forecasting of river discharge: the case study of the Himalayan mountainous river. Earth Sci Inf 18:242

Ravindiran G, Hayder G, Kanagarathinam K, Alagumalai A, Sonne C (2023) Air quality prediction by machine learning models: a predictive study on the Indian coastal city of Visakhapatnam. Chemosphere 338:139518

Ravindra K, Rattan P, Mor S, Aggarwal AN (2019) Generalized additive models: building evidence of air pollution, climate change and human health. Environ Int 132:104987

Reynolds SA, Beery S, Burgess N, Burgman M, Butchart SHM, Cooke SJ, Coomes D, Danielsen F, Di Minin E, Durán AP, Gassert F, Hinsley A, Jaffer S, Jones JPG, Li BV, Mac Aodha O, Madhavapeddy A, O'Donnell SAL, Oxbury WM, Peck L, Pettorelli N, Rodríguez JP, Shuckburgh E, Strassburg B, Yamashita H, Miao Z, Sutherland WJ (2025) The potential for AI to revolutionize conservation: a horizon scan. Trends Ecol Evol 40(2):191–207

Ribeiro MT, Singh S, Guestrin C (2018) Anchors: high-precision model-agnostic explanations. In: Proceedings of the thirty-second AAAI conference on artificial intelligence, New Orleans, LA, USA, 2–7 February

Robnik-Šikonja M, Kononenko I (2008) Explaining classifications for individual instances. IEEE Trans Knowl Data Eng 20:589–600

Roll I, Wylie R (2016) Evolution and revolution in artificial intelligence in education. Int J Artif Intell Educ 26:582–599

Rutenberg I, Gwagwa A, Omino M (2021) Use and impact of Artificial Intelligence on climate change adaptation in Africa. In: Oguge N et al (eds) African handbook of climate change adaptation. Springer International Publishing, Cham, pp 1107–1126

Ryan M, Stahl BC (2021) Artificial intelligence ethics guidelines for developers and users: clarifying their content and normative implications. J Inf Commun Ethics Soc 19(1):61–86

Samek W, Montavon G, Vedaldi A, Hansen LK, Müller KR (eds) (2019) Explainable AI: interpreting, explaining and visualizing deep learning. Lecture notes in computer science. Springer International Publishing, Cham

Selvakumar P, Sivashanmugam P (2018) Multi-hydrolytic biocatalyst from organic solid waste and its application in municipal waste activated sludge pre-treatment towards energy recovery. Process Saf Environ Prot 117:1–10

Shang Z, Deng T, He J, Duan X (2019) A novel model for hourly PM2.5 concentration prediction based on CART and EELM. Sci Total Environ 651:3043–3052

Sharma S, Sato K, Gautam BP (2023) A methodological literature review of acoustic wildlife monitoring using artificial intelligence tools and techniques. Sustainability 15(9):7128

Shobanke M, Bhatt M, Shittu E (2025) Advancements and future outlook of artificial intelligence in energy and climate change modeling. Adv Appl Energy 17:100211

Silvestro D, Goria S, Sterner T et al (2022) Improving biodiversity protection through artificial intelligence. Nat Sustain 5:415–424

Singapore Government (2020) Model artificial intelligence governance framework (2nd edn). Available at: https://www.pdpc.gov.sg/-/media/files/pdpc/pdf-files/resource-for-organisation/ai/sgmodelaigovframework2.pdf. Accessed 4 Apr 2025

Smuha NA (2021) Beyond a human rights-based approach to AI governance: promise, pitfalls, plea. Philos Technol 34:91–104

Sokol K, Santos-Rodríguez R, Flach PA (2019) FAT forensics: a python toolbox for algorithmic fairness, accountability and transparency. CoRR, arXiv:1909.05167

Soleymani AR, Moradi M (2018) Performance and modeling of UV/persulfate/Ce(IV) process as a dual oxidant photochemical treatment system: kinetic study and operating cost estimation. Chem Eng J 347:243–251

Soori M, Arezoo B, Dastres R (2023) Artificial intelligence, machine learning and deep learning in advanced robotics: a review. Cognit Robot 3:54–70

Srinivasan AV, de Boer M (2020) Improving trust in data and algorithms in the medium of AI. Maandblad Voor Accountancy en Bedrijfseconomie 94(3–4):147–160

Tang Y, Huang Y, Wang H, Wang C, Guo Q, Yao W (2018) Framework for artificial intelligence analysis in large-scale power grids based on digital simulation. CSEE J Power Energy Syst 4:459–468

Tang Y, Sun Y, Han Z, Soomro S, Wu Q, Tan B, Hu C (2023) Flood forecasting based on machine learning pattern recognition and dynamic migration of parameters. J Hydrol Reg Stud 47:101406

Trocin C, Mikalef P, Papamitsiou Z, Conboy K (2021) Responsible AI for digital health: a synthesis and a research agenda. Inf Syst Front:1–19

Tschopp M, Ruef M (2019) Artificial intelligence – is it worth the risk? Titanium Trust Report. Available at: https://www.researchgate.net/publication/336849066_Artificial_Intelligence_-_Is_it_worth_the_risk.Zugegriffenam. Accessed 4 Apr 2025

Vakkuri V, Kemell KK, Abrahamsson P (2019) AI ethics in industry: a research framework. arXiv preprint. Available at: https://arxiv.org/abs/1910.12695. Accessed 4 Apr 2025

Vanijjirattikhan R, Khomsay S, Kitbutrawat N, Khomsay K, Supakchukul U, Udomsuk S, Suwatthikul J, Oumtrakul N, Anusart K (2022) AI-based acoustic leak detection in water distribution systems. Results Eng 15:100557

Verma S, Dickerson JP, Hines K (2020) Counterfactual explanations for machine learning: a review. CoRR, arXiv:2010.10596

Vilone G, Longo L (2021) Notions of explainability and evaluation approaches for explainable artificial intelligence. Inf Fusion 76:89–106

Wachter S, Mittelstadt B, Russell C (2018) Counterfactual explanations without opening the black box: automated decisions and the GDPR. Harv J Law Technol 31:841–887

Wang C, Ye Z, Yu Y, Gong W (2018) Estimation of bus emission models for different fuel types of buses under real conditions. Sci Total Environ 640–641:965–972

Wei S, Yuwei W, Chongchong Z (2018) Forecasting CO2 emissions in Hebei, China, through moth-flame optimization based on the random forest and extreme learning machine. Environ Sci Pollut Res 25(29):28985–28997

Wikle CK, Datta A, Hari BV, Boone EL, Sahoo I, Kavila I, Castruccio S, Simmons SJ, Burr WS, Chang W (2023) An illustration of model agnostic explainability methods applied to environmental data. Environmetrics 34(1):e2772

Winfield AFT, Jirotka M (2018) Ethical governance is essential to building trust in robotics and artificial intelligence systems. Philos Trans R Soc A Math Phys Eng Sci 376:20180085

Woodward JF (2003) Making things happen: a theory of causal explanation. Oxford University Press, New York

Xu C, Xue Z (2024) Applications and challenges of artificial intelligence in the field of disaster prevention, reduction, and relief. Nat Hazard Res 4(1):169–172

Zhao Y, Xiao F, Wang S (2013) An intelligent chiller fault detection and diagnosis methodology using Bayesian belief network. Energ Buildings 57:278–288

Zheng G, Li X, Zhang RH, Liu B (2020) Purely satellite data-driven deep learning forecast of complicated tropical instability waves. Sci Adv 6(29):1482

Artificial Intelligence and Environmental Decision Support Systems

11

Daria Gribkova and Yulia Milshina

Abstract

The environment, as a dynamic system influenced by anthropogenic factors, requires evidence-based decision-making to ensure sustainable resource management and mitigate adverse effects. This chapter explores the transformative role of artificial intelligence (AI) in environmental decision support systems (EDSS), focusing on its potential to address complex environmental challenges. EDSS, characterized by their interactivity and ability to process unstructured data, have evolved significantly since their inception in the late twentieth century. However, their adoption has been hindered by mismatches between system outputs and user expectations, as well as a lack of trust and standardized evaluation criteria. The integration of AI technologies, such as machine learning, predictive analytics, and computer vision, has revolutionized the functionality of EDSS, enabling real-time data analysis, improved forecasting, and enhanced decision-making processes. This chapter highlights the symbiotic relationship between AI advancements and urban development, demonstrating how AI-driven tools optimize resource allocation, improve strategic planning, and contribute to sustainable growth. Case studies illustrate the application of AI-based EDSS in areas such as climate risk prediction, disaster management, and environmental impact assessments for infrastructure projects.

By examining the drivers and barriers to EDSS implementation, this chapter underscores the importance of interdisciplinary approaches and robust AI models in addressing global environmental challenges. The findings emphasize that AI-powered EDSS are essential for navigating the complexities of modern environmental management and achieving long-term sustainability goals.

D. Gribkova (✉) · Y. Milshina
HSE University, Institute for Statistical Studies and Economics of Knowledge, Moscow, Russia
e-mail: dgribkova@hse.ru; ymilshina@hse.ru

B. Thokchom et al. (eds.), *Artificial Intelligence Enabled Real Time Environmental Monitoring*, https://doi.org/10.1007/978-981-95-3767-9_11

Keywords

Artificial intelligence · Environmental decision support systems · Environmental issues · Forecasting · Decision support · Evidence-based decision-making

11.1 Introduction

The environment is a complex system of physical, chemical, and biological factors, of living and nonliving elements and relationships, in which all organisms that inhabit the planet are immersed (Buongiorno and Chiaramonte 2023). It is a dynamic system, with its own internal laws of development and interaction between its components, and susceptible to the influence of anthropogenic factors. The progress of human society has always been, and still is, dependent on the environment. However, its resources are not unlimited. Careful planning of resource use and environmental interventions (including those related to construction, agriculture, logging, etc.), balanced and evidence-based decision-making are becoming increasingly important.

The environment and ecosystems are one of the most important topics in today's scientific publications (Scinapse 2024); previously perceived as isolated, sociocultural, biophysical and economic components of the system are studied in a complex. An interdisciplinary approach is becoming more common in research and governance of the ecological environment; as the availability of data on the system and technologies for analyzing them increases, so does the complexity of the system representation. This determines the need for appropriate decision support tools.

In the context of deep uncertainty (Aleskerov 2023) and wild card events, tightening environmental regulations, and the growing importance of sustainable development and circular economy principles, decision-makers are faced with the challenge of making complex decisions that will have a strong impact and long-term consequences. The level of uncertainty in environmental systems makes forecasting in this area challenging (Guillaume et al. 2015). The decision-making process is becoming more complex, creating a need for tools such as decision Support systems.

Environmental decision support systems (EDSS) are intelligent (computer-based) information systems with user-friendly interfaces that supports one or more phases of decision-making, improve the time taken to make decisions in environmental management, as well as the consistency and quality of the decisions, expressed in terms of metrics that are characteristic of the area of application (Haagsma and Johanns 1994; Cortés et al. 2000; Matthies et al. 2007). Such systems are characterized by interactivity, flexibility, the ability to work with unstructured or poorly structured data, and support one or more phases of decision-making (intelligence, defining a model of a system, identifying alternatives, etc.). EDSS do not simply provide models, estimates, or options, but help structure complex problems, improve their understanding, and increase the transparency of decision-making processes (McIntosh et al. 2011). They are designed to help (not replace) managers at different levels of management to identify interactions and causalities and make effective, reasonable, justifiable, legally, and morally correct decisions.

The impact of such systems becomes particularly noticeable when working with large amounts of data. Environmental management is challenging due to the complexity and scale of natural systems, as well as the intricate and impactful interplay of social, economic, resource, and political subsystems, which are often misaligned. The role of digital technologies, especially artificial intelligence (AI), in improving the accuracy, speed, and efficiency of such systems is growing. It is necessary to look closely at two aspects that have proven to be interlinked—increased attention to environmental and climate issues, and the growing role of cutting-edge technologies, in this case AI, which is associated with potential breakthroughs in many areas of human activity, including those related to the environment.

11.2 Literature Review

The field of environmental management includes closely related activities such as continuous environmental monitoring, forecasting and analysis, disaster management, post-incident mitigation and recovery planning, decisions impact assessment, and environmental impact assessment of major infrastructure or industrial projects, that is, planning for environmentally sound man-made industrial products and processes (Filip and Moisil 2000).

Experts note that the development of decision support tools is associated with a change in the perception of the world and different spheres of human activity and life—from clearly delineated disciplines to complex systems consisting of many interconnected elements and processes with their own development dynamics, with the emergence of new, larger-scale and global problems, with changes in socio-technical systems (Schot and Steinmueller 2018), and with the inadequacy of applied management and decision-making tools to current conditions (Borrás and Serger 2023). While the dynamics of some components of change can be predicted with relative certainty, their combination creates situations of deep uncertainty (Yakobson 2023). Accordingly, approaches to decision-making and management of such systems had to change.

Environmental decision support systems (in the sense that they correspond to the level of development of the computer and visualization technologies at the time) began to appear in the late 1970s and 1980s (Rossman 1980; Bell 1985; Johnson 1986) and have been used for site assessment and environmental management (Massei et al. 2014), waste collection route planning and hazardous waste tracking, water quality and wastewater treatment (Panagopoulos et al. 2012; Castillo et al. 2017; Jayanthi et al. 2022) and other resource management, agricultural impact assessment (Ascough II et al. 2002), energy planning (Kumar et al. 2017; Trivyza et al. 2018), etc. Despite a long history of development, such technologies have not been widely adopted and have not delivered the expected results (Wrisberg et al. 2002; French and Geldermann 2005; Newman et al. 2017). Experts note that although many models have been developed, very few have come to market (Poch et al. 2017) and the evidence base for the effectiveness of such systems is very limited (Hamouda et al. 2009; Reiter et al. 2018; Zasada et al. 2017). The reasons for the lack of demand are mainly related to the mismatch between the views of decision-makers and the results produced by the systems (Mysiak et al. 2005), the lack of trust (Boschetti et al. 2012) or belief that the proposed solutions are not worth the cost of purchasing and implementing decision support systems, a long time lag between a decision being made using the system and visible results, and the difficulty of establishing cause-and-effect relationships (Matthews et al. 2011).

Such systems can be understood as a wide range of tools that support decision-making in the field of environmental management or in activities that have a direct or indirect impact on the environment. There are also no clearly defined criteria for evaluating such systems. Walling and Vaneeckhaute (2020) point out that there is no generally accepted guide or standard for the production of EDSS. In reviewing previous attempts to formulate criteria for the assessment of EDSS, authors list the problems faced by developers of such systems and propose solutions, noting that despite the scope of their work, it cannot be considered comprehensive. They also note that such systems have unique features and that evaluation criteria must be built into the design process to ensure high system quality.

Merritt et al. (2017), based on an analysis of 15 water resources modeling projects, identified 33 factors that influence the success of such systems, relating to project participants and stakeholder interactions, project management, contextual conditions, model development, etc. They concluded that good relationships between the development team and end users, reliable model validation, and the availability of highly skilled developers were associated with a successful outcome.

With the development of technologies, including observation and sensor technologies, and the growth of computer processing power, the capabilities and functionality of decision support systems have expanded, as has the range of their applications and the scope of topics, which have become much more focused and sophisticated at the same time. The greater availability of technology, its relative ease of use, and the evolution of methodology have enabled scientists from a variety of disciplines to develop their models for making predictions about environmental issues.

For example, Agyemang et al. (2025) have conducted research to develop an adaptive framework for developing a decision support system for a multimodel assessment of the combined effects of health and environmental efforts in the food system using data on food production, water consumption, and energy use. The development of the scientists allows exploration of different intervention scenarios, taking into account short-, medium-, and long-term goals, to create resilient and sustainable food systems for the future.

One group of work is represented by studies of the possibilities of using systems in various aspects of the transport sector and related fields (energy systems, energy storage technology, pollution monitoring, etc.) (Ma et al. 2025; Suganya et al. 2024; Li et al. 2025). Alrashdi et al. (2025) presented the single-valued neutrosophic sets structure for ranking alternative vehicle technologies with overcoming information uncertainty. Schulte and Voß (2015) propose an integrated decision support approach to balance free-floating car-sharing systems and evaluate user-oriented relocation strategies. They suggest a regression and neural network approach to model car2go demand data to support efficient relocation plans. Furthermore, the authors illustrate how demand forecasts and real-time data from the system provider can be used in an integrated decision support approach to guide the application of relocation strategies derived from the empirical analysis of the case study (Schulte and Voß 2015).

A brief review of research papers has shown that the range of potential applications for such systems is widening and that narrowly focused themes are emerging that cover a wide range of factors. EDSS are faced with complex tasks and increasing expectations of functionality. Social, economic, technological, and environmental processes are considered in a complex way, which requires that relevant factors are taken into account in the development of systems.

The chapter is structured in the following way: it begins with a brief overview of the environmental challenges and how modern technologies are being used to address them, then considers the role of AI in EDSS, identifies the drivers and barriers to the use of such systems, provides the examples of practical EDSS applications, and identifies the prospects and difficulties for future development.

11.3 AI-Based Approaches to Addressing Environmental Challenges

11.3.1 Environmental Challenges

There is an increasing focus on improving our understanding of environmental processes. The range of interrelated challenges facing the Earth and its natural systems—long-term and profound changes in weather patterns, damage to the biosphere, local air pollution, the spread of pollutants in water and the health of the oceans, energy use, and global climate change—is growing at international and national, governmental, corporate, and individual levels. Sustainable development and environmental issues, on which people's well-being depends, are increasingly the focus of governments, academia, and civil society.

WEF ranks extreme weather events, critical changes to Earth systems, biodiversity loss and ecosystem collapse, natural resources shortages, and pollution as the most serious short- and long-term development risks (WEF 2025). Rising global average temperatures and air and water pollution have a significant impact on human life and well-being, threatening environmental sustainability and the socio-economic development of countries. Negative environmental processes exacerbate problems with long-term consequences at the intersection of social, economic, and environmental spheres: reduction of productive resources, negative health impacts, conflicts over water and land resources, and forced migration of populations. Environmental risks are playing an increasingly important role in business and governance risk calculations (UNDP 2024). Environmental change, geopolitical conflicts, and health crises are inextricably linked.

Between 2010 and 2020, the death rate from floods, droughts, and cyclones will be 15 times higher in regions extremely vulnerable to climate change, where some 3.3–3.6 billion people live (UN 2023). Between 1961–1990 and 2014–2023, an increase in the number of days with extreme precipitation affected 61% of the world's land area (Romanello et al. 2024). Irrational use of natural resources leads to the depletion of productive resources. Excessive use of fertilizers hurts soil and water quality and therefore agriculture. Soil degradation, land desertification, and water scarcity threaten the activities and well-being of more than one billion farms, and thus the food and health security, labor productivity, and economic stability of countries. Average annual economic losses from weather-related extreme events will increase by 23% between 2010–2014 and 2019–2023, to US$227 billion (Ibid.).

The need for accurate and reliable instruments to ensure the adoption of rational and effective decisions is determined by environmental processes and phenomena and their impact on related areas.

11.3.2 Environmental Management Solutions

The role of innovative solutions and modern technologies in ecosystem assessment, modeling, simulation, forecasting, planning and decision-making, environmental management, and protection is constantly growing. In science, government, and business, the idea of artificial intelligence is widely accepted as a technology capable of radically transforming information, financial, transport, medical, scientific, logistics, and other sectors. UNEP estimates that seven technologies—AI, digital access, fast internet, cloud, Internet of Things, extended reality, and blockchain—could impact 103 of the 169 Sustainable Development Goal targets (UNEP).

The use of AI technologies (computer vision, natural language processing, predictive analytics, etc.) in combination with other technologies—cloud computing, IoT, advanced software, data mining, and automation tools—has great transformative potential for management and data-driven decision-making. Today, intelligent decision and management support technologies (Talakauskas et al. 2024) based on machine learning, neural networks, and AI are among the most in-demand in business and government.

For instance, data mining is used to transform large amounts of data into useful, actionable, interpretable, and decision supporting information. Computer vision collects and processes continuously updated image and video data from the environment. There is an increasing focus on developing more robust and efficient models that can process large amounts of real-world data and provide real-time situational awareness. AI, particularly Generative (Gen AI) and Agentic AI, is used to automate routine tasks, streamline operations, support decision-making, and speed up decision-making by reducing the time between data collection and analysis. AI-powered tools can facilitate communication, provide real-time translation, break down language barriers, and facilitate global connectivity.

Predictive modeling enables the prediction of physical climate risk events: floods, hail, windstorms, fire hazards, and permafrost degradation. Using current and historical data sets with characteristics that influence the realization of a risk event, the algorithm calculates the probability of the occurrence of these risks for the forecast period based on scenario analysis. Such models help improve the quality of strategic planning and adaptation to climate change in regions and companies and contribute to improving decision-making systems to protect the lives, health, and property of citizens, preserve biodiversity, ensure food security, and reduce greenhouse gas emissions.

Machine learning and artificial intelligence are increasingly being used to make decisions that affect people's lives and well-being, resource consumption, and the functioning of infrastructure. However, experts note that the current state of development of these technologies allows them to ask questions of "what," "when," and "where," rather than "why," "what if," and "how" with confidence (Camps-Valls et al. 2025). Progress is linked to Explainable AI, which will not only explain the logic of the algorithms but also improve understanding of the environmental processes themselves. Explainable AI aims to provide transparency into how the system acts and makes decisions (Coussement et al. 2024). Understanding why a system can make certain decisions is as important as the accuracy of the decisions themselves, the interpretability, and the transparency of the results. This increases the analytical validity of the decisions made.

However, a qualitatively new stage in the development of AI for decision-making is associated with agentic systems. Agentic AI functions include autonomous decision-making with a focus on long-term goals and action execution. Agentic AI can set goals and strategies to achieve them and adapt its approach based on real-time environmental data. The functional foundations of the Agentic AI framework are decision algorithms and reinforcement learning. Unlike Gen AI, which mimics the creative human thought process, Agentic AI uses reinforcement learning to gradually improve its decision-making capabilities (E&ICT Academy 2025).

At the same time, its wider implementation is not without challenges—there are several operational, technical, ethical, and cultural barriers to implementing advanced technologies, including AI—lack of or poor quality of data, high energy consumption, cost of training language models, old technical base incompatible with advanced technologies, and lack of specialists with the necessary skills. Ethical and legal issues need to be addressed, including those related to the limits of using AI to support decision-making. What is the balance between the benefits of increased automation and the need for human oversight? What happens if AI makes a mistake in such a sensitive area, where people's lives and well-being depend on the decision? Who is responsible if AI gets it wrong? Some experts point out that intelligent technologies look for the most efficient way, not the most ethical way, which is natural for them (Security Intelligence 2025). In addition, legislation is lagging behind the pace of technological development, including that of artificial intelligence.

AI is expected to bring about radical transformations in all spheres of human life and activity. Some believe that AI's ability to predict climate events will be crucial in the fight against environmental degradation (Basso and Li 2025). AI is associated with optimizing energy systems, supporting biodiversity conservation, and making more efficient use of resources. Key application areas include dynamic energy management, renewable energy, climate modeling, environmental monitoring, and the implementation of circular economy principles (Fingreen 2025). Nie and Liu (2025) highlight the benefits of integrating AI with traditional environmental modeling and management methods but conclude that LLMs should augment, not replace, human expertise, with the ultimate responsibility for environmental decisions remaining with humans.

11.4 Environmental Decision Support Systems and AI Functions

The term decision support system (DSS) was coined in the early 1970s when managerial decision support systems began to appear in the literature (Gorry and Scott Morton 1971; Filip and Moisil 2000). The concept, which was initially rather vague and focused on the aspects of "decision" and "support," began to be associated with information systems as technologies for collecting, accumulating, and analyzing data developed. Efforts to develop technologies to support environmental management organizations in their search for solutions to complex problems led to the development of what has been termed environmental DSS or EDSS (Guariso and Werthner 1989; Rizzoli and Young 1997).

AI systems are already being used to varying degrees to process data in areas such as energy, construction, environmental monitoring emergency preparedness, and others (Fig. 11.1). But their potential is far from exhausted.

In the energy sector, AI systems analyze vast amounts of real-time data, including historical energy consumption, weather conditions, and operational trends, to dynamically optimize energy consumption, forecast the load manage the smart grid,

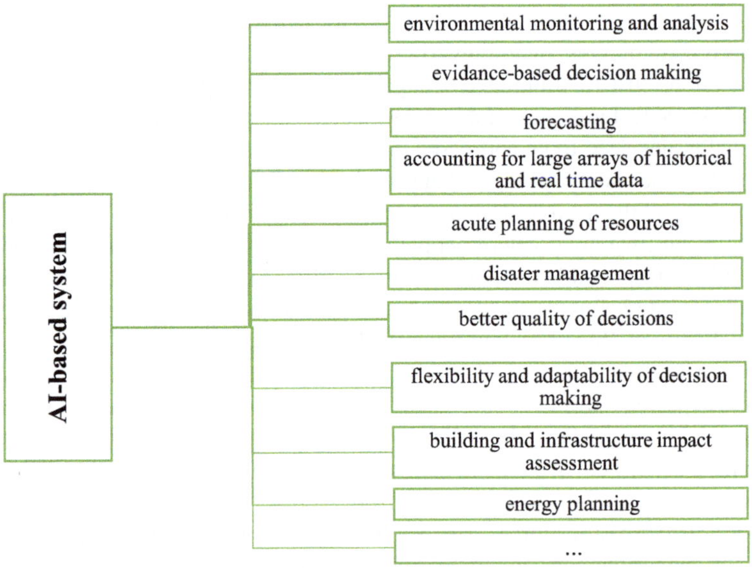

Fig. 11.1 Some functions and effects of AI-based systems use. (Source: authors)

and make optimal decisions. This could potentially enable proactive regulation of energy consumption at the household, neighborhood, city, or industry level. Smart grids balance supply and demand, ensuring a more efficient distribution of energy. In industrial settings, AI-based solutions can identify inefficient energy use and optimize equipment performance, and therefore its energy consumption.

The use of AI in energy management can improve the reliability of energy systems and the efficiency of renewable energy generation, which depends on variable weather conditions. AI systems can also help determine the best location for renewable energy generation capacity (wind turbines and solar panels) (Bilgili et al. 2024; Wind Systems 2023), less reliable sources of energy than gas or coal, based on the analysis of factors such as the number of sunny days and hours, sunlight exposure, wind patterns, and demand centers. For example, Google Cloud and Total have developed the Solar Mapper, a tool designed to speed up the installation of solar panels on private rooftops. The AI platform makes it possible to assess the potential of solar power generation technology for specific homes (ESG Today 2020).

As for construction, AI analysis is applicable in the planning of works, including large road and infrastructure projects, energy and water infrastructure, housing and utility facilities, as well as in assessing the long-term effects of climate policy implementation and its adaptation. In Russia, for example, various AI technologies are used in the construction industry, from forecasting to monitoring hazardous situations on site. Large construction companies (Balfour Beatty, Hochtief, Kiewit, Square Engineering, etc.) are collecting big data on projects and work phases, including environmental conditions, to use AI to optimize the processes involved (Robok 2023; Hochtief 2020; Crunching the Numbers; The Future of Commerce 2025). In the design and construction of buildings and infrastructure, the role of AI systems that can model and accurately predict climate processes is crucial as this is important for considering the sustainability of future structures and calculating the type and amount of materials used. AI helps estimate the quantity and cost of resources, track the location and use of equipment, control the quality of work, ensure the safety of people on site, and prevent risks to the environment. AI can generate different design options for buildings and structures, taking into account location, climate conditions, regulatory requirements, and other factors. In 2024, the Russian Federal Agency for Technical Regulation and Metrology approved standards regulating the use of AI in construction and road equipment, as well as the use of AI in the design of new composite materials (Federal Agency for Technical Regulation and Metrology 2024).

The use of AI, namely predictive analytics, helps identify many problems through the analysis of accumulated statistical data. It enables the modeling of the impact of a large number of different factors on the construction process, making decisions more rational and evidence-based. Data for the predictive analytics system can be collected using the Building Information Modeling (BIM) tools and help make optimal decisions more accurately (Construction newspaper 2024). The larger and higher the quality of the sample of such data about the object, the more accurate and correct the final forecast.

AI-based decision-making tools are particularly valuable when planning construction in special climatic conditions. For example, scientists at the St. Petersburg National Research University of Information Technologies, Mechanics, and Optics (ITMO University) have taught AI to design industrial and port infrastructure for the Arctic zone. It allows a detailed plan of a future facility to be created in a matter of minutes, taking into account building regulations, terrain features, and harsh climatic conditions. To make a plan of the required site, it is enough to select an area of 10–100 hectares on the map and indicate the connection points to existing transport communications and the desired indicators. Data on wind, sea waves, ice characteristics, and hydrometeorological hazards are automatically imported from a 3D model of the atmosphere–ice–ocean–soil, including potential climate change. The result is an interactive 3D scene for the user. The program will reduce the labor intensity and time frame for project development and allow the creation of complex architectural solutions that take into account the characteristics of the natural environment (ITMO News 2023).

In environmental monitoring, image recognition and sensor-based monitoring are used to track endangered species, detect illegal activities, identify illegal landfills, and monitor forest health and deforestation in real time. For example, AI systems analyze data from drones and camera traps to identify wildlife populations and movement patterns, providing information to conservationists. In marine ecosystems, AI is helping to monitor the health of coral reefs and track the migration of aquatic species, predicting environmental threats such as habitat destruction or climate-related changes and enabling early intervention to protect ecosystems. In this way, AI is improving the effectiveness of measures to protect biodiversity and conserve ecosystems that are vital to maintaining the planet's ecological balance.

Emergency preparedness could be significantly improved by AI solutions. By integrating vast datasets from satellites, weather stations, and environmental sensors, AI systems provide detailed information on changing weather conditions, temperature fluctuations, and the likelihood of extreme events that could threaten stability, economic security, and biodiversity loss. Such models can enable private and public organizations to develop proactive actions and response measures.

Another promising area for the application of AI in decision-making is the development of logistics in difficult natural and climatic conditions. For example, the innovative company Morintech and the Lomonosov Moscow State University Marine Research Center have developed a system for ensemble forecasting of ice conditions along the Northern Sea Route based on AI technologies (Ink 2023). Climate change increases the need for such systems to improve efficiency and unlock the economic potential of the Arctic. Reliable hydro-meteorological support, including forecasts of ice, weather, and oceanographic data, is necessary for the safety of shipping and the minimization of environmental risks. The system can increase the accuracy of ice condition forecasts, reduce the risk of accidents and spills critical to Arctic shipping, resource extraction, and fishing, optimize ship routes, diminish fuel consumption and greenhouse gas emissions, and reduce the risk of collisions with ice fields. The solution can potentially be used by organizations in industries such as transport, logistics, construction, industry, and the oil and gas sector.

However, it remains to be seen how accurate, transparent, and reliable AI models can be created for analyzing extreme natural events and, more importantly, for decision-making. Despite the many proposed ML algorithms for predicting extreme weather events, there are no universal algorithms; the specifics of most are determined by the characteristics of a particular region or use case (Camps-Valls et al. 2025).

11.5 Drivers and Barriers for AI-Based Environmental Decision Support Systems

The use of AI in decision support systems makes it possible to increase efficiency, simplify the decision-making process, and reduce the risk of human error. AI is expanding the capabilities of decision support systems, providing more opportunities for real-time information processing, improved comprehensive data analysis and risk assessment, more accurate biodiversity and ecosystem monitoring and pollution detection, spatial analysis, modeling, trends forecasting, scenario building, increased computational accuracy and speed of governance decision making, and preparation of response measures and then correlation of policy implementation with changes in environmental markers. There is great promise in the development of Generative AI, which can be used to read, translate, summarize, analyze, and fill the gaps in large volumes of unstructured data.

In the field of environmental research, the application of AI is expanding. AI can be used in climate modeling and biodiversity monitoring to develop proactive measures to reduce negative impacts on the environment. Climate and environmental modeling are needed to develop scenarios and make informed governance decisions for timely intervention. AI increases the accuracy and efficiency of calculations. Machine learning algorithms can analyze historical climate data to identify patterns and accurately predict future climate conditions. For example, deep learning models have been used to improve temperature and precipitation forecasts by assimilating data from various sources such as satellite sensors, weather stations, field capacities, camera traps, drones, and ocean buoys (Answer This 2024).

AI ensures the adaptability of decision support systems by processing relevant real-time data. This is important for predicting extreme events or simply atypical weather processes for a particular area. The information obtained is important for timely warning, notification, and preparation. Forecasts prepared using AI are more detailed, the speed of modeling intervention options increases, and it allows for calculating probable consequences and assessing the effects of various climate intervention strategies, for instance, carbon emission reduction policies. Thus, the use of AI increases the reliability of climate forecasts and chances of a quick response to natural disasters, supports the development of targeted and effective climate policy, and sustainable management of natural resources, and also facilitates prompt enforcement of environmental regulations.

Despite the benefits of the use of AI-based EDSS, there are still a number of barriers to overcome. They are related to the quality of data, the level of sophistication of the systems already in use with which the EDSS must be integrated and compatible, the dynamics of the regulatory environment, and the energy intensity of AI technologies.

Barriers to AI adoption include fragmented, unstructured, incomplete, duplicated, and error-prone data. Historical data may be limited or unavailable for some geographies or environmental variables. To create value, data must be standardized, organized, and manageable. Data quality is measured by accuracy, completeness, coherence, timeliness, validity, uniqueness, and consistency. Other metrics are also important—error rate, duplicate records, and data time-to-value, or the time it takes to process and analyze data for decision-making. As early as 2000, Cortés et al. (2000) raised the issue of the quality of the information available—the amount of data is huge and heterogeneous, coming from many different disciplines or sources (Cortés et al. 2000). The question is still relevant a quarter of a century later.

Training AI models on large and poorly controlled datasets creates vulnerabilities such as data poisoning (introducing malicious data into training sets), false data injection (sending false data that causes equipment to malfunction), and overreliance on artificial intelligence to analyze results can lead to incorrect interpretations and conclusions.

As data plays a key role in the application of AI to environmental decision-making, it is crucial to ensure that it is of appropriate quality. Poor quality data can lead to incorrect decisions, which, given the extreme sensitivity of this area, can have extremely negative consequences. Data development, collection, analysis, and dissemination processes must be of high quality and holistic. Timeliness in the use of data is essential. Accurate but outdated data can also lead to mistaken decisions. Modern technology can shorten the time between data collection, processing, and use. The speed of big data technologies gives people time to prepare and prevent irreversible changes. The value of the resulting data can be enhanced by ensuring a continuous flow of high-quality and timely data from national, international, and private big data sources and human-generated digital data. The data cycle should be consistent with the decision-making cycle (Data Revolution.org 2014).

Du et al. (2024) use a SWOT (strengths, weaknesses, opportunities, threats) approach to analyze the use of AI in decision-making in spatial planning, which involves multiple stakeholders. On the one hand, AI facilitates stakeholder involvement, but, on the other hand, the lack of data limits the operation of AI systems. The situation has changed little over the decades, even with the development of AI technologies and their proliferation in decision-making systems—data in different formats is generated in huge volumes, and its processing and proper storage require a lot of time and financial resources. Data bias, including that from social networks and crowdsourcing platforms, creates difficulties. An additional hurdle is the complexity of environmental data.

There are concerns about the environmental footprint of AI: the extraction of raw materials, rare earths, and minerals, which go into manufacturing the hardware used for AI, data center construction and operation, energy consumption, greenhouse gas emissions, water used for cooling, and also waste and e-waste (UN 2025).

The use of AI, particularly for decision-making, is associated with high energy consumption, which contradicts the idea of reducing the resources spent. The amount of energy and time required to train an AI model can have a significant impact on environmental quality, primarily due to its high carbon footprint (Konya and Nematzadeh 2024). Thus, one request to AI consumes ten times more energy than a standard Google search query. By 2026, the energy consumption of data centers will double compared to 2022 levels (Womble Bond Dickinson 2025). In the United States, for example, by 2030 the share of energy consumed by data centers may increase to 25% of all generated energy (4% now) (Wall Street Journal 2024). In this case, even if AI improves decision-making on energy efficiency, it could become one of the main drivers of energy demand. So, there is growing importance of collaboration between environmentalists and AI experts in developing more energy-efficient AI models for decision-making (Konya and Nematzadeh 2024).

There are costs associated with integrating AI into decision support systems, the amount of which depends on the scale of the system, data storage and management facilities, the complexity of the AI model, and the requirements of the infrastructure on which the model is deployed.

The use of AI in decision-making systems will require coordination with regulatory requirements. The regulatory framework for the use of AI is still evolving, and this process is likely to continue as technologies continue to improve while legislation fails to keep pace. Current regulations, for example, EU AI Act (EUR-lex 2024), focus on the fact that it is not always possible to understand why a particular AI decision has been made. The AI Act does not define specific cases of the use of AI in environmental decision-making; it considers the impact of AI on the environment in terms of the energy intensity of the technology. However, the Act defines four categories of risks associated with the use of AI: unacceptable, high, limited, and minimal. For example, the unacceptable risks potentially relevant to the environment include malicious manipulation and deception, and exploitation of vulnerabilities using AI. High risks include the security of critical infrastructure (e.g., transport), which is related to the life and health of citizens.

Given the scale and seriousness of the potential consequences of mistaken or intentionally spoiled environmental decisions, the requirements for the safety of such systems will constantly increase. There are some factors to consider that will allow effective use of the EDSS' potential (Fig. 11.2).

11.6 Use Cases

11.6.1 FireAId Project

The application of AI-based environmental decision-making systems is a pressing issue for both government, research, and commercial sectors. Initiatives are being launched at international, state, regional, and corporate levels to harness the

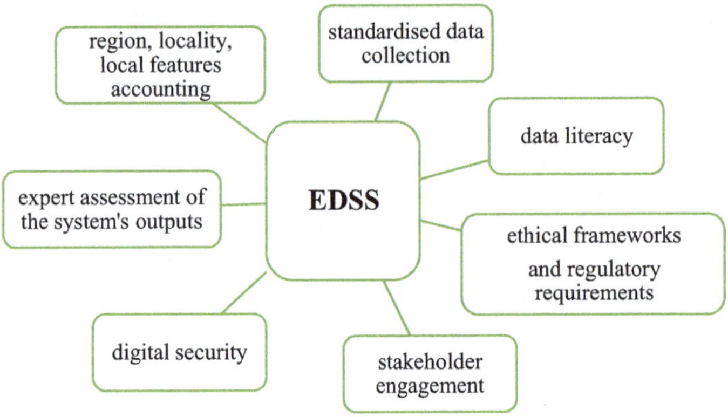

Fig. 11.2 Factors that ensure realization of EDSS potential. (Source: authors based on Ncube and Ngulube 2024)

potential of AI in decision support. The following are examples of the use of AI in decision support systems.

Globally, wildfires are becoming an increasingly complex problem due to both human factors and climate change. Today, in addition to massive amounts of historical fire data, experts have high-quality data collected in real time by the Internet of Things devices, sensors, drones, satellites, and ground cameras, or even from social media. Huge sets of relevant data, the high computing power of modern computers, and sophisticated AI models make it possible to develop better algorithms for fire detection and spread. WEF is implementing the *FireAId Project* (WEF 2023). One of the areas of work is related to the development of a dynamic risk map, optimized resource allocation, and proposals for response to forest fires. Efforts are aimed at building global cooperation in the field of fire management, coordinating the actions of its participants to develop globally scalable fire management support systems. The collected data also makes it possible to model forest fires in a specific location and to work out scenarios of the response of all services involved in fire situations.

One of the countries where the FireAId Project was implemented was Turkey (Koç Holding 2023). The technology was used as a decision-making tool to allocate and move resources between regions, plan fire patrol teams and firefighting maintenance, and make decisions regarding the chains of resource movements from regions with a low probability of fire to regions with a high probability of fire. Initially, the Turkish Ministry of Agriculture and Forestry used a wildfire risk assessment tool based only on meteorological parameters. However, domain experts and fieldworkers stressed that meteorological parameters alone are not sufficient for assessing wildfire risks. Energy transmission lines, vegetation, topographical characteristics, etc., had to be taken into account. Therefore, there was an urgent and fundamental need for a risk map that would take into account static and dynamic parameters (WEF 2023).

Daily experts checked the meteorological risk map in the morning and identified regions at risk. Based on this, responsible organizations in the region were informed of the risks, and areas were planned for patrolling. In the event of a fire, the fire departments responded. If necessary, the teams were able to call for help from neighboring regions without compromising the preparedness of those regions.

11.6.2 Air Quality Monitoring System City Air

The *Russian air quality monitoring system City Air* allows the identification of industrial emission sources and thus the prediction of environmental risks (City Air). The system can be used to model the dispersion of passive pollutants from sources, calculate the transport of chemically active pollutants in the air, and make forecasts of pollutant concentrations for 72 h. The models used take into account the local topography and allow a high degree of prediction accuracy to be achieved. The modeling is based on a forecast of the wind speed field and data on the main emission sources. Data from City Air's air pollution monitoring stations and a range of data from other available sources are used to train and refine the physical models for a given area. The following are examples of the use of AI in decision support systems.

11.6.3 Soil and Water Assessment Tool Virtual Extension Assistant

In late 2024, Texas A&M AgriLife Research and IBM released an AI tool based on big data sets for assessing soil and water conditions and making data-driven decisions—*Soil and Water Assessment Tool Virtual Extension Assistant* (SWAT VEXA). It allows users to make decisions in a variety of scenarios, including soil erosion, pollution control, and disaster risk. The system democratizes access to knowledge and speeds up decision-making (Phys.org 2025).

11.6.4 Severstal AI System

Severstal, one of the world's most efficient metals and mining companies, uses AI to collect and analyze data, predict and prevent negative environmental impacts, and optimize the use of natural resources. For example, the system analyses data from CCTV cameras located on different sides of the coke oven battery. If a coke oven gas emission is detected, the system registers the gassing and immediately alerts process personnel with an audible signal. The average time to eliminate gas emissions during the model's industrial operation was reduced by almost two times. The solution also automatically generates a daily report on the duration of gassing, which allows the condition of the unit to be monitored and new coke oven gas emissions to be prevented.

Thanks to the implementation of the model from November 2021 to March 2022, more than 100,000 cubic meters of coke oven gas have been prevented from escaping into the atmosphere, which meets the overall goal of reducing fugitive emissions (Severstal). In addition, controlling the gassing of the coke oven battery helps extend the life of the unit. Finally, the technological discipline of the personnel in this area has improved since the implementation: violations are now identified and corrected much more quickly.

11.7 Key Results and Outlook

Despite its long history, the application of AI systems to environmental decision-making is at an early stage and faces similar challenges to the application of AI in other fields: data quality and organization, availability of powerful solutions to handle large amounts of data, applicability of modern hardware and software solutions, and ethical and accountability issues.

AI is good at solving logical and probabilistic problems, but not at making justifications or taking into account the probabilities of unpredictable events (Chen et al. 2025). Despite the benefits of using AI in decision-making, it will not replace humans in the foreseeable future. Intelligent systems are effective at performing repetitive and data-driven tasks with fixed parameters. However, they struggle with tasks that require strategic planning, emotional intelligence, or fine-tuning. When it comes to understanding context, AI cannot replace humans.

Finding solutions to data-related problems requires collaboration between researchers, developers, suppliers of technology solutions, and government representatives in establishing data exchange mechanisms, creating data collection and storage infrastructure, and developing standardized formats (Answer This 2024). There should be no doubt about the reliability and quality of the data used in environmental decision-making systems, that is, the source of the data should be authoritative. This is a complex task that is being addressed at different levels of governance. For instance, the UN Biodiversity Lab is a repository of spatial data collected by satellites and national and international research groups and used to create maps consisting of 400 layers of the world's data on nature, climate change, and sustainable development (UN Biodiversity Lab). The goals of the system are to democratize access to verified spatial data and analytical tools as a digital public good, support decision-makers in the use of spatial data, and enable stakeholders to use spatial data for national monitoring and reporting. Another example of an attempt to create reliable databases is the International Methane Emissions Observatory (IMO) launched by UNEP in 2021 (International Methane Emissions Observatory).

The issues of data storage and processing are important. Tools for ensuring data quality are also needed, allowing to find solutions of duplication, manual correction of errors, and storage of data. The ability to quickly process large volumes of data in different formats and transform them into valuable information determines the demand for AI solutions. At the same time, the dynamics of the trend will depend on the development of the technology itself, in particular, on the ability of AI to analyze data based not on probabilistic relationships (probabilistic language), but on cause-and-effect relationships (WEF 2024). Causal AI is focused on understanding the cause-and-effect relationships of phenomena and processes, rather than recognizing data patterns (Garter 2023). It allows us to understand how decisions are made, whereas traditional AI models do not provide the ability to determine what caused a particular decision. Causal models are explainable, which means they increase user involvement in the decision-making process.

Another promising direction is focusing AI models on solving acute environmental problems, for example, monitoring water quality (Yang et al. 2022), managing the use of natural resources, and urban planning (Urban et al. 2021). More comprehensive and effective solutions can be obtained through the integration of AI, GIS, remote sensing, big data analytics, edge computing, and wireless communications.

Some of the most pressing issues surrounding the use of AI in decision support systems relate to privacy, security, and safety. The constantly evolving regulatory environment sets increasingly stringent requirements for the collection, storage, and distribution of data. This determines the growing attention to the creation of reliable data governance frameworks.

The use of AI in decision support systems in the field of ecology has great potential, related to both the speed of decision-making and the quality of decisions. However, there are barriers. Further research should focus on assessing and enhancing the potential of AI-based systems in environmental data collection, modeling, decision-making, and environmental management. As AI models become more complex, they lose transparency, posing challenging questions for engineering teams to answer when building and deploying AI agents. That is why the development of more interpretable models is becoming an increasingly important issue. Another question is related to this aspect—researchers note the importance of developing socially responsible and ethical approaches to the use of AI.

Acknowledgments This chapter is based on a study funded by the Basic Research Program of HSE University.

References

Agyemang P, Kwofie EM, Baum JI, Wang D, Kwofie EA (2025) Environmental-health convergence: a deep learning-oriented decision support system for catalyzing sustainable healthy food systems. Environ Model Softw 185. https://doi.org/10.1016/j.envsoft.2024.106309

Aleskerov F (2023) Decision-making under deep uncertainty. Available at https://www.hse.ru/en/news/research/879136988.html. Accessed 3 Mar 2025

Alrashdi I, Ali AM, Sallam KM, Abdel-Basset M (2025) Intelligent decision support framework for assessment of alternative vehicle technologies in transportation system: a sustainable approach toward environmental remedy. Sustain Futures 9. https://doi.org/10.1016/j.sftr.2025.100472

Answer This (2024) AI in environmental research. Available at https://answerthis.io/blog/ai-in-environmental-research. Accessed 10 Mar 2025

Ascough JC II, Rector HD, Hoag DL, Mcmaster GS, Vandenberg BC, Shaffer MJ, Weltz MA, Ahuja LR (2002) Multicriteria spatial decision support systems for agriculture: overview, applications, and future research directions. Environmental modeling international conference proceedings. In: Rizzoli AE, Jakeman AJ (eds) Integrated assessment and decision support proceedings of the 1st biennial meeting of the IEMSS, vol 3, Lugano, pp 175–180

Basso M, Li C (2025) How AI can move from hype to global solutions. WEF. Available at https://www.weforum.org/stories/2025/01/ai-transformation-industries-responsible-innovation/. Accessed 10 Mar 2025

Bell PC (1985) Visual interactive modeling as an operations research technique. Interfaces 15(4):26–33. Available at https://www.jstor.org/stable/25060718?seq=1. Accessed 10 Mar 2025

Bilgili A, Arda T, Kilic B (2024) Explainability in wind farm planning: a machine learning framework for automatic site selection of wind farms. Energy Convers Manag 309. https://doi.org/10.1016/j.enconman.2024.118441

Borrás S (2023) The governance of science, technology, and innovation policy missions for grand challenges in Europe //. The Tokyo Foundation for Policy Research. Available at https://www.tokyofoundation.org/research/detail.php?id=936. Accessed 11 Mar 2025

Boschetti F, Fulton EA, Bradbury RH, Symons J (2012) What is model, why people don't trust them and why they should. In: Raupach MR, McMichael A, Finnigan J, Manderson L, Walker B (eds) Negotiating our future: living scenarios for Australia to 2050, vol 2. Australian Academy of Science, pp 107–118. Available at https://www.per.marine.csiro.au/staff/Fabio.Boschetti/papers/Model2050.pdf. Accessed 10 Mar 2025

Buongiorno F, Chiaramonte X (2023) Environment. In: Wallenhorst N, Wulf C (eds) Handbook of the anthropocene. Springer, Cham. https://doi.org/10.1007/978-3-031-25910-4_8

Camps-Valls G, Fernández-Torres MÁ, Cohrs KH et al (2025) Artificial intelligence for modeling and understanding extreme weather and climate events. Nat Commun 16(1919). https://doi.org/10.1038/s41467-025-56573-8

Castillo A, Vall P, Garrido-Baserba M, Comas J, Poch M (2017) Selection of industrial (food, drink and milk sector) wastewater treatment technologies: a multi-criteria assessment. J Clean Prod 143:180–190. https://doi.org/10.1016/j.jclepro.2016.12.132

Chen Y et al (2025) A manager and an AI walk into a bar: does ChatGPT make biased decisions like we do? Manuf Serv Oper Manag 27(2):354–368. https://doi.org/10.1287/msom.2023.0279

City Air. Available at https://cityair.ru/ru/. Accessed 18 Mar 2025

Construction newspaper (2024) Artificial intelligence in construction. Available at https://stroygaz.ru/publication/technologies/iskusstvennyy-intellekt-v-stroitelstve/. Accessed 17 Feb 2025

Cortés U, Sànchez-Marrè M, Ceccaroni L, Roda IR, Poch M (2000) Artificial Intelligence and environmental decision support Systems. Appl Intell 13:77–91. https://doi.org/10.1023/A:1008331413864

Coussement K, Abedin MZ, Kraus M, Maldonado S, Topuz K (2024) Explainable AI for enhanced decision-making. Decis Support Syst 184. https://doi.org/10.1016/j.dss.2024.114276

Crunching the Numbers. Available at https://newsroom.kiewit.com/kieways/crunching-the-numbers/. Accessed 11 Feb 2025

Data Revolution.org (2014) A world that counts. Mobilizing the data revolution for sustainable development. Available at https://www.undatarevolution.org/wp-content/uploads/2014/12/A--World-That-Counts2.pdf. Accessed 10 Feb 2025

Du J, Ye X, Jankowski P, Sanchez TW, Mai G (2024) Artificial intelligence enabled participatory planning: a review. Int J Urban Sci 28(2):183–210. https://doi.org/10.1080/12265934.2023.2262427

E&ICT Academy (2025) Agentic AI vs. Generative AI: key differences and use cases in 2025. Available at https://eicta.iitk.ac.in/knowledge-hub/artificial-intelligence/agentic-ai-vs-generative-ai/. Accessed 13 Feb 2025

ESG Today (2020) Total and Google Cloud Bring AI Solutions to Home-Based Solar Energy Deployment. Available at https://www.esgtoday.com/total-and-google-cloud-bring-ai-solutions-to-home-based-solar-energy-deployment/. Accessed 13 Feb 2025

EUR-lex (2024) Regulation (EU) 2024/1689 of the European Parliament and of the Council of 13 June 2024 laying down harmonised rules on artificial intelligence and amending Regulations (EC) No 300/2008, (EU) No 167/2013, (EU) No 168/2013, (EU) 2018/858, (EU) 2018/1139 and (EU) 2019/2144 and Directives 2014/90/EU, (EU) 2016/797 and (EU) 2020/1828 (Artificial Intelligence Act) (Text with EEA relevance). Available at https://eur-lex.europa.eu/legal-content/EN/TXT/?uri=CELEX%3A32024R1689. Accessed 11 Feb 2025

Federal Agency for Technical Regulation and Metrology (2024) Artificial intelligence technologies in construction and road equipment. Terms and definitions. Available at https://pravdaosro.ru/wp-content/uploads/2024/12/gost-r-71750-2024.pdf. Accessed 01 Feb 2025

Filip F-G, Moisil I (2000) Decision support for environmental management. Environ Syst 3. Available at https://www.eolss.net/sample-chapters/c09/E4-20-04-00.pdf. Accessed 01 Feb 2025

Fingreen AI (2025) How AI is transforming sustainability in 2025: key trends to watch. Available at https://fingreen.ai/blog/ai-transforming-sustainability-2025-key-trends/#:~:text=In%20 2025%2C%20AI%20is%20driving,wildlife%20monitoring%2C%20and%20circular%20 economies. Accessed 01 Feb 2025

French S, Geldermann J (2005) The varied contexts of environmental decision problems and their implications for decision support. Environ Sci Policy 8:378–391. https://doi.org/10.1016/j.envsci.2005.04.008

Garter (2023) What's new in Artificial Intelligence from the 2023 Gartner Hype Cycle. Available at https://www.gartner.com/en/articles/what-s-new-in-artificial-intelligence-from-the-2023-gartner-hype-cycle. Accessed 01 Feb 2025

Gorry GA, Scott Morton MS (1971) A framework for management information systems. Sloan School working paper. Sloan Manag Rev 13(1) Available at https://dspace.mit.edu/bitstream/handle/1721.1/47936/frameworkformana00gorr.pdf?sequence=1&isAllowed=y. Accessed 03 Feb 2025

Guariso G, Werthner H (1989) Environmental decision support systems. Ellis Horwood-Wiley, New York. Available at https://books.google.ru/books/about/Environmental_Decision_Support_Systems.html?id=UyROAQAAIAAJ&redir_esc=y. Accessed 18 Feb 2025

Guillaume JHA, Kummu M, Rasanen TA, Jakeman AJ (2015) Prediction under uncertainty as a boundary problem: a general formulation using iterative closed question modeling. Environ Model Softw 70:97–112. https://doi.org/10.1016/j.envsoft.2015.04.004

Haagsma IG, Johanns RD (1994) Decision support systems: an integrated and distributed approach. Environ Syst II:205–212

Hamouda MA, Anderson WB, Huck PM (2009) Decision support systems in water and wastewater treatment process selection and design: a review. Water Sci Technol 60(7):1757–1770. https://doi.org/10.2166/wst.2009.538

Hochtief (2020) HOCHTIEF and the digital construction site. Available at https://www.hochtief.com/careers/inside-hochtief/hochtief-and-the-digital-construction-site. Accessed 08 Feb 2025

Koç Holding (2023) Davos Summit celebrates the success of Koç Holding's FireAId initiative, an AI-driven project to predict and fight wildfires. Available at https://www.koc.com.tr/media-center/latest-news/2023/davos-celebrates-success-of-koc-holdings-fireaid-initiative. Accessed 14 Feb 2025

Ink (2023) The AI market in Russia has grown by 17% in a year, but implementing the technology is still not easy. How does the government help with this?. Available at https://incrussia.ru/specials/ai-tech-growth/. Accessed 12 Feb 2025

International Methane Emissions Observatory. Available at https://www.unep.org/topics/energy/methane/international-methane-emissions-observatory. Accessed 11 Feb 2025

ITMO News (2023) Artificial Intelligence to design arctic infrastructure. Available at https://news.itmo.ru/ru/news/13331/ Accessed 01 Feb 2025

Jayanthi M, Kumaran M, Vijayakumar S, Duraisamy M, Anand PR, Samynathan M, Thirumurthy S, Kabiraj S, Vasagam KPK, Panigrahi A, Muralidhar M (2022) Integration of land and water resources, environmental characteristics, and aquaculture policy regulations into site selection using GIS based spatial decision support system. Mar Policy 136. https://doi.org/10.1016/j.marpol.2021.104946

Johnson LE (1986) Water resource management decision support systems. J Water Resour Plan Manag 112(3):308–325. https://doi.org/10.1061/(ASCE)0733-9496(1986)112:3(308)

Konya A, Nematzadeh P (2024) Recent applications of AI to environmental disciplines: a review. Sci Total Environ 906. https://doi.org/10.1016/j.scitotenv.2023.167705

Kumar A, Sah B, Singh AR, Deng Y, He X, Kumar P, Bansal R (2017) A review of multi-criteria decision making (MCDM) towards sustainable renewable energy development. Renew Sust Energ Rev 69:596–609. https://doi.org/10.1016/j.rser.2016.11.191

Li M, Tang Y, Wu K, Cheng H (2025) Autonomous vehicle pollution monitoring: an innovative solution for policy and environmental management. Transp Res Part D: Transp Environ 139. https://doi.org/10.1016/j.trd.2024.104542

Ma S-C, Fan Y, Yao X, Fang H, Xu C (2025) Cost and environmental impacts assessment of electric vehicles and power systems synergy: the role of vehicle-to-grid technology. Environ Impact Assess Rev 114. https://doi.org/10.1016/j.eiar.2025.107899

Massei G, Rocchi L, Paolotti L, Greco S, Boggia A (2014) Decision support systems for environmental management: a case study on wastewater from agriculture. J Environ Manag 146:491–504. https://doi.org/10.1016/j.jenvman.2014.08.012

Matthews KB, Rivington M, Blackstock K, McCrum G, Buchan K, Miller DG (2011) Raising the bar?—the challenges of evaluating the outcomes of environmental modeling and software. Environ Model Softw 26(3):247–257. https://doi.org/10.1016/j.envsoft.2010.03.031

Matthies M, Giupponi C, Ostendorf B (2007) Environmental decision support systems: current issues, methods and tools. Environ Model Softw 22(2):123–127. https://doi.org/10.1016/j.envsoft.2005.09.005

McIntosh BS, Ascough IIJC et al (2011) Environmental decision support systems (EDSS) development e challenges and best practices. Environ Model Softw 26:1389–1402. https://doi.org/10.1016/j.envsoft.2011.09.009

Merritt WS, Fu B, Ticehurst JL, El Sawah S, Vigiak O, Roberts AM, Dyer F, Pollino CA, Guillaume JH, Croke BF (2017) Realizing modeling outcomes: a synthesis of success factors and their use in a retrospective analysis of 15 Australian water resource projects. Environ Model Softw 94:63–72. https://doi.org/10.1016/j.envsoft.2017.03.021

Mysiak J, Giupponi C, Rosato P (2005) Towards the development of a decision support system for water resource management. Environ Model Softw 20:203–214. https://doi.org/10.1016/j.envsoft.2003.12.019

Ncube MM, Ngulube P (2024) Enhancing environmental decision-making: a systematic review of data analytics applications in monitoring and management. Discov Sustain 5:290. https://doi.org/10.1007/s43621-024-00510-0

Newman JP, Maier HR, Riddell GA, Zecchin AC, Daniell JE, Schaefer AM, van Delden H, Khazai B, O'Flaherty MJ, Newland CP (2017) Review of literature on decision support systems for natural hazard risk reduction: current status and future research directions. Environ Model Softw 96:378–409. https://doi.org/10.1016/j.envsoft.2017.06.042

Nie Q, Liu T (2025) Large language models: tools for new environmental decision-making. J Environ Manag 375. https://doi.org/10.1016/j.jenvman.2025.124373

Panagopoulos GP, Bathrellos GD, Skilodimou HD et al (2012) Mapping Urban water demands using multi-criteria analysis and GIS. Water Resour Manag 26:1347–1363. https://doi.org/10.1007/s11269-011-9962-3

Phys.org (2025) Generative AI for decision-making in agricultural and natural resources production, and management. Available at https://phys.org/news/2025-02-generative-ai-decision-agricultural-natural.html Accessed 11 Mar 2025

Poch M, Comas J, Cortés U, Sànchez-Marrè M, Rodríguez-Roda I (2017) Crossing the Death Valley to transfer environmental decision support systems to the water market. Global Chall 1. https://doi.org/10.1002/gch2.201700009

Reiter D, Meyer W, Parrott L, Baker D, Grace P (2018) Increasing the effectiveness of environmental decision support systems: lessons from climate change adaptation projects in Canada and Australia. Reg Environ Chang 18:1173–1184. https://doi.org/10.1007/s10113-017-1255-9

Rizzoli AE, Young WJ (1997) Delivering environmental decision support systems: software tools and techniques. Environ Model Softw 12(2–3):237–249. https://doi.org/10.1016/S1364-8152(97)00016-9

Robok (2023) AI-driven roadworks with Balfour Beatty & National Highways.. Available at https://robok.ai/f/ai-driven-roadworks-with-balfour-beatty-national-highways. Accessed 11 Mar 2025
Romanello M, Walawender M et al (2024) The 2024 report of the lancet countdown on health and climate change: facing record-breaking threats from delayed action. Lancet 404(10465):1847–1896. Available at https://www.thelancet.com/journals/lancet/article/PIIS0140-6736(24)01822-1/fulltext. Accessed 01 Mar 2025
Rossman LA (1980) Synthesis of waste treatment systems by implicit enumeration. J Water Pollut Control Fed 52(1):148–160. Available at https://www.scopus.com/record/display.uri?eid=2-s2.0-0018864133&origin=inward&txGid=a87b120b36fca18db6cc926e437b66b5 Accessed 07 Mar 2025
Schot J, Steinmueller WE (2018) Three frames for innovation policy: R&D, systems of innovation and transformative change. Res Policy 47(9):1554–1567. https://doi.org/10.1016/j.respol.2018.08.011
Schulte F, Voß S (2015) Decision support for environmental-friendly vehicle relocations in free-floating car sharing Systems: the case of Car2go. Procedia CIRP 30:275–280. https://doi.org/10.1016/j.procir.2015.02.090
Scinapse. Research Trends Insights (2024) Are we witnessing an inflection point in global research? Why research in 2025 will be a turning point for science.. Available at https://insights.pluto.im/global-research-in-2025/. Accessed 07 March 2025
Security Intelligence (2025) AI decision-making: where do businesses draw the line?. Available at https://securityintelligence.com/articles/ai-decision-making-where-do-businesses-draw-the-line/. Accessed 06 Mar 2025
Severstal. Monitoring leaks of raw coke gas through doors and fittings at the top of coke oven battery, No. 6. https://it.severstal.com/severstal-digital/kontrol-utechek-syrogo-koksovogo-gaza//
Suganya R, Leo Joseph LMI, Kollem S (2024) Understanding lithium-ion battery management systems in electric vehicles: environmental and health impacts, comparative study, and future trends: a review. Results Eng 24. https://doi.org/10.1016/j.rineng.2024.103047
Talakauskas DS, Dem'yanova AV, Abashkin VL (2024) Artificial Intelligence. Series of information and analytical materials "Artificial Intelligence", №11 // ISSEK HSE University. Available at https://issek.hse.ru/mirror/pubs/share/986408357.pdf. Accessed 06 Mar 2025
The Future of Commerce (2025) Construction trends 2025: a tech reckoning, workforce development, and greener building.. Available at https://www.the-future-of-commerce.com/2025/01/29/construction-trends-2025/. Accessed 11 Mar 2025
Trivyza NL, Rentizelas A, Theotokatos G (2018) A novel multi-objective decision support method for ship energy systems synthesis to enhance sustainability. Energy Convers Manag 168:128–149. https://doi.org/10.1016/j.enconman.2018.04.020
UN (2023) Goal 13: take urgent action to combat climate change and its impacts.. Available at https://www.un.org/sustainabledevelopment/climate-change/. Accessed 11 Mar 2025
UN (2025) AI and the environment: risks and potentials.. Available at https://unric.org/en/ai-and-the-environment-risks-and-potentials/. Accessed 11 Mar 2025
UN Biodiversity Lab. Available at https://unbiodiversitylab.org/en/. Accessed 19 Mar 2025
UNDP (2024) 2024 UNDP trends report. The landscape of development.. Available at https://www.undp.org/future-development/publications/2024-undp-trends-report-landscape-development Accessed 21 Mar 2025
UNEP. Digitalization for sustainability. Available at https://www.unep.org/topics/digital-transformations/digitalization-sustainability. Accessed 29 Mar 2025
Urban A, Hick D, Noennig JR, Kammer D (2021) With a little help from AI. Int J Urban Plan Smart Cities 2(2):19–33. https://doi.org/10.4018/IJUPSC.2021070102
Wall Street Journal (2024) Artificial intelligence's 'Insatiable' energy needs not sustainable, arm CEO says.. Available at https://www.wsj.com/tech/ai/artificial-intelligences-insatiable-energy-needs-not-sustainable-arm-ceo-says-a11218c9. Accessed 29 Mar 2025
Walling E, Vaneeckhaute C (2020) Developing successful environmental decision support systems: challenges and best practices. J Environ Manag 264. https://doi.org/10.1016/j.jenvman.2020.110513

WEF (2023) The next frontier in fighting wildfires: FireAId pilot and scaling. Insight report. Available at https://www3.weforum.org/docs/WEF_The_Next_Frontier_Wildfire_Fighting_2023.pdf. Accessed 29 Mar 2025

WEF (2024) Causal AI: the revolution uncovering the 'why' of decision-making. Available at https://www.weforum.org/agenda/2024/04/causal-ai-decision-making/. Accessed 21 Mar 2025

WEF (2025) Global Risks Report 2025. Available at https://www.weforum.org/publications/global-risks-report-2025/. Accessed 17 Mar 2025

Wind Systems (2023) AI applications in wind-energy systems. Available at https://www.windsystemsmag.com/ai-applications-in-wind-energy-systems/. Accessed 17 Mar 2025

Womble Bond Dickinson (2025) Energy suppliers embrace Artificial Intelligence (AI). Available at https://discover.wbd-uk.com/reports/global-energy-transition-2025/artificial-intelligence. Accessed 17 Mar 2025

Wrisberg N, Udo de Haes HA, Triebswetter U, Eder P, Clift R (2002) Analytical tools for environmental design and management in a systems perspective. Kluwer Academic Publishers, Dordrecht

Yakobson LI (2023) Strategic planning in conditions of uncertainty: adaptability for the sake of sustainability. J New Econ Assoc 4(61):230–236. https://doi.org/10.31737/22212264_2023_4_230-236

Yang L, Driscol J, Sarigai S, Wu Q, Lippitt CD, Morgan M (2022) Towards synoptic water monitoring systems: a review of AI methods for automating water body detection and water quality monitoring using remote sensing. Sensors 22(6):48. https://doi.org/10.3390/s22062416

Zasada I, Piorr A, Novo P, Villanueva AJ, Valánszki I (2017) What do we know about decision support systems for landscape and environmental management? A review and expert survey within EU research projects. Environ Model Softw 98:63–74. https://doi.org/10.1016/j.envsoft.2017.09.012

Artificial Intelligence for Urban Planning and Building Smart Cities

12

Alina Demekhina and Yulia Milshina

Abstract

Today profound transformation takes place in urban development, primarily driven by the advanced information and communication technologies (ICTs). The term "smart city" first emerged in the 1990s as a result of urbanization and the digital revolution. At that time, many people moved to cities, raised their living standards, and actively participated in the network economy. As a result, the concept of a smart city ceased to be an element of techno-utopian discourse and began to present smart technologies as a rational tool for solving various urban problems, forming a more open, flexible, and participatory format of the city. With the rapid development of AI, more tasks are transferred to smart algorithms. At its core, a smart city uses the power of AI to revolutionize urban areas, representing a shift toward more algorithmic, automated, and predictive models of city services.

This chapter seeks to clarify the multifaceted applications of AI in the areas of smart cities, highlighting its key role in shaping and driving urban development. The symbiotic relationship between AI and smart cities is explored in detail, demonstrating how AI technologies play a significant role in optimizing resource allocation, improving decision-making processes, and ultimately enhancing the quality of life of citizens. The chapter highlights the interrelationship between progress in fundamental and applied research on AI technologies and urban development, highlighting how concerted efforts to develop AI capabilities can propel cities toward a future characterized by sustainable growth.

A. Demekhina · Y. Milshina (✉)
HSE University, Institute for Statistical Studies and Economics of Knowledge, Moscow, Russia
e-mail: ademekhina@hse.ru; ymilshina@hse.ru

B. Thokchom et al. (eds.), *Artificial Intelligence Enabled Real Time Environmental Monitoring*, https://doi.org/10.1007/978-981-95-3767-9_12

Keywords

Smart cities · Urban area · Artificial intelligence · Machine learning · Big data · Urban planning · Urban development · Sustainable development

12.1 Core Topics and Considerations

Rapid urbanization presents complex challenges for cities, ranging from managing dense populations and optimizing resource allocation to addressing environmental sustainability. In response to the growing challenges of urbanization, the concept of smart cities, which rely heavily on ICTs, has been introduced in both developed and developing countries (Aguilera et al. 2016; Galindo 2014). AI emerged in the 1950s as a game-changing technology but developed slowly as a consequence of immature complementary technologies. Now, it is deployed at a scale of cities and megacities (Yigitcanlar et al. 2020). The various computational technologies employed are tailored to specific functions, and these technologies play an increasingly critical role in public service delivery (The OECD Observatory of Public Sector Innovation 2023). The majority of smart city initiatives and technologies are focused on generating data about city dynamics (Kar et al. 2017). By employing AI methods, cities can harness and analyse this data to make the best possible decisions (Liu et al. 2019). The further prospective benefits of AI for cities are still being explored in the context of smart cities—that are enabled by community, technology, and policy to enhance productivity and livability (Desouza et al. 2015; Batty 2018).

Smart cities are designated as the "catalysts" for economic growth, which are grounded on a technology-powered base. By 2026, smart cities are projected to generate an economic benefit of $20 trillion on a global scale (Cyber Defense Magazine 2019). The primary socioeconomic and technological trends in building AI-powered smart cities are citizen-centric design (focusing on citizen needs), data-driven decisions (with the application of AI-powered analytics), and sustainable development (with a focus on gaining city resource efficiency).

AI is based on key technological branches: machine learning (ML), neural networks, and big data processing. ML is a method that allows computers to learn based on incoming information. The model is built on a large volume of input data, analyzes it, and forms predictions or decisions. All this happens without the need for explicit programming at each stage. Another technology, neural networks, are multilevel systems that can automatically learn and adapt. Their use allows for solving problems related to image, speech, and text recognition. Another essential part of AI is the big data processing. Modern cities generate amounts of information—from traffic and resource consumption to population information. Being able to process a large amount of received data, AI can make the most of them to solve complex problems in an urban environment.

Traditionally, AI meant algorithms that do not involve calculations according to some pre-known formula, but rather search for solutions in conditions of great uncertainty and incomplete data. A more complex version involves the ability to analyze and independently model specific processes and solve problems that require professional skills and knowledge. The advanced AI technologies to date are combined with the concept of *artificial general intelligence (AGI)*. AGI can be defined as a synthetic highly autonomous intelligence that performs a wide range of tasks and has good generalization ability across different contexts when performing heterogeneous tasks. In other words, AI can perform tasks as successfully as a human does (McKinsey 2024; Mitchell 2024).

The integration of AI with the IoT has given rise to a transformative paradigm known as *artificial intelligence of things (AIoT)*. This convergence has numerous points of application especially in the areas of smart cities. The technical components of the AIoT framework embody a robust structure that seamlessly integrates the computing and analytical power of AI with the ubiquitous connectivity and plentiful city data generated by the IoT (Al-Turjman et al. 2021; Bronner et al. 2021).

The following AI technologies, or subtechnologies, are employed in several ways to create application solutions for data-intensive smart cities: natural language generation (NLG), text analytics and natural language processing (NLP), image recognition, speech recognition, emotion recognition, ML platforms, DL platforms, virtual agents, decision management, robotic process automation, biometrics, cyber defense, content creation, and marketing automation. The combination of these technologies makes it possible to form the so-called "urban super brain."

The literature on AI in smart cities is worth to be observed using a bibliometric analysis approach. An example of the results of such approach, using the Scopus and Web of Science bibliographic databases, is given next. These databases provide an extensive variety of scientific articles and research papers with broad availability and thematic coverage across multiple scientific disciplines. The bibliometric research began by focusing on publications that included the terms "smart city" and "artificial intelligence" (research query for Web of Science was: TS = "smart city" AND "artificial intelligence"; and research query for Scopus was: TITLE-ABS-KEY ("smart city" AND "artificial intelligence")). The search focused on items published between 2010 and 2023. A significant number of publications appeared in both databases between 2018 and 2023. Before this period, references to AI in the context of smart cities were rare and constituted "emerging topics" (Fig. 12.1).

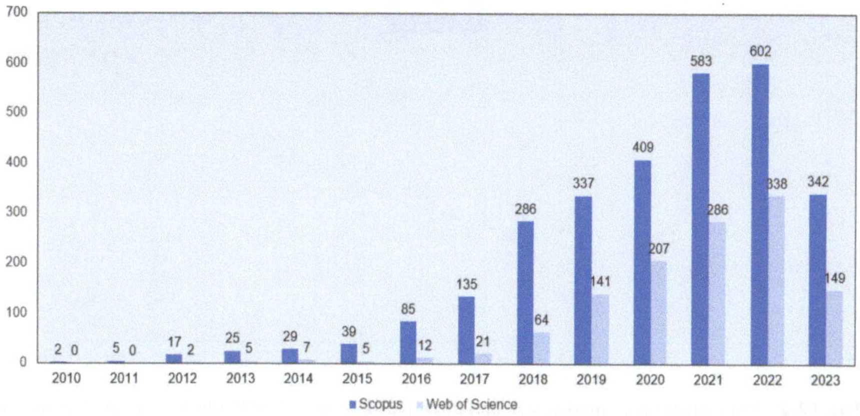

Fig. 12.1 Bibliometric analysis of literature on AI in smart cities. (Source: Authors based on Szpilko et al. (2023))

The long-term and lasting impact of the COVID-19 pandemic and environmental crisis highlights the challenges that must be addressed in urban intelligent technology implementation. Governments are starting to tackle the problems posed by fast urbanization. All technological advances carry potential threats and unintended consequences relating to cities and government agencies, compounded by the critical role of public services and the sensitive nature of personal data. For example, 66% of 167 cities surveyed by ESI ThoughtLab had invested in AI, and 80% would do this within the next 3 years (ESI ThoughtLab 2021). A study by S&P Global showed that public sector representatives emphasized the need to sustain robust security measures to maintain the integrity of smart city initiatives (S&P Global 2024). Managing these measures requires a proactive approach, including proper system design and protection for secure smart city data usage. Despite much enthusiasm for smart city initiatives, cautious adoption is warranted to avoid foreseeable "pitfalls."

The global AI in smart cities market size was valued at USD 45.6 billion in 2024 and is projected to grow to USD 208.3 billion by 2033, expanding at a compound annual growth rate (CAGR) of 18.4% from 2024 to 2033 (Fig. 12.2). The AI market for smart cities has been experiencing rapid growth, driven by advancements in technology and the growing demand for sustainable urban solutions that are becoming essential for managing the complexity of urban environments.

Currently, the key concept is smart sustainable cities (SSC), for the implementation of which two strategies are proposed—the introduction of AI-powered functions into existing cities using the "add-on" method (brownfield) (examples include Moscow, Dublin, Sao Paulo), and the construction of smart cities as new territorial units (greenfield) (examples include Songdo, Republic of Korea; Toyota Woven City, Japan; Masdar, UAE).

This chapter consists of the following parts: the overview of AI applications in smart cities in various urban management systems (including digital governance,

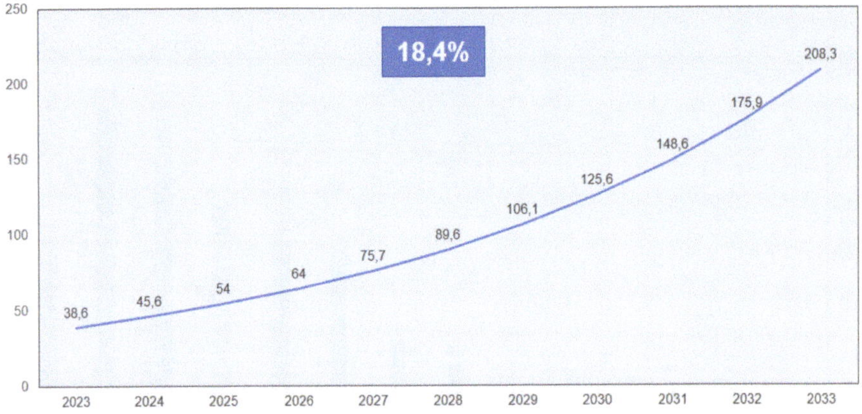

Fig. 12.2 AI in smart cities market size and growth 2024–2033, USD billion. (Source: Cervicorn Consulting (2025))

urban planning, urban growth monitoring, etc.) and sectors (transport and logistics, energy, water, waste management, etc.); key obstacles and challenges of AI implementation in smart cities; directions of future development of AI solutions for smart urban planning and management.

12.2 Application of AI in Smart Cities and Urban Planning

With a focus on initiatives to improve the quality of life of citizens, intelligent decision-making, and combat climate change, the areas of AI application that are highly applicable in various areas of design, construction, and maintenance of smart cities, as well as depending on the function that the urban infrastructure serves, are shown next. The visual landscape of smart city areas that are of research interest is shown in Fig. 12.3.

12.2.1 Smart City Management

Megacities are experiencing significant growth: currently, over half of the global population—approximately 4.4 billion individuals—resides in urban areas. This trend is projected to persist, with the urban population anticipated to more than double by 2050, resulting in nearly 70% of people living in cities.[1] The challenges posed by overloaded infrastructure, traffic congestion, environmental issues, accidents, and other urban difficulties necessitate that city officials not only engage in long-term planning but also respond swiftly to immediate concerns. Traditional management strategies are increasingly inadequate for handling this demanding workload, thereby highlighting the importance of digitalization.

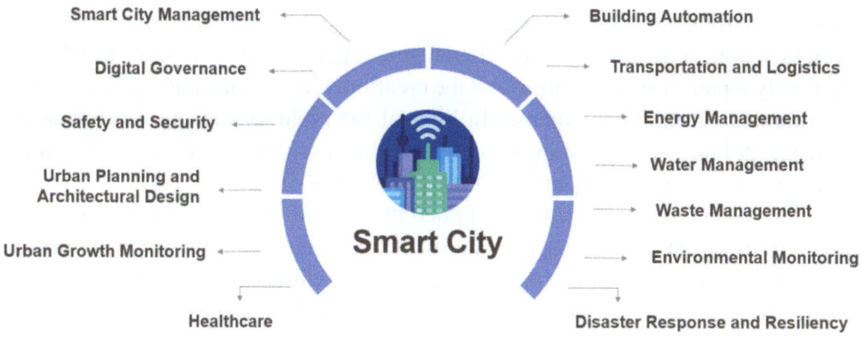

Fig. 12.3 Areas of a smart city visualization. (Source: Authors)

[1] World Bank Group. Urban Development Overview. URL: https://www.worldbank.org/en/topic/urbandevelopment/overview

Digital ecosystems of urban services that aggregate and analyze data from many different sources make it possible to repeatedly increase the efficiency of urban management, and the level of security in the city, and actively involve citizens in the decision-making process.

The main elements of intellectual city management include operating systems, city management panels, transport systems, energy networks, connected sound sensors, video surveillance systems, Wi-Fi traffic monitoring systems, dynamic lighting systems, sharing platforms, and more. The most important issues of these elements are concentrated on data safety, protection confidentiality, and guarantees of the inclusion of citizens as a proactive side.

At the same time, flexible modeling is provided by *city information modeling* (CIM)—dynamic three-dimensional city models or platforms, combining the geospatial data and data of urban facilities, with using AI, ML, IoT, virtual and mixed reality to optimize processes and forecast city needs. For example, CIM-based automated systems offer solutions to improve transport infrastructure, energy consumption, and resource management. The creation of interactive and realistic models of cities is possible thanks to the development of IoT.

The expansion of the use of *digital twins*, which creates digital copies of physical buildings, allows monitoring in real time, tests various scenarios for object management, and optimizes building construction operations. The concept of a digital twin for a city represents a groundbreaking approach to urban management. This technology facilitates not only forecasting and development planning but also the rapid resolution of day-to-day issues. Integrating data into a cohesive system enables real-time analysis, allowing for the identification of even minor problems that might be overlooked by conventional methods. For instance, it can pinpoint the location of accidents, road surface damage, or unauthorized waste dumping as they occur. The prompt detection of such issues empowers city services to act immediately, thereby reducing any inconvenience for residents and avoiding potential repercussions.

Moreover, a digital twin aids in scenario modeling, transportation route optimization, environmental enhancement, and ensuring transparency in processes. It addresses both broad and specific challenges, ranging from emission reduction to the timely repair of street lighting and the creation of pedestrian and cyclist-friendly environments at minimal expense. In light of the rapid expansion of urban areas, these technologies are becoming essential for the development of smart cities of the future, enhancing the safety and comfort of residents' lives.

Digital twins are currently being applied in countries that are actively engaged in digital transformation. For instance, in Shanghai, a city digital replica encompasses over 100,000 components of the urban landscape, facilitating the planning of emergency response strategies (Shanghai Daily 2022). In Zurich, architects leverage this technology to evaluate the effects of new constructions on transportation and lighting systems (ETH Zurich 2023). In Rennes, France, the implementation of a digital twin has led to a 70% reduction in the construction duration of metro stations, while in Kunming, China, it has enhanced transportation management, resulting in a 20% decrease in traffic congestion. In New South Wales, Australia, this technology aids in project planning by providing a comprehensive map that includes 22 million trees and 540,000 buildings. The World Economic Forum reports that such systems can enhance management efficiency and lower costs by 35%, contributing to improved environmental conditions in cities like New York (WEF 2022). These examples illustrate the significant potential of digital models in optimizing urban processes at various levels, enhancing urban practices, and preventing design flaws in city planning.

12.2.2 Digital Governance

The concept of "smart governance" seeks to govern effectively and improve urban outcomes through technological innovations. Smart governance is applied to improve administrative services and focuses on *citizen-oriented digital platforms*, and AI is revolutionizing administrative operational processes by enabling government-to-citizens (G2C) and government-to-business (G2B) services, improving multilingual communication for various social groups. Despite limitations and not yet reaching its full potential, various trends in the use of AI in public administration are beginning to be applied worldwide. The countries or regions that are approaching AI in smart governance are the United States, China, the European Union, India, Japan, and Brazil (IBM 2024).

AI applications used in smart governance include data analytics tools and virtual assistants. Leveraging AI-driven platforms, city governments can streamline operations and enhance citizen engagement. For example, Bloomberg has created a new digital platform City AI Connect,[2] designed to be a "one-stop shop" for local government leaders to collaborate with colleagues across cities to brainstorm, develop, and test new features of generative AI. AI-powered virtual assistants and NLP-powered chatbots help citizens to access information and services. AI-powered virtual assistants are already widespread in many industries, and with their help, cities will evolve into digital, customer-centric platforms as "city brains" to orchestrate city activities, and provide event correlation, root-cause analysis, predictive analytics, incident management, and operational intelligence.

NLP-powered chatbots are transforming routine processes, offering citizens day-and-night access to services, and streamlining workflows for city staff. In cities across Germany, Austria, and Switzerland, innovative chatbot solutions are setting new standards for "responsive" city governance. According to one study, the key outcomes of AI-chatbot deployment in the US government services are optimized workloads, improved communication, shorter wait times, and a broader range of use cases (Chen and Gascó 2024). Chatbot implementations still face familiar obstacles such as data selection and accessibility. Research has revealed an obvious but important fact: understanding the nuances of residents' language preferences requires constantly updating the chatbot's knowledge base. According to some estimates, generative AI will not replace the current use of search-based chatbots anytime soon because organizations want to prioritize consistency in the responses and

[2] Johns Hopkins University. City AI Connect. URL: https://cityaiconnect.jhu.edu

reactions chatbots convey to the public to reduce legal risks (Government Technology 2024).

A primary example of deploying AI for smart governance is Singapore, which focuses on ways to expand AI use cases within sectors such as healthcare and education and has launched more than 100 solutions using generative AI. This initiative is a joint project of the Ministry of Digital Development and Information, Digital Industry Singapore, Smart Nation Group, and Google Cloud (The Straits Times 2024). As part of this initiative, the group of MIT researchers are working on human–machine interactions and are continuing to build on AI initiatives in Singapore. This collaboration hopes to ensure that AI technologies are integrated into the city's infrastructure and services (MIT 2023).

12.2.3 Safety and Security

Public sector and personal data are increasingly becoming objects of cyberattacks. Crimes and offenses are becoming high-tech and long-lasting, challenging the stability of the socioeconomic ecosystem. To secure the frontiers of growth, security operations need to be spearheaded by reliable and affordable technological solutions.

A productive AI-powered public safety framework consists of three main components:

- Data management
- Advanced analytics
- Response optimization

Each plays an important role in creating a proactive urban security approach. AI-powered public security systems rely on a vast array of data types that are persistently integrated into an analytics system, with sources such as real-time street camera footage, emergency call recordings, social media data, environmental parameters, traffic patterns of urban vehicles, weather and climate data, and historical crime data. Predictive policing systems analyze the collected data using a variety of advanced techniques: pattern recognition, behavioral analysis, temporal–spatial correlation, risk assessment, and anomaly detection systems. AI further assists public security by optimizing response strategies through resource allocation, real-time patrol route planning, emergency response optimization, interagency coordination, and city community engagement.

For example, Singapore's Smart Nation[3] initiative includes a comprehensive approach to urban security that integrates AI-powered predictive policing algorithms with smart city ICT infrastructure. This system demonstrates how predictive policing can be part of a multicomponent ecosystem of hyperconnected urban services. As a result, Singapore has reported a 35% decrease in street crime and a 50% increase in the productivity of police patrol force distribution.

[3] Smart Nation Singapore. URL: https://www.smartnation.gov.sg

One of the most noteworthy AI applications is *facial recognition technology (FRT)*. Facial recognition as a key technology bolsters smart cities with advanced security, integrating transport, cyber, border control, and crime prevention seamlessly. In the past decade, FRT has gained momentum, contesting to be one of the most reliable and affordable technologies for modern city economies. With a focused trend toward technology and AI advancements, FRT offers one-of-a-kind security solutions that cater to the diverse infrastructural requirements of a smart city (Shao et al. 2021). FRT surveillance at border checkpoints enables rapid background checks, verification of legal requirements for entry into smart city areas, and verification of whether individuals are subject to cross-border restrictions. Legal investigations are increasingly becoming technology-driven in terms of video monitoring, lie detection, global database integration, and intelligent alert systems. However, these processes are not fully automated and rely on manual commands. This leaves room for human error, which can lead to false accusations and delays in response. Multifactor authentication, implemented using FRT, allows for identity verification and re-verification, preventing access violations to the city's critical infrastructure and protecting virtual borders and resources to secure smart infrastructures.

AI use cases also include, for example, forecasting payment transactions through online transactions, identifying abnormal behavior with computer vision tools, and monitoring migration flows within the city at different times to plan for the overload of urban infrastructure. When considering the "blue-sky" vision for AI within public safety, it is expected that AI tools have the great potential to reshape public safety by automating data collection and analysis, with the generation of "judgement" points and conclusions, which are critical to making high-impact, high-priority decisions.

12.2.4 Urban Planning and Architectural Design

Urban planning is a multifaceted field that encompasses both technical and political aspects, concentrating on land utilization, the design of constructed environments, transportation systems, infrastructure development, population distribution, and the accessibility of facilities. The examples of AI-powered urban planning practices break the patterns of traditional urban planning. These examples are clear cases of building more "renewable living" and making optimal cultural and social structures for data-saturated smart cities.

It is easy to understand that smart cities do not have to be fresh start projects. Using AI tools to improve existing cities that carry hundreds of years of human history is the most challenging mission in urban planning. The example is the Barcelona Smart City project,[4] which revitalizes public spaces by integrating AI technology, thereby improving the city's livability. In addition, Barcelona has embraced open data initiatives, encouraging public sector agencies to employ the numerous urban

[4] Barcelona Smart City. URL: https://www.barcelona.cat/infobarcelona/en/tema/smart-city

planning tools and real-time information centers that continuously improve the city's operations and services.

Urban planning practices are directly related to measuring the quality of urban spaces and urban condition assessment. The use of AI in this area is gaining popularity: researchers conduct experiments with satellite imagery to quantify urban sprawl and calculate the required infrastructure facilities for deployment. ML tools create large-scale maps of poverty, wealth, and income in developing countries, with the aim to calculate an objective measure of urban decay. Examples include the urban mapping practices to determine the location of physical attributes of decay (potholes, trash, tents, broken windows, dilapidated facades, utility markings, etc.) by training ML models and further urban blight indicating to measure changes in the prevalence of urban blight over time in San Francisco and Mexico City (Vallebueno and Lee 2023).

Harnessing AI-powered generative design has significantly expanded the solution space, facilitating a fusion between traditional and innovative methods. Generative design is based on the usage of computational capabilities in order to optimize the search for feasible solutions to various design problems. AI algorithms favor the analysis of large data sets, or heterogeneous collections of data elements, providing novel findings to an otherwise cumbersome process, especially when the list of factors a designer or engineer should consider. Through the expansion of designers' reference frameworks, the key objective of AI-driven generative design is to deliver entirely innovative alternatives that differ from conventional models. Generative AI-based solutions used in urban planning allow assessing the potential of land plots, taking into account environmental and social factors, and predicting the impact of future projects on the urban environment. In this regard, AI tools are applicable in the work of city governments.

PromeAI, DeepBlocks, Gendo, and Autodesk Forma are just a few examples of such generators. For example, the NVIDIA's AI System helps in integrating styles and forms in architectural sketches and assists in optimizing the shape of a building in accordance with environmental parameters (NVIDIA 2023). Zaha Hadid Architects has implemented generative AI tools in its design process, significantly increasing productivity and efficiency. The firm has reduced rendering times and improved design innovation by using platforms such as Midjourney and Gendo for text-to-image generation and custom AI models.

Traditionally, urban planning has operated under a "top-down" approach. However, the advent of generative AI is transforming this paradigm by facilitating greater citizen involvement and engagement, thereby simplifying the often complex bureaucratic processes. Residents will be able to utilize AI-driven tools, such as conversational chatbots, to navigate intricate zoning and building codes, as well as to interpret sometimes conflicting engineering standards and planning regulations. Additionally, companies and governmental bodies will be able to utilize AI and immersive VR to create visual representations of prospective urban landscapes, enabling citizens to grasp and actively participate in the planning process. Nonetheless, there are apprehensions that the emphasis on modeling and algorithms in smart cities may overlook certain qualitative aspects of urban life that are challenging to measure (Fortune 2024).

12.2.5 Urban Growth Monitoring

Urban population growth and the trend toward the development of urban agglomerations provide new perspectives for exploring and monitoring urban expansion. AI helps urban planners make data-driven and data-informed decisions about urban growth and makes it practicable to produce detailed forecasts, which facilitates the determination of the most effective strategies for infrastructure planning. The main types of ML models used for urban expansion prediction include shallow learning (e.g., random forest, support vector machine) and DL architectures, taking into account the influence of diverse variables such as demographics, environmental factors, and land use parameters.

Using GIS frameworks with multilayer perceptron (a type of ANN consisting of multiple layers of neurons) to simulate a wide array of urban growth scenarios for informed expansion of urban infrastructure. For example, by achieving 93.8% accuracy in urban growth prediction in Kathmandu Valley in Nepal, it became possible to optimize spatial patterns for future urban growth under three scenarios: baseline, environment-friendly, and resource-efficient sustainable development. Spatial patterns have provided insights into urban growth trajectories up to 2040 (Bharti and Biswas 2024).

Geospatial big data is currently attracting great attention and is at the forefront worldwide, and Google Earth Engine[5] is currently the platform for big data processing for remote sensing. Overall, Google Earth Engine has already proven its reliability in achieving the purpose of urban expansion assessment. Google Earth Engines and ML algorithms have been widely used by researchers for urban change detection and prediction, for example, in Singapore (Sidhu et al. 2018), Greenville County (the United States) (Zurqani et al. 2019), City of Kigali (Rwanda) (Mugiraneza et al. 2020), Rondônia State (Brazil) (Floreano and de Moraes 2021), Datong City (China) (Xue 2021), Munneru River Basin (India) (Loukika et al. 2021), and Raiganj Municipality (India) (Saha et al. 2023).

12.2.6 Healthcare

Healthcare is a vital part of any urban system and ensuring it is accessible, efficient, and of high quality is critical to the well-being of the community. With the rapid advancement of technology, new solutions are emerging that can revolutionize urban healthcare systems. The fundamental technologies that underpin smart city healthcare services are IoT devices, wearables, telemedicine, electronic health records, mobile health (mHealth), and biometric sensors, which are most promising to be managed by AI algorithms (Mohammadzadeh et al. 2023).

In recent years, with the development of AI and IoT, the digitalization of healthcare has entered a new phase. Smart devices and wearable technologies allow

[5] Google Earth Engine. A Planetary-Scale Platform for Earth Science Data & Analysis. URL: https://earthengine.google.com

real-time monitoring of health status, and ML algorithms help in the diagnosis and prognosis of diseases. The integration of various healthcare systems and the exchange of data between institutions are becoming a global standard, which improves the quality of service and increases the efficiency of healthcare organizations. The concept of *4P Medicine* is developing, which assumes the functioning of the healthcare system on the principles of personalization, predictability, prevention, and participation (Flores et al. 2013). 4P Medicine takes into account the characteristics of each patient, helps identify and reduce individual risks of disease development through prevention and early diagnosis, and also provides the opportunity for independent health management, thereby setting the landscape of healthcare of the future.

One notable approach is the telemedicine services that allow patients to receive medical consultations from remote locations (Burrell 2023). Additionally, smart city infrastructure is used to improve transportation options and provide easier and more convenient access to healthcare facilities for people. A demonstrative example is the use of portable technology (wearables) by healthcare providers to continuously monitor patient health and identify potential health issues at the early stages. This approach to preventing early illness will greatly contribute to patient outcomes and general health improvements. AI algorithms can foresee and follow the spread of irresistible infections by investigating information from an administration, medical services, and different sources. Therefore, AI can play a crucial part of worldwide general well-being as a solution for combatting epidemics and pandemics (Sowmitha et al. 2022).

Generative AI algorithms can be used to model the flow of patients and medical staff, optimizing the use of space in a hospital. For example, algorithms can analyze patient data to determine which areas are being used the most and then allocate space accordingly (Munavalli et al. 2021; Varnosfaderani and Forouzanfar 2024).

Another promising concept that is implemented with AI is *ambient assisted living (AAL)* (Guerra et al. 2023), which focuses on using technology to improve the life quality and well-being of people with limited mobility or the elderly, both indoors and outdoors, which is suitable for the implementation of the so-called "smart silver city" format. It aims to give people the opportunity to maintain their independence and autonomy while ensuring their safety and providing assistance when needed. Human activity recognition is widely regarded as the most popular methodology in the field of ambient assisted living, namely the automatic detection and classification of activities using sensor systems, wearable and fixed sensors and algorithms ranging from simple threshold methods to more advanced DL approaches.

The urban smart healthcare sector is making incremental progress as it learns to utilize the appropriate big data frameworks and models. IoT-enabled healthcare devices generate volumes of data (big IoT data) that can be analyzed to better guide healthcare delivery. The key component of the successful functioning of this system is the capacity to manage data while taking care of all the basic ethical aspects of security, such as data confidentiality, authorization, and authentication. In the near future, it is conceivable that smart hospitals will integrate into a broader, interconnected ecosystem that supports the achievement of both population health and preventive healthcare objectives.

12.2.7 Building Automation

Smart buildings are considered one of the main elements of the building environment in a smart city. The first definition of a smart building was published in 1989 by the US Intelligent Building Institute, and a smart building was described as "providing an efficient environment through optimized structures, systems, services and management, and the relationships between them" (Ghaffarianhoseini et al. 2016). Smart cities, whose operational efficiency largely depends on buildings, are the most noticeable trend in creating a holistic environment of the future (Apanaviciene et al. 2020; Albino et al. 2015).

The recent active implementation of AI in building automation systems has led to new opportunities in terms of using modern programming tools and languages, convenient visualization for different user groups, and has opened the way to the rapid processing of large volumes of data from engineering equipment. These methods for managing building engineering equipment facilitate the evolution from an *intelligent building* to a *cognitive building*. AI-based applications can be divided into conditional categories, such as optimized management of facility equipment, infrastructure and areas; load management; preventive maintenance; improving the efficiency of personnel use; provision of paid services; improving the efficiency of sensor use, etc.

The capabilities of AI and ML technologies allow for intelligent operation of buildings, which is technically achieved by such widely used algorithms as multi-layer perceptrons (MLP), generative adversarial networks (GAN), convolutional neural networks (CNN), recurrent neural networks (RNN), networks with long short-term memory (LSTM), and networks with radial basis functions (RBFN). Predictive maintenance and AI-based analytics help minimize damage to systems and buildings, reduce downtime, and optimize the overall performance of the building ecosystem.

Advanced sensors, AI, and IoT capabilities will optimize space and resources based on the number of people in a building at a given time. Data aggregated by sensors and detectors can be used to enable or disable HVAC systems (heating, ventilation, and air-conditioning) for maintaining good indoor air quality (IAQ). Occupant movement data can help identify underutilized space or, conversely, space that requires expansion, which is especially important for the organization of flexible common spaces after the COVID-19 pandemic. The Internet of Things is also driving developments in the field of fire safety beyond standard alarms and emergency lighting. For example, sensors, fire alarms, wireless communications, and AI will help determine the location of fires faster and more accurately.

The major tech companies develop applications to transform the industry, every process, and every product. Schneider Electric has pioneered open innovation platforms for buildings that enable rapid connectivity of IoT devices. The company offers a wide range of products and services, including building automation and industrial safety systems. Honeywell has leveraged its vast talent pool to develop innovative solutions. Their Honeywell Vector Space Sense has revolutionized the use of building space by analyzing data from multiple sources, enabling efficient space optimization. Cisco is a recognized leader in networking equipment, high-tech products, services, and telecommunications equipment, offering solutions for intelligent buildings through optimized lighting, building automation, and IoT technologies.

Startup platform Enertiv[6] uses neural networks to monitor energy consumption at sites and in buildings. AI analyzes the data and suggests measures to reduce energy costs, which can reduce energy costs by 20–30%. Startups like 75F[7] and Verdigris Technologies[8] are leading the way in making energy efficiency an exciting, user-friendly, and installer-friendly approach to saving energy, combined with occupant-oriented services.

The Edge[9] is a building in Amsterdam, the Netherlands, and is often called one of the smartest and greenest buildings in the world, and uses AI to create a highly efficient and personalized work environment. AI systems manage lighting, climate control, and energy consumption by analyzing data from a network of sensors, optimizing energy consumption and improving occupant comfort. At 45 Broadway, the implementation of the BrainBox AI[10] system resulted in a 15.8% reduction in HVAC energy consumption, resulting in savings of over $42,000 and a significant reduction in the building's carbon footprint. The AI optimizes temperature settings by analyzing real-time weather and occupancy data. Living Tomorrow[11] is an innovation campus in Vilvoorde, Belgium, highlighting various AI applications in building management and everyday life. The campus features smart home automation systems, AI-powered energy management, and intelligent security solutions. Living Tomorrow demonstrates how AI can create more efficient, sustainable, and user-friendly building environments, providing a glimpse into the future of smart living spaces.

The end user will switch from systems and devices that need to be selected and configured independently to complex integrated end-to-end solutions. This will change the business strategies of market players: they will focus on the creation and maintenance of smart building ecosystems. AI is expected to play a key role in the creation of smart cities, where buildings and infrastructure will be integrated into a single network controlled by AI. This will help significantly improve the quality of life and sustainability of cities. Such ecosystems will optimize energy consumption: for example, automatically adjust lighting and heating depending on the presence of people and weather conditions. Neural networks will be used to develop and manage construction materials that can change their properties depending on operating conditions. This will allow for the creation of more durable and sustainable buildings.

---

[6] Enertiv. URL: https://www.enertiv.com

[7] 75F. URL: https://www.75f.io

[8] Verdigris. URL: https://www.verdigris.co.

[9] The Edge. URL: https://edge.tech/portfolio/the-edge

[10] BrainBox AI. URL: https://brainboxai.com

[11] Living Tomorrow. URL: https://www.livingtomorrow.com

12.2.8 Transportation and Logistics

Urban transportation is undergoing a major shift, thanks to AI. From optimizing traffic routes to predicting mobility patterns, AI is reshaping how vehicles move through cities. Rapid urban population growth has increased the demand for efficient transportation systems and sustainable infrastructure. The rise of smart cities powered by AI is transforming urban infrastructure optimization and transportation planning.

Here is how AI is making a difference in urban transportation:

- Predictive route planning: AI-powered predictive tools analyze historical data to anticipate peak traffic times and suggest alternative routes.
- Route optimization: AI-powered route optimization tools analyze traffic conditions, weather patterns to determine the fastest and most efficient routes, with real-time adaptability and dynamic route adjustments (rerouting).
- Traffic management: AI-driven traffic management systems analyze real-time data, reducing unnecessary road traffic and urban congestion.

AI-powered solutions, including intelligent traffic management, real-time data analytics, and autonomous mobility, improve transportation efficiency, reduce congestion, and promote environmental sustainability. Additionally, AI-powered predictive maintenance of roads and transportation networks minimizes infrastructure disruptions and ensures operational efficiency. Autonomous and connected vehicles, a critical component of AI-powered urban mobility, contribute to improved transportation efficiency. AI enables vehicle-to-vehicle (V2V) and vehicle-to-infrastructure (V2I) communications, improving traffic coordination and reducing accident risks.

AIoT services enhance the efficiency of transportation systems, which is a critical aspect in the implementation of advanced AIoT solutions aimed at planning smart transportation. By integrating advanced technologies such as AI, ML, DL, IoT, edge computing, fog computing, and cloud computing, as well as advanced communications, this framework creates the foundation for prompt and proactive decision-making based on real-time data analytics, addressing varied urban traffic flow issues (Jagatheesaperumal et al. 2024).

The core AI-based solution that has been around for a few years is *autonomous vehicles (AV)*—connected self-driving cars and trucks. While some companies, like Tesla, that are poised to play on the cutting edge have already released products, most are holding off. Building a high computing system with real-time adapting algorithms to a changing landscape requires a great deal of effort. For example, Waabi is a startup that uses generative AI to power large-scale autonomous vehicles, including the long-haul trucking sector. The company is using NVIDIA DRIVE Thor (Waabi 2024)—a centralized in-vehicle computer, to bring to market a safe, generative AI-powered solution for autonomous trucking. Waabi's AI model works similarly to text-to-image models like DALL-E or OpenAI's Sora. It takes the LiDAR data point clouds that visualize a 3D map of the car's surroundings and breaks them down into chunks, similar to how image generators break down photographs into pixels. Then it is possible to predict how all the LiDAR data points will move based on its training data. Doing this continuously allows it to generate predictions 5–10 s into the future. As a result, it is possible to suppose that generative AI is a predictive technology. This model could be even more forceful if it gathered data not just from the car's IoT sensors, but also from the urban environment itself.

Neural networks as a crucial component of AI models can explore and search for parking spaces, which solves a significant problem in megacities that causes drivers to waste time and money, as well as contribute to air pollution and global warming (Kemik et al. 2024). Neural networks can also be used to improve the efficiency of autonomous smart traffic management (ASTM) systems, reduce traffic congestion, and improve traffic flow speed (Goenawan and Soo 2024). Moscow Transport implements services based on neural networks, including systems with elements of unmanned driving, a smart route network of ground transport with an update function based on predictive analytics, dynamic redistribution of transport on individual routes in order to reduce waiting times at stops, and online selection of the optimal charging station for electric buses. It is expected that Moscow AI-based transport services will be implemented in ecosystem services of Russian tech companies (Sber, VTB, T-Bank, MTS, and Yandex).

Urban logistics has also become AI-powered. AI is at the heart of modern urban logistics, using data-driven algorithms and edge computing to improve efficiency. Logistics companies leverage AI to optimize everything—from warehouse management to last-mile delivery. By analyzing vast amounts of data, AI can identify delivery trends, predict challenges, offer shared delivery models for multiple businesses, reduce the number of vehicles on the road, and enable self-driving delivery robots and drones. Accelerating change in AI-driven urban logistics is inseparable from additional end-to-end technologies. For example, the researchers proposed an innovative approach to optimizing urban logistics using a multilayer architecture that seamlessly integrates AI, IoT technologies, and autonomous vehicles (Mohsen 2024).

The application of AI possesses strategic potential for digital sovereignty and ensuring long-term city development. However, for the successful implementation of AI projects in urban transport, a few challenges need to be solved. First, it is necessary to ensure access to high-quality data that can be applied to train AI algorithms. Additionally, it is extremely important to develop infrastructure with a data-centric approach in the design of transportation systems. Furthermore, road safety is expected to reach new heights as we get closer to realizing the vision of full AV (USA National Highway Traffic Safety Administration 2023).

12.2.9 Energy Management

Cities account for 75% of carbon emissions despite occupying only 3% of the Earth's surface.[12] There is an increasing need to optimize energy resource allocation, accurately forecast energy demand, and reduce carbon footprints, with the latter being critical. Smart energy development includes smart meter networks and AI energy management systems that help optimize energy consumption, reduce costs, and improve energy efficiency toward the so-called "net-zero emissions journey."

Some of the tasks performed by AI for energy management include

- Forecasting energy demand based on historical data and current consumption patterns, optimizing energy distribution accordingly.
- Regulating heating, cooling, and lighting based on occupancy and consumption patterns, resulting in significant energy savings and a reduced carbon footprint.
- Facilitating the integration of renewable energy sources by predicting fluctuations in energy production and consumption, and managing the operation of a mix of renewable and nonrenewable energy.

Forecasting the *building energy efficiency* is a key strategy for urban energy management. For example, a semi-supervised DL method can be used for this purpose to predict the energy use intensity of buildings (EUI). In addition, it is acceptable and appreciable to analyze the urban space in terms of carbon dioxide emissions. Sustainable energy efficiency of buildings focuses not only on analyzing and minimizing the energy needs of buildings but also on reducing CO_2 emissions. For this purpose, spatial modeling of urban carbon emissions fits to reduce carbon emissions.

Cities are implementing AI-powered energy management systems that use ML or DL algorithms to predict energy supply and demand by integrating data from weather forecasts and energy consumption sensors. The system dynamically adjusts energy distribution and storage, using battery storage systems during excess production and redistributing it during peak demand. Cities like Oslo are prime examples of sustainability in action. Oslo is committed to renewable energy, with an impressive 98% of its electricity coming from hydropower (Sweco 2024).

Cities actively collect real-time energy consumption data from meters and sensors distributed throughout the city infrastructure. This data is analyzed using AI and Big Data to identify trends and identify areas where energy can be saved. Blockchain-based *peer-to-peer electricity transactions* are projected to play a significant role in the operation of future energy infrastructures. However, on-demand energy usage is a major challenge for households seeking cost-effectiveness. To solve this challenge, the ML-based predictive electricity trading framework was created to support distributed energy resources in real time, daily monitoring, and scheduling. The framework includes two modules: fuel trading and smart contracting with predictive analytics

[12] Cities—United Nations Sustainable Development Action 2015. URL: https://www.un.org/sustainabledevelopment/cities

components. In particular, the smart contracting module enables peers to track energy consumption in real time, manage electricity trading, model rewards, and irreversible records of electricity trading transactions (Tang et al. 2022).

The building and construction sector stands as one of the largest consumers of energy in the world, contributing 25–40% of global CO_2 emissions. In China, for example, the construction sector is among the top three energy-consuming sectors, representing about 22% of carbon emissions from energy-related sectors. China has set forth ambitious carbon neutrality goals, aiming for carbon peaking by 2030 and carbon neutrality by 2060. In this regard, there was an attempt to forecast the energy consumption of urban buildings in the central part of Shanghai by developing a robust method for predicting the energy consumption of buildings at the city scale. More than 18,000 city blocks were collected and designated as the simulation and prediction datasets. This methodology offers a viable approach to estimating the building energy consumption at the city scale, especially when data availability is limited, and can be effectively integrated with workflows that use remote sensing and AI to identify the carbon footprints of buildings (Jiang et al. 2024).

There are still some controversies in the evaluations regarding the impact of AI development on urban energy efficiency. In concrete terms, excessively fast update repetition rates and unwarranted treatment of outdated AI equipment may enhance energy consumption and hinder the balanced development of urban ecosystems (Zhuk 2023). In addition, there are a number of threats to success, such as high-cost infrastructure and investment demands required to deploy smart technologies. These technologies are based on a robust ICT infrastructure and are often hampered by a variety of issues, including data protection directives and technical requirements.

12.2.10 Water Management

Most developing countries face water-related challenges in water management and water pollution, which causes health problems. Moreover, uncertain rainfall, frequent cyclones, and droughts are becoming commonplace. Maintaining water availability and water quality balance analysis is important in water collection and delivery systems as they determine the relationship between storage capacities, reuse demand, and supply reliability. Designs must, in particular, be efficient not only at full capacity but also in situations with limited service, such as suboptimal, limited supplies, and unexpected conditions. However, over-reliance on supply-driven urban water supply is increasingly being discovered and criticized, and they are considered wasteful and expensive supply-side solutions. This usually results in improved living standards for all population groups, while the same technology, on the other hand, causes environmental degradation. Basic AI techniques (knowledge-based systems, genetic algorithms, fuzzy inference systems) and advanced AI techniques (ML, DL, and RL) are common tools for developing expert AI-powered water management services (Jayakumar et al. 2024).

The potential of AI in water management in cities is widely recognized by both the research and practice communities, with a growing number of applications demonstrating solutions to water supply, stormwater, and wastewater problems. AI is also recognized as having a fundamental role in the development of *urban water infrastructure (UWI)*. AI is expected to play a key role in the transition of UWI to the future, linked to the development of a circular economy, greening of urban spaces, decarbonization, and automation. Efficient and safe water management is at the intersection of food, energy, environmental, and urban issues.

For example, the City of Tucson in Arizona, USA, implemented AI to expand the water loss control program and optimize the centralized management of the city water system, which consists of more than 4600 miles of distribution water pipes. The city government implemented Voda.ai project,[13] which detects patterns in historical pipe failure data and evaluates weather, land use, and more to develop unerring forecasts for pipe bursts and other failures. The AI then calculates a Probability of Failure (LoF) and Consequence of Failure (CoF) score for each pipe segment: from these scores, the technology generates a quarterly risk exposure score, allowing utilities to focus their resources on the most critical sections of pipe as a tangible asset.

Data democratization, widespread adoption of no-code platforms, and the introduction of advanced infrastructure will expand the horizons of AI implementation in water management in cities. Significant research efforts are also needed to develop and apply bio-inspired learning algorithms to improve the understanding of water processes and the water cycle. The implementation of cyber-safe water policies based on AI requires collaboration among various discipline representatives. Thus, integrating AI for sustainable water management practices will create a future that effectively addresses the problem of responsible water use for city residents.

12.2.11 Waste Management

As waste generation rapidly outpaces the abilities of traditional management approaches, the need for innovation has become more vital. Research confirms that the environmental and social impacts of poor waste management are severe, with landfills releasing large amounts of methane, a potent greenhouse gas that significantly contributes to global warming. By 2050, the world is expected to generate 3.4 billion tons of waste annually, up significantly from 2 billion tons in 2018 (World Bank 2018).

Materials that could be recycled or composted are constantly ending up in landfills. With waste generation projected to increase, the problem could get worse (World Bank 2022). AI for solid waste sorting and processing becomes increasingly essential in waste management. Developed countries, including Austria, Germany, New Zealand, the United States, the United Kingdom, Japan, Singapore, Switzerland, South Korea, and Canada, have already started implementing AI technologies to extend resource utilization and recycling possibilities throughout the solid waste management cycle (Soni et al. 2019).

[13] Voda.AI. URL: https://voda.ai

Smart waste management solutions with AI and IoT technologies have already demonstrated their enormous potential in many cities around the world. For example, smart bins equipped with IoT sensors can detect when they are almost full and automatically signal collection services. This reduces the likelihood of overflow and ensures that collection resources are only deployed when necessary (Addas et al. 2024). Another promising application of AI and the IoT in waste management is automated sorting systems, which play a crucial role in recycling. Waste collected from cities is transported to a materials recovery facility (MRF), with subsequent sorting. This sorting process is inefficient because incorrectly classified waste materials contaminate the recyclables. Additionally, sorting is largely manual and labor-intensive as people rush to scan and sort recyclables on a fast-moving conveyor belt. In this regard, AI is transforming waste management with automated sorting systems that use robotics and ML algorithms for waste classification (Sutikno et al. 2024).

Robotic AI sorting systems are employed to enhance waste logistics by effectively identifying and categorizing various waste types, assessing the capacity of containers and garbage trucks, and devising optimal routes for transporting municipal solid waste from collection sites to recycling facilities. To address these objectives, AI leverages data, including waste type, morphological composition before and after sorting, volume, weight, transportation method, capacity of waste processing plants, length of transport routes, and other interrelated variables. The implementation of AI in waste logistics can lead to a reduction in transportation distances by as much as 36.8%, a decrease in costs by up to 13.4%, and time savings of up to 28.2%. Furthermore, AI can achieve waste identification and sorting accuracy ranging from 72.8% to 99.9%. When integrated with chemical analysis, AI enhances waste pyrolysis, carbon footprint evaluation, and energy conversion processes (Fang et al. 2023).

To illustrate the solution landscape, it is appropriate to itemize a few of the successfully commercialized cases. AMP Robotics,[14] a key player in AI-based sorting, has introduced powerful inkjet systems that can make thousands of picks per minute on conveyor belts moving at 600 feet/min. In addition to requiring a fraction of the manual effort, these systems require minimal downtime, significantly increasing the throughput of waste processing plants. The RecycleOS[15] platform can sort items with over 95% accuracy. AI algorithms collect and analyze precise data on the shape, size, weight, material, packaging type, product cost, and even brand information of each recyclable material that passes through the plant. These systems adapt to new waste types over time, ensuring that sorting processes are adaptive even as the composition of the waste changes. Greyparrot,[16] which has developed the AI-powered waste recognition system, currently services about 50 waste recycling facilities in Europe. The system works like this: special smart cameras are installed above the conveyor belts at waste recycling plants, which send images in real time to the startup's software. It, in turn, analyzes the objects using AI. Then the software classifies different types of waste and sends them for appropriate recycling. In

[14] AMP Smart Sortation. URL: https://ampsortation.com

[15] RecycleOS. URL: https://www.everestlabs.ai/recycleos

[16] GreyParrot. URL: https://www.greyparrot.ai

addition, the Greyparrot neural network can calculate the approximate volumes and ratios of different types of waste.

A novel advancement in this field is the implementation of *AI-driven delta robots*, which exhibit remarkably high operational speeds of up to 150 actions per minute. These robots utilize suction cups to extract waste from the flow. Despite this productivity constraint, each robot can effectively replace two human workers. Delta robots have demonstrated superior performance in quality control tasks following traditional sorting methods, which tend to have a certain percentage of blockages at the output, as well as in the sorting of separately collected waste.

Technological solutions described above have promising potential in smart waste management, encouraging to improvement of the system of separate collection reducing the number of landfills for waste disposal, and improving the state of the environment. These technologies can achieve full automation of waste collection, increase the efficiency and speed of processing, and reduce the human factor.

12.2.12 Environmental Monitoring

The integration of artificial intelligence (AI) into environmental monitoring has become a game-changer, especially in the field of air quality management. The synergy between AI, sensors, and the Internet of Things (IoT) is revolutionizing the way we predict, manage, and reduce air pollution. Here are the most functioning points of using AI in environmental monitoring and air pollution management:

- Real-time air quality monitoring: IoT and AI devices equipped with air quality sensors can be strategically placed throughout the city to provide real-time data on air quality and detect pollution hotspots.
- Combating air and noise pollution: By integrating transportation, energy and waste disposal systems, cities can streamline urban operations and mitigate carbon footprint.
- Predictive analytics for air quality management: predictive analytics involves correlating air quality data with other relevant factors, such as traffic patterns, industrial operations, and meteorological conditions.
- Predictive modeling based on historical data: AI models use algorithms to forecast future air quality conditions based on anticipated changes (for instance, during rush hours or under specific weather conditions).

AI-powered sensor systems integrate data from multiple sources, including particulate matter (PM) sensors, gas sensors, and other environmental monitoring devices. Sensor networks that are part of IoT systems collect real-time data on various pollutants such as particulate matter, nitrite, sulfur dioxide, and other hazardous substances. The data is then processed using ML and DL algorithms that can simultaneously process multiple variables such as meteorological data, emission levels, and operating conditions. This holistic analysis provides the most detailed picture of air quality that traditional methods cannot provide (Particulas 2024).

ML and DL algorithms play a key role in analyzing the large volumes of data generated by these sensor networks. These algorithms can detect complex nonlinear relationships between variables, improving the accuracy of air quality predictions. An illustrative case of an IoT-integrated environmental toxicology model involves the usage of ANN to assess and forecast air quality in real time, issuing alerts upon the detection of harmful pollutants (Popescu et al. 2024). ML algorithms and edge computing are also employed to evaluate the potential effects of these substances on both human health and the environment. For instance, e-Nose algorithms, which are based on olfactory principles, analyze data gathered from sensors to identify hazardous chemicals by recognizing their distinct chemical signatures. These algorithms can implement a range of methodologies, including pattern recognition, ANN, and fuzzy logic. A significant benefit of e-Nose technologies is their capacity to detect dangerous chemicals in real time, facilitating prompt action in response to potential hazards. Furthermore, e-Nose technologies have diverse applications, such as monitoring air quality in urban settings (Jeong and Choi 2022).

Integration of AI with IoT and Big Data technologies is essential for effective air quality monitoring. IoT sensors distributed across different locations can collect real-time data on operations and emissions. This data is then stored and processed using Big Data technologies, providing a robust foundation for AI models to analyze and predict air quality. For example, Kunak AIR air quality monitoring stations use IoT sensor networks to collect real-time data on key pollutants. This data is analyzed using advanced ML and DL algorithms, providing accurate and timely information. Kunak AIR Cloud software provides easy data visualization and analysis, helping professionals make more informed decisions.[17]

Real-time monitoring and forecasting are two of the most significant benefits of AI in air quality management. AI systems can continuously monitor air quality, providing immediate alerts when pollutant levels exceed safe thresholds. This real-time data is invaluable for air quality management in municipalities. A partnership between NASA and the US State Department is a prime example of how AI can improve air quality forecasting. Using satellites and other data, NASA creates AI-powered air pollution forecasts. These forecasts integrate global model outputs, and satellite air quality observations into advanced ML algorithms, significantly improving the existing monitoring program (American Foreign Service Association 2021).

12.2.13 Disaster Response and Resiliency

Cities are increasingly focusing on resilience as a critical aspect. Resilience here refers to a city's ability to respond in a timely and effective manner to potentially disruptive events, such as natural disasters or pandemics. As a result, cities are creating "urban predictive operations centers" to increase their resilience. These intelligence centers bring together data, technology, and human expertise from multiple

[17] Kunak. URL: https://kunakair.com

departments, such as police and fire departments, highway and transportation departments, hospitals, and homeland security agencies. They transform the data into information that different city departments can use to respond quickly and effectively to emergencies. These centers leverage the power of the Internet of Things, big data analytics, and AI to identify patterns that are vital in situations such as an infectious disease outbreak. Growing financial support from city governments, as well as grants from large companies that seek to fulfill corporate social responsibility obligations and contribute to solving humanitarian crises, is contributing to the growth of the global market for digital emergency response services.

Emergency management includes a set of measures for their prevention, response, and elimination. In recent years, there has been a surge in AI integration, providing increased flexibility for emergency management in cities:

- Pre-emergency technologies include AI and ML for early identification of natural hazards and environmental risks, and early warning systems for natural disasters (hurricanes, earthquakes, floods, forest fires, etc.).
- Postemergency response technologies include unmanned vehicles, satellite navigation, which help quickly assess the situation and coordinate efforts, and remote sensing to assess damage and prioritize recovery activities.

AI creates advanced capabilities for predicting man-made and natural disasters, monitoring humanitarian crises, and optimizing responses to emergencies of various natures. AI algorithms can analyze large volumes of meteorological, geological, spatial, and other data to predict natural disasters long before they occur. For achieving the goals of predicting disasters, a combination of AI and complementary technologies is used. For example, AiDASH, a company that develops SaaS applications, launched the Climate Risk Intelligence System (CRIS),[18] an AI-based platform that uses satellite imagery and meteorological data to improve preparedness for natural disasters and extreme effects of climate change, plan for emergency response, and identify and map the risks of loss or destruction of valuable material assets in the event of an emergency. In addition, recently, an end-to-end DL model has been presented that directly links satellite captures and large-scale city damage maps, with the main objective of assisting near real-time decision-making in postshock emergency responses. This model seamlessly integrates high-resolution satellite imagery, geospatial data, geographic information systems (GIS), and DL methods. A case study has shown that using GIS data enables automated sub-image extraction with exceptional correctness (Braik and Koliou 2024).

LLM offers significant advantages in urban disaster simulation and can efficiently process datasets, including historical records of disaster events, topography, meteorological conditions, etc., with extracting worthwhile parameters to provide robust data support for simulations. In addition, LLM helps in simulating advanced city scenarios: by learning from extensive data, LLM can understand complex

[18] AiDASH. CRIS Climate Risk Intelligence System. URL: https://www.aidash.com/climate-risk-intelligence-system

relationships within scenarios such as building collapses, fire spread, or weather cataclysms, resulting in more exact simulation outcomes. LLM can produce response plans based on the existing disaster conditions, providing powerful support for decision-makers (Chen et al. 2024).

In the face of AI's potentially revolutionary impact on pre- and postdisaster response, it is critical to acknowledge the significant challenges that are immanently expected in software development. These challenges include deployment safety: LLM, particularly deep neural networks, have already demonstrated exceptional performance in supervised benchmarks, but their security and reliability in real-life cases raise legitimate concerns, particularly in safety-critical scenarios such as post-hazard response (Sun et al. 2024). In addition, there is a significant increase in the use of explainable AI (XAI) methods for disaster risk management, which highlights the growing need for transparency and interpretability (Ghaffarian et al. 2023).

12.3 Key Challenges of AI Implementation in Smart Cities

In the quickly evolving realm of urban development, the incorporation of AI into smart city projects represents a significant advancement while simultaneously presenting intricate challenges. As cities globally aim to enhance efficiency and livability, AI provides remarkable prospects for transforming urban management and elevating the quality of life for residents. Nonetheless, this technological advancement is accompanied by various obstacles, especially regarding privacy, cybersecurity, and ethical considerations.

Lack of proper infrastructure	Developing economies struggle with inadequate infrastructure, which hampers the deployment of smart technologies. This includes not only hard infrastructure but also digital frameworks necessary for data gathering, processing, storage, management, and analysis.
Inadequate funding, financing, and budgeting	Financial constraints are a significant barrier to the adoption of AI solutions. A key step in any smart city financing effort is developing a complete strategic plan to capitalize on the strong points of projects. Main financial tools: national innovation-driven development strategy funds, expanded ministry grants and awards, municipal bonds, public–private partnership funds, local government special bonds, and policy bank loans.
Cybersecurity and information security risks	Hyper-connected cities acquire risks of cyber threats. Ensuring robust cybersecurity measures is a key step in protecting sensitive information and maintaining public trust. The practical implementation of smart city projects poses several challenges, especially when it comes to the requirements of digital ethics and digital sovereignty in data processing. Smart city actors must maintain a fine line between data processing for creating and adding value for society, and simultaneously protecting privacy.
Environmental risks	Ongoing increases in rates of urbanization have led to undesirable effects on the environment. Smart city solutions must prioritize sustainability and ESG claims in risk mitigation, with possible nature-first smart urban planning.
Trust issues	Building public trust in AI technologies is pivotal. Engaging communities in the development process can enhance acceptance and increase the cognition-based part of trust toward AI systems and applications.

12.4 Directions for Future Development

Projections indicate a substantial rise in the number of smart cities globally by 2040, fueled by urbanization and technology advancements. Data-intensive cities like Singapore, Tokyo, London, and Moscow are expected to lead in smart city initiatives, establishing benchmarks for others through their AI infrastructure and innovative urban management strategies. This domain is experiencing significant growth, overcoming various challenges, with emerging trends and innovations poised to transform the landscape of smart cities. It is assumed that further enhanced development of smart cities and smart city areas will be determined by the following core technological innovations:

The rise of AIoT	AIoT is poised to become one of the most defining smart city trends in the near future. As the number of interconnected IoT devices increases, urban areas will produce substantial volumes of data, which AI will analyze in real time. For example, traffic management systems will leverage AI to interpret data collected from sensors and cameras, allowing for the prediction of traffic patterns and the dynamic adjustment of traffic light schedules. Similarly, air quality monitoring systems will enable cities to respond to environmental concerns by detecting pollution levels and suggesting real-time actions to improve air quality.
AI-powered city digital twins	The emerging technological trend with growing potential is the development of comprehensive city digital twins that integrate real-time data from multiple sources. Powered by ML algorithms, such digital twins will allow urban planners and managers to simulate and test multiscale urban scenarios, and optimize resource allocation and infrastructure planning, which is essential for building automation, architectural design, and improving disaster preparedness.
Human–AI collaboration in urban management	The future of smart cities does not lie in AI systems operating autonomously but in effective collaboration between human decision-makers and AI technologies. This trend toward human-AI collaboration is manifested through participatory AI: involving citizens in the design and implementation of AI systems through public consultations, workshops, and citizen science initiatives. By developing this collaboration, cities can harness the analytical power of AI while maintaining human judgment, empathy, and ethical considerations in urban management, healthcare, and digital governance.
Advanced edge computing and distributed AI	The significant trend of AI-powered smart cities with forthcoming impact is moving toward advanced edge computing and distributed AI architectures. This approach involves processing data closer to its source—at the "edge" of the network—instead of sending all the information to centralized cloud servers, and facilitates decentralized data processing on IoT devices. For smart cities, this solution offers several benefits: reduced latency for real-time applications (e.g., in traffic management, emergency response); increased privacy and security by minimizing data transmission and storage; increased resilience to network failures, reduced power consumption, and reduced bandwidth requirements, which is essential in resource management, waste recycling, and environmental monitoring.

12.5 Conclusion: Navigating the Future of AI-Powered Smart Cities

The future of smart cities globally appears bright, driven by technological advancements that are set to reshape urban landscapes. AI is expected to assume a pivotal role in urban management, enhancing various systems, including transportation, healthcare, and resource allocation. The integration of IoT devices with AI sensors is anticipated to streamline city operations, fostering predictive capabilities and greater stability. These innovations promise to transform urban living by facilitating quicker emergency responses and optimizing energy usage.

For organizations, businesses, and governments involved in the development of smart cities, prioritizing inclusivity through knowledge sharing and access to AI will enhance collaborative efforts and amplify their impact. Ideally, AI will serve as a democratizing force, allowing individuals to access, comprehend, and visualize data from their personal computers, thereby enabling more informed decision-making over time.

The path forward for building AI-powered smart cities in both developed and developing economies is fraught with challenges, but it also holds great potential. By addressing the infrastructure gap, securing funding, and building public trust, countries can harness the capabilities of AI to empower urban planners for informed decisions and create smarter urban environments.

Developing AI-driven smart cities necessitates addressing various technical hurdles alongside a range of ethical considerations. As AI systems grow more intricate, adaptable AI methodologies will be crucial for understanding decision-making processes and establishing trust in software solutions.

Ultimately, the success of AI in smart cities will depend on the ability to balance technological progress with universally accepted human and social values. An ongoing dialogue between technologists, policymakers, and citizens will be needed to guarantee that AI-driven urban systems support the diverse needs and aspirations of city communities.

Acknowledgments The chapter is based on the study funded by the Basic Research Program of the HSE University.

References

Addas A et al (2024) Waste management 2.0 leveraging internet of things for an efficient and eco-friendly smart city solution. PLoS One 19(7):e0307608

Aguilera D et al (2016) Citizen-centric data services for smarter cities. Futur Gener Comput Syst 76:234–247

Albino V et al (2015) Smart cities: definitions, dimensions, performance, and initiatives. J Urban Technol 22(1):3–21

Al-Turjman F et al (2021) Intelligence of things: AI-IoT based critical-applications and innovations. Springer, 234 p

American Foreign Service Association (2021) AI air quality forecasting: a state-NASA partnership. URL: https://afsa.org/ai-air-quality-forecasting-state-nasa-partnership
Apanaviciene R et al (2020) Smart building integration into a smart city (SBISC): development of a new evaluation framework. Energies 13(9):2190
Batty M (2018) Artificial intelligence and smart cities. Environ Plan B Urban Anal City Sci 45(1):3–6
Bharti P, Biswas A (2024) Predicting urban growth of Kathmandu Valley using artificial intelligence. J Geovis Spat Anal 8:40
Braik AM, Koliou M (2024) Automated building damage assessment and large-scale mapping by integrating satellite imagery, GIS, and deep learning. Comput Aided Civ Inf Eng 39(15):2389–2404
Bronner W et al (2021) Sustainable AIoT: how artificial intelligence and the internet of things affect profit, people, and planet. Springer Books. In: Oliver Gassmann O, Ferrandina F (eds) Connected business. Springer, pp 137–154
Burrell DN (2023) Dynamic evaluation approaches to telehealth technologies and artificial intelligence (AI) telemedicine applications in healthcare and biotechnology organizations. Merits 3(4):700–721
Cervicorn Consulting (2025) AI in smart cities market size and growth 2024 to 2033. URL: https://www.cervicornconsulting.com/artificial-intelligence-in-smart-cities-market
Chen T, Gascó M (2024) Uncovering the results of AI chatbot use in the public sector: evidence from US state governments. Public Perform Manag Rev:1–26. https://doi.org/10.1080/15309576.2024.2389864
Chen DB et al (2024) AI-based strategies for urban disaster emergency response management. Eng Technol Trends 2(3):11–15
Cyber Defense Magazine (2019) Smart cities predicted to deliver more than USD 20 trillion in additional economic benefits by 2026. URL: https://www.cyberdefensemagazine.com/smart-cities-predicted-to-deliver-more-than-usd-20-trillion-in-additional-economic-benefits-by-2026
Desouza KC et al (2015) Local government 2035: strategic trends and implications of new technologies. Issues Technol Innov 27:1–27
ESI ThoughtLab (2021) Smart city solutions for a riskier world. URL: https://thoughtlabgroup.com/wp-content/uploads/2021/03/Smart-City-Solutions-eBook-.pdf
ETH Zurich (2023) Virtual design for real-world buildings. URL: https://baug.ethz.ch/en/news-and-events/news/2023/10/virtual-design-for-real-world-buildings.html
Fang B et al (2023) Artificial intelligence for waste management in smart cities: a review. Environ Chem Lett 21:1959–1989
Floreano IX, de Moraes LAF (2021) Land use/land cover (LULC) analysis (2009-2019) with Google Earth Engine and 2030 prediction using Markov-CA in the Rondônia State, Brazil. Environ Monit Assess 193(4):239
Flores M et al (2013) P4 medicine: how systems medicine will transform the healthcare sector and society. Pers Med 10(6):565–576
Fortune (2024) AI and the future of cities. URL: https://fortune.com/2024/12/02/ai-and-the-future-of-cities
Galindo F (2014) Methods for law and ICT: an approach for the development of smart cities. In: Conference paper. AI approaches to the complexity of legal systems, pp 26–40
Ghaffarian S et al (2023) Explainable artificial intelligence in disaster risk management: achievements and prospective futures. Int J Disaster Risk Reduct 98:104123
Ghaffarianhoseini A et al (2016) What is an intelligent building? Analysis of recent interpretations from an international perspective. Archit Sci Rev 59(5):338–357
Goenawan CR, Soo HD (2024) ASTM: autonomous smart traffic management system using artificial intelligence CNN and LSTM. arXiv preprint arXiv:241010929
Government Technology (2024) How are chatbots being used in government agencies? URL: https://www.govtech.com/artificial-intelligence/how-are-chatbots-being-used-in-government-agencies

Guerra BMV et al (2023) Ambient assisted living for frail people through human activity recognition: state-of-the-art, challenges and future directions. Front Neurosci 17:1256682
IBM (2024) AI in government. URL: https://www.ibm.com/think/topics/ai-in-government
Jagatheesaperumal SK et al (2024) Artificial intelligence of things for smart cities: advanced solutions for enhancing transportation safety. Comput Urban Sci 4:10
Jayakumar D et al (2024) Unlocking the potential of artificial intelligence for sustainable water management focusing operational applications. Water 16(22):3328
Jeong J, Choi J (2022) Artificial intelligence-based toxicity prediction of environmental chemicals: future directions for chemical management applications. Environ Sci Technol 56(12):7532–7543
Jiang Q et al (2024) Predicting building energy consumption in urban neighborhoods using machine learning algorithms. Front Urban Rural Plan 2:6
Kar AK et al (2017) Advances in smart cities: smarter people, governance, and solutions. CRC Press, 229 p
Kemik H et al (2024) Smart solutions for mega-cities: utilizing long short-term memory and multi-head attention in parking prediction. ISPRS Int J Geo Inf 13(12):449
Liu Y et al (2019) Intelligent edge computing for IoT-based energy management in smart cities. IEEE Netw 33(2):111–117
Loukika KN, Keesara VR, Sridhar V (2021) Analysis of land use and land cover using machine learning algorithms on Google Earth Engine for Munneru River Basin, India. Sustainability 13(24):13758
McKinsey (2024) What is artificial general intelligence (AGI)? URL: https://www.mckinsey.com/featured-insights/mckinsey-explainers/what-is-artificial-general-intelligence-agi
MIT (2023) SMART launches research group to advance AI, automation, and the future of work. URL: https://news.mit.edu/2023/smart-launches-m3s-research-group-advance-ai-automation-future-work-0823
Mitchell M (2024) Debates on the nature of artificial general intelligence. Science 383(6689):eado7069
Mohammadzadeh Z et al (2023) Smart city healthcare delivery innovations: a systematic review of essential technologies and indicators for developing nations. BMC Health Serv Res 23:1180
Mohsen BM (2024) AI-driven optimization of urban logistics in smart cities: integrating autonomous vehicles and IoT for efficient delivery systems. Sustainability 16(24):11265
Mugiraneza T et al (2020) Continuous monitoring of urban land cover change trajectories with Landsat time series and LandTrendr-Google Earth Engine cloud computing. Remote Sens 12(18):2883
Munavalli JR et al (2021) Real-time capacity management and patient flow optimization in hospitals using AI methods. In: Masmoudi M, Jarboui B, Siarry P (eds) Artificial intelligence and data mining in healthcare. Springer, Cham
NVIDIA (2023) The future of generative AI in architectural design practices. URL: https://resources.nvidia.com/en-us-general-aec/enterprise-56
Particulas (2024) The power of artificial intelligence in air quality monitoring and forecasting in the mining and industrial sectors. URL: https://particulas.cl/2024/05/the-power-of-artificial-intelligence-in-air-quality-monitoring-and-forecasting-in-the-mining-and-industrial-sectors
Popescu SM et al (2024) Artificial intelligence and IoT driven technologies for environmental pollution monitoring and management. Front Environ Sci 12:1336088
S&P Global (2024) The rise of AI-powered smart cities. URL: https://www.spglobal.com/en/research-insights/special-reports/ai-smart-cities
Saha S, Sarkar D, Mondal P (2023) Urban expansion monitoring using machine learning algorithms on Google Earth Engine platform and cellular automata model: a case study of Raiganj Municipality, West Bengal, India. In: Rahman A, Sen Roy S, Talukdar S, Shahfahad (eds) Advancements in urban environmental studies. GIScience and geo-environmental modelling. Springer, Cham, pp 43–55
Shanghai Daily (2022) Shanghai harnessing "digital twin" technology to improve city management. URL: https://www.shine.cn/news/metro/2202151932

Shao X et al (2021) How does facial recognition as an urban safety technology affect firm performance? The moderating role of the home country's government subsidies. Saf Sci 143:105434
Sidhu N et al (2018) Using Google Earth Engine to detect land cover change: Singapore as a use case. Eur J Remote Sens 51(1):486–500
Soni U et al (2019) Forecasting municipal solid waste generation using artificial intelligence models—a case study in India. SN Appl Sci 1:162
Sowmitha R et al (2022) Artificial intelligence in smart cities and healthcare. EAI Endorsed Trans Smart Cities 6(3):e5
Sun L et al (2024) Convergence of AI and urban emergency responses: emerging pathway toward resilient and equitable communities. Appl Sci 14(17):7949
Sutikno T et al (2024) Artificial intelligence technologies in urban smart waste management. High-Tech Innov Ser 1:55–66
Sweco (2024) Leading the (Nor)way in electric power generation. URL: https://www.sweco.co.uk/blog/leading-the-norway-in-electric-power-generation
Szpilko D et al (2023) Artificial intelligence in the smart city—a literature review. Eng Manag Prod Serv 15(4):53–75
Tang Z et al (2022) Machine learning assisted energy optimization in smart grid for smart city applications. J Interconnect Netw 22(3):2144006
The OECD Observatory of Public Sector Innovation (2023) Artificial intelligence in the public sector. URL: https://oecd-opsi.org/work-areas/ai
The Straits Times (2024) More than 100 solutions developed as part of generative AI initiative. URL: https://www.straitstimes.com/tech/tech-news/more-than-100-solutions-developed-as-part-of-generative-ai-initiative
USA National Highway Traffic Safety Administration (2023) Automated vehicles for safety. URL: https://www.nhtsa.gov/vehicle-safety/automated-vehicles-safety
Vallebueno A, Lee YS (2023) Measuring urban quality and change through the detection of physical attributes of decay. Sci Rep 13(1):17316
Varnosfaderani SM, Forouzanfar M (2024) The role of AI in hospitals and clinics: transforming healthcare in the 21st century. Bioengineering (Basel) 11(4):337
Waabi (2024) Waabi to bring the first generative AI-powered trucking solution, built on NVIDIA DRIVE Thor, to market. URL: https://waabi.ai/nvidia-drivethor
WEF (2022) Digital twin cities: framework and global practices. URL: https://www3.weforum.org/docs/WEF_Global_Digital_Twin_Cities_Framework_and_Practice_2022.pdf
World Bank (2018) What a waste 2.0: a global snapshot of solid waste management to 2050. URL: https://openknowledge.worldbank.org/entities/publication/d3f9d45e-115f-559b-b14f-28552410e90a
World Bank (2022) Solid waste management. URL: https://www.worldbank.org/en/topic/urbandevelopment/brief/solid-waste-management
Xue M (2021) Expansion and evolution of a typical resource-based mining city in transition using the Google Earth Engine: a case study of Datong, China. Remote Sens 13(20):4045
Yigitcanlar T et al (2020) Contributions and risks of artificial intelligence (AI) in building smarter cities: insights from a systematic review of the literature. Energies 13(6):1473
Zhuk A (2023) Artificial intelligence impact on the environment: hidden ecological costs and ethical-legal issues. J Digit Technol Law 1(4):932–954
Zurqani HA et al (2019) Mapping urbanization trends in a forested landscape using Google Earth Engine. Remote Sens Earth Syst Sci 2:173–182

MIX
Papier aus verantwortungsvollen Quellen
Paper from responsible sources
FSC® C105338

If you have any concerns about our products,
you can contact us on
ProductSafety@springernature.com

In case Publisher is established outside the EU,
the EU authorized representative is:
Springer Nature Customer Service Center GmbH
Europaplatz 3, 69115 Heidelberg, Germany

Printed by Libri Plureos GmbH
in Hamburg, Germany